教育部科学技术研究重点项目（No.211189）
甘肃省高校学科带头人扶持项目（11ZX-04）　　资助出版
天水师范学院重点学科
天水师范学院"青蓝"人才工程

卟啉自组装膜电化学

Electrochemistry of porphyrin on self-assembled monolayers

左国防　卢小泉　编著

兰州大学出版社

图书在版编目（CIP）数据

卟啉自组装膜电化学/左国防，卢小泉编著. —兰州：兰州大学出版社，2011.7
ISBN 978-7-311-03714-7

Ⅰ.①卟… Ⅱ.①左…②卢… Ⅲ.①单分子膜—电化学—研究 Ⅳ.①O646

中国版本图书馆CIP数据核字（2011）第145327号

策划编辑　陈红升
责任编辑　郝可伟　陈红升
封面设计　刘　杰

书　　名	卟啉自组装膜电化学
作　　者	左国防　卢小泉　编著
出版发行	兰州大学出版社　（地址：兰州市天水南路222号　730000）
电　　话	0931-8912613（总编办公室）　0931-8617156（营销中心） 0931-8914298（读者服务部）
网　　址	http://www.onbook.com.cn
电子信箱	press@lzu.edu.cn
印　　刷	兰州德辉印刷有限责任公司
开　　本	787mm×1092mm　1/16
印　　张	15.75
字　　数	380千
版　　次	2011年9月第1版
印　　次	2011年9月第1次印刷
书　　号	ISBN 978-7-311-03714-7
定　　价	28.00元

（图书若有破损、缺页、掉页可随时与本社联系）

前 言

自组装单分子膜是超分子化学的一个重要分支,经过三十余年的发展,它已成为凝聚态物理、材料科学、合成化学、结构化学、微电子学及生物化学等研究领域的交叉性前沿课题,是近代化学发展的一个更高层次。

组织有序、定向、密集、完好的自组装单分子膜可在分子水平上达到控制电极界面微结构的要求。与传统的修饰电极修饰方法相比,自组装膜具有更好的稳定性。特别是它所特有的明晰的微结构,为电化学研究提供了一个重要的实验平台,借此可探索电极表面分子微结构和宏观电化学响应之间的关系,并可以研究表面功能基团的反应性和分子识别问题。可以认为,自组装膜化学修饰电极是单分子膜化学修饰电极发展的最高形式,是研究有关表面和界面各种复杂现象的理想模型体系。

卟啉及其衍生物是血红素、细胞色素和叶绿素等生命活性物质的核心部分,已被广泛地应用于光电转换、催化、导电、铁磁体、载氧体、分子电子器件和医学等领域。研究表明,卟啉自组装单分子膜是最接近天然生物膜的理想模型,被广泛应用于自组装膜领域的研究。卟啉类自组装膜在电化学领域显示出巨大的优势,主要表现在:利用自组装膜的卟啉分子设计和结构可控的特点,在分子水平上预先设计膜结构,获得特殊功能,达到模拟生物膜的目的,是研究界面电子转移的理想模型;卟啉作为一种光电活性物质,将其自组装于基底表面可以有效地研究其光电性质,用于模拟植物体内光合作用过程和能量传递,在光电分子器件的设计和合成上有潜在的应用;卟啉及金属卟啉分子在均相和异相化学反应中是良好的催化剂,可用于构建各种不同的电化学生物传感器,在生命系统以及在仿生学中发挥着良好的功能,备受研究者的关注;卟啉化合物由于结构的特殊性,通过在卟啉环周边进行化学修饰,引入特定官能团,使之与底物产生多重相互作用,可进行分子大小、形状、功能基团和手性异构体的识别;另外,利用卟啉作为分子机能材料对未来情报信息的表示、传递、识别,超导材料的制备,有机电致发光以及太阳能电池的开发研究已成为国内外十分活跃的研究领域。

本书将自组装单分子膜和卟啉化合物有机结合,系统介绍了卟啉自组装膜的发展、特点、表征及其应用,尤其对卟啉自组装膜的电化学研究方法及其应用研究进展进行了全面阐述。全书共分十二章,其中第一章到第五章系统介绍了自组装单分子膜的发展、特点、研究方法、界面电化学行为、性质及其应用研究进展;第六章到第九章介绍了卟啉及其衍生物的合成及应用;第十章到第十二章系统介绍了卟啉自组装膜的制备、表征、电化学研究方法及其应用研究。

全书在左国防博士学位论文基础上增补了大量本领域研究的最新成果,由左国防博士和卢小泉教授共同完成,卢小泉教授对全书进行了全面审读,并在内容体系上提出了建设性意见。

兰州大学出版社陈红升老师和郝可伟老师在本书出版中进行了认真的审阅和加工,李艳红博士和张建斌老师为本书出版收集、整理和翻译了部分资料,在此表示深切的谢意。

在本书编写过程中,编者参考了相关研究领域中的大量国内外文献资料,多为国际学术刊物上近期公开发表的有一定影响的论著,分列在各章中,以供读者进一步查阅。

限于编者水平,以及本领域研究的不断扩展和深入,书中肯定存在不足之处,敬请读者不吝指正,将深表感谢。

编者

2011年5月

目 录

第一章 自组装单分子膜 ... 1
第一节 自组装单分子膜简介 ... 1
1. 自组装膜的类型 ... 1
2. 自组装膜的制备 ... 3
3. 自组装膜的结构 ... 3
4. 自组装膜的特点 ... 4
5. 自组装膜的表征 ... 5

第二节 自组装单分子膜修饰电极的电极过程动力学 ... 5
1. Langmuir等温吸附 ... 6
2. Frunkin等温吸附 ... 6
3. Temkin等温吸附 ... 7

第三节 自组装成膜过程动力学研究 ... 7
1. 自组装成膜过程动力学研究概述 ... 7
2. 分子自组装的基本原理 ... 9
3. 硫醇分子在不同单晶Au表面自组装的模型系统 ... 11

第四节 自组装膜长程电子转移研究 ... 12
1. 自组装膜的电子转移理论 ... 12
2. 电子传递距离对电子转移速率的影响 ... 14
3. 膜表面分子的设计和状态对长程电子转移的影响 ... 14
4. 硫醇自组装膜末端基团对其电荷输运特性的影响 ... 16

第五节 自组装单分子膜的性质和电化学行为 ... 17
1. 双电层结构和电容 ... 17
2. 自组装膜的稳定性和致密度 ... 18
3. 自组装膜的微结构和覆盖度 ... 18
4. 自组装膜的针孔型缺陷 ... 18
5. 自组装膜表面的酸碱性 ... 20

参考文献 ... 21

第二章 自组装膜研究方法 ... 31
第一节 自组装膜电化学研究方法 ... 31
1. 循环伏安法 ... 31

2. 交流阻抗法 … 36
  3. 光谱电化学方法 … 36
  4. 扫描电化学显微镜 … 38
  5. 电化学石英晶体微天平 … 40
  6. 电化学发光技术 … 41
 第二节 自组装膜其他研究方法 … 41
  1. 自组装膜计算机模拟研究方法 … 41
  2. 自组装膜非电化学研究方法 … 43
 参考文献 … 45

第三章 氧化还原自组装膜界面电子转移研究 … 52
 第一节 氧化还原自组装膜电子传递研究的电化学方法 … 52
  1. 循环伏安法 … 53
  2. 检测K_{ET}的Laviron法 … 54
  3. Marcus密度态理论检测K_{ET} … 54
  4. 计时安培(CA)法 … 56
  5. 交互电流伏安(ACV)法 … 56
  6. 电化学交流阻抗(EIS)法 … 57
  7. 扫描电化学显微镜(SECM) … 58
  8. 间接激光—诱导温度跳跃(ILIT)法 … 60
 第二节 自组装膜上K_{ET}电化学测量的氧化还原体系 … 60
  1. 金属茂 … 60
  2. 钌—胺络合物 … 61
  3. 铁—氮杂络合物 … 61
  4. 富勒烯(C_{60}) … 62
  5. 溶液中的氧化还原物质 … 62
  6. 金属簇 … 63
  7. 金属蛋白 … 64
 第三节 电子传递动力学的外球效应 … 66
  1. 溶剂及离子对效应 … 66
  2. 质子参与的电子传递(PCET) … 67
 第四节 氧化还原自组装单层膜的结构 … 69
  1. 自组装膜的结构形式 … 69
  2. 自组装膜缺陷的电化学分析 … 69
  3. 自组装膜缺陷对电子传递测量的影响 … 70

  4.自组装膜的形式 70
  5.自组装膜形成之后氧化还原中心的结合 71
 第五节 氧化还原自组装膜烷基链结构 72
  1.烷基链的结构 73
  2.功能基团 73
  3.共轭烷基链 74
  4.肽键 74
  5.核酸桥联 75
 第六节 氧化还原自组装膜电子传递研究中的电极材料 76
  1.电极材料的影响——理论和实验 76
  2.电极材料 77
 第七节 自组装结构的模型体系研究 79
  1.早期的工作——分子动力学 79
  2.在表面上的自组装膜模型 80
  3.尾端基团效应 80
  4.自组装膜表面Fc的模型研究 81
 参考文献 82

第四章 基于硅基自组装膜的分子电子器件研究

 第一节 分子电子器件的出现及其目前的地位 99
  1.以Au为代表的金属基底的烷基硫醇单层膜 99
  2.有机分子经由自组装过程结合于SiO_2或Si基底上的单层膜 99
 第二节 基于硅基自组装膜的电子传导 100
  1.理论模型 101
  2.通过脂肪短链的电子传导机制 102
  3.电子传导性质与距离间的关系 102
  4.基于自组装膜的介电衰减 102
 第三节 基于硅基自组装膜的分子电子装置 103
  1.分子二极管 103
  2.分子共振隧穿二极管(MRTD) 104
  3.分子记忆 105
  4.分子晶体管 105
 参考文献 106

第五章 自组装膜应用研究 108

 第一节 具有光功能的自组装膜 108

1. 自组装单分子膜光诱导电子传递 ……………………………………………… 108
　　2. 自组装膜覆盖的纳米簇光诱导电子传递 …………………………………… 111
　　3. 自组装膜上的光致异构化 …………………………………………………… 111
　　4. 自组装膜的化学发光 ………………………………………………………… 112
　　5. 自组装膜光形图案化 ………………………………………………………… 114
　第二节　自组装膜在金属防护中的应用 …………………………………………… 115
　　1. 自组装膜在金防护中的应用 ………………………………………………… 115
　　2. 自组装膜在银防护中的应用 ………………………………………………… 115
　　3. 自组装膜在铜防护中的应用 ………………………………………………… 116
　　4. 自组装膜在钢铁防护中的应用 ……………………………………………… 117
　　5. 自组装膜在其他材料防护中的应用 ………………………………………… 117
　第三节　自组装膜在电分析化学中的应用 ………………………………………… 118
　　1. 自组装膜在电分析化学中的应用 …………………………………………… 118
　　2. 自组装膜与生物传感技术 …………………………………………………… 119
　第四节　自组装膜的其他应用 ……………………………………………………… 121
　　1. 自组装单分子膜的摩擦学行为研究 ………………………………………… 121
　　2. 微触点技术 …………………………………………………………………… 121
　　3. 电寻址固定 …………………………………………………………………… 122
　　4. 分布阻隔技术 ………………………………………………………………… 123
　　5. 自组装膜在生物分子器件中的应用 ………………………………………… 123
　参考文献 ……………………………………………………………………………… 123

第六章　卟啉及金属卟啉的合成及其应用概述 …………………………………… 135
　第一节　卟啉的合成 ………………………………………………………………… 135
　　1. Adler-Longo法 ………………………………………………………………… 135
　　2. Lindsey法 ……………………………………………………………………… 136
　　3. 卟啉合成的模块法 …………………………………………………………… 136
　　4. 由2-位取代吡咯合成卟啉 …………………………………………………… 137
　　5. 由线形四吡咯出发合成卟啉 ………………………………………………… 138
　　6. 微波激励法 …………………………………………………………………… 138
　第二节　金属卟啉的合成 …………………………………………………………… 138
　　1. 金属卟啉合成的基本反应过程 ……………………………………………… 138
　　2. 金属卟啉合成的主要方法 …………………………………………………… 139
　第三节　卟啉及金属卟啉应用概述 ………………………………………………… 141
　　1. 卟啉在材料化学中的应用 …………………………………………………… 141

    2. 卟啉在医学中的应用 ………………………………………………………… 141
    3. 卟啉在生物化学中的应用 ……………………………………………………… 142
    4. 卟啉在分析化学中的应用 ……………………………………………………… 142
    5. 卟啉在能源方面的应用 ………………………………………………………… 143
    6. 卟啉在地球化学中的应用 ……………………………………………………… 143
    参考文献 …………………………………………………………………………… 144

第七章 卟啉非线性光学材料研究 ………………………………………………… 151
第一节 非线性光学理论 ……………………………………………………………… 151
    1. 非线性光学原理 ………………………………………………………………… 152
    2. 非线性光限幅效应与反饱和吸收 ……………………………………………… 153
    3. 非线性光限幅材料 ……………………………………………………………… 154
第二节 卟啉非线性光学材料的分子设计 …………………………………………… 155
    1. 卟啉非线性光学材料的分子设计 ……………………………………………… 155
    2. 卟啉的二阶非线性光学效应 …………………………………………………… 156
    3. 卟啉的三阶非线性光学效应 …………………………………………………… 156
第三节 卟啉光限幅材料研究进展 …………………………………………………… 158
    参考文献 …………………………………………………………………………… 159

第八章 卟啉传感器及金属卟啉模拟生物酶研究 ………………………………… 163
第一节 卟啉传感器 …………………………………………………………………… 163
    1. 卟啉氧传感器 …………………………………………………………………… 163
    2. 卟啉二氧化碳传感器 …………………………………………………………… 164
    3. 卟啉氯化氢传感器 ……………………………………………………………… 164
    4. 卟啉氮氧化物传感器 …………………………………………………………… 165
    5. 卟啉生物传感器 ………………………………………………………………… 165
    6. 其他卟啉传感器 ………………………………………………………………… 165
第二节 金属卟啉模拟生物酶研究 …………………………………………………… 166
    1. 金属卟啉模拟血红素酶研究 …………………………………………………… 166
    2. 金属卟啉模拟细胞色素P450酶研究 …………………………………………… 167
    参考文献 …………………………………………………………………………… 171

第九章 卟啉超分子化合物 …………………………………………………………… 175
第一节 卟啉超分子化合物的组装合成 ……………………………………………… 175
    1. 共价键构筑卟啉聚合物 ………………………………………………………… 175
    2. 非共价键构筑卟啉聚合物 ……………………………………………………… 177
    3. 卟啉功能材料的最新研究进展 ………………………………………………… 178

第二节 金属卟啉超分子催化剂与分子识别 ... 179
1. 金属卟啉超分子催化剂 ... 179
2. 金属卟啉与分子识别 ... 180

第三节 卟啉超分子化合物在分子器件中的应用 ... 181
1. 卟啉分子导线 ... 181
2. 卟啉分子开关 ... 182
3. 光合作用中的卟啉分子器件 ... 183
4. 其他卟啉电子器件 ... 184

第四节 卟啉超分子在光学领域中的应用 ... 185

第五节 卟啉类超分子的其他应用 ... 186

参考文献 ... 187

第十章 卟啉自组装膜概述 ... 194

第一节 卟啉自组装膜的制备方法 ... 194
1. 直接法 ... 194
2. 轴向配位法 ... 196
3. 其他方法 ... 199

第二节 卟啉自组装膜的电化学表征技术 ... 199
1. 循环伏安法 ... 200
2. 电化学交流阻抗法 ... 200
3. 扫描电化学显微镜法 ... 200

第三节 卟啉自组装膜研究概述 ... 202
1. 卟啉自组装膜成膜过程动力学研究 ... 202
2. 卟啉自组装膜长程电子转移研究 ... 203

参考文献 ... 204

第十一章 卟啉自组装膜电子传递性质研究 ... 212

第一节 卟啉自组装膜与分子信息存储 ... 212
1. 卟啉自组装膜分子信息存储的装置设计 ... 212
2. 卟啉自组装膜分子信息存储的影响因素 ... 213
3. 卟啉自组装膜分子信息存储研究现状 ... 217

第二节 卟啉自组装膜电子性质研究 ... 220
1. 卟啉自组装膜界面电子转移研究 ... 220
2. 金属卟啉自组装膜对分子氧催化还原的电化学行为研究 ... 222
3. 卟啉自组装膜与分子识别 ... 222

参考文献 ... 223

第十二章 卟啉光诱导电子转移和能量传递以及卟啉自组装膜光电转换研究 ········· 230

第一节 卟啉光诱导电子转移和能量传递 ·· 230
1. 金属卟啉及其配合物的光诱导性质 ·· 230
2. 周边基团修饰卟啉衍生物的光诱导性质 ·· 230
3. 卟啉和富勒烯组装体的光诱导性质 ·· 231
4. 卟啉自组装体的光诱导性质 ··· 231
5. 卟啉—纳米薄膜层的光诱导性质 ··· 232

第二节 卟啉自组装膜光电转换研究 ··· 233
1. 卟啉自组装膜光电转换研究 ··· 233
2. 卟啉—C_{60}以及二茂铁—卟啉—C_{60}复合系统自组装膜光电转换研究 ········· 234
参考文献 ·· 238

第一章 自组装单分子膜

自组装单分子膜是超分子化学的一个重要分支,经过三十余年的发展,它已成为凝聚态物理、材料科学、合成化学、结构化学、微电子学及生物化学等研究领域的交叉性前沿课题,是近代化学发展的一个更高层次。

1946 年,Zisman 提出了在洁净的金属表面通过表面活性剂的吸附(即自组装)制备单分子膜的方法[1],这拉开了自组装单分子膜研究的序幕。然而,对自组装现象的研究真正兴起于 20 世纪 80 年代。1980 年,Sagiv[2] 报道了十八烷基三氯硅烷在硅片上形成的自组装膜。而后,在 1983 年 Nuzzo 和 Allara 通过从稀溶液中吸附二烷基二硫化物在金表面形成自组装单分子膜[3],成功地解决了制备自组装单分子膜的两个主要问题:一是避免了对水敏感的烷基三氯硅烷;二是使用了较理想的成膜表面——金表面。他们的工作无疑具有开创性的意义。从此,几种制备单分子膜的体系逐渐发展和完善起来[4]。在过去的三十年中,自组装膜在构造、表征、理论及应用研究无论在深度还是在广度方面均取得了长足的进步,受到研究者的普遍关注,对其的研究方兴未艾。

第一节 自组装单分子膜简介

组织有序、定向、密集、完好的自组装单分子膜可在分子水平上达到控制电极界面微结构的要求。与传统的修饰电极修饰方法(如共价键合法、吸附法、欠电位沉积法、聚合物薄膜法以及气相沉积法等)相比,自组装膜具有更好的稳定性。特别是它所特有的明晰的微结构,为电化学研究提供了一个重要的实验平台,借以可探索电极表面分子微结构和宏观电化学响应之间的关系,并可以研究表面功能基团的反应性和分子识别问题。可以认为,自组装膜化学修饰电极是单分子膜化学修饰电极发展的最高形式,是研究有关表面和界面各种复杂现象的理想模型体系。

1. 自组装膜的类型

常见自组装膜的类型有有机硫化物在金属表面形成的自组装膜、表面硅烷化形成的有机硅烷自组装膜以及脂肪酸在金属氧化物表面强吸附形成的自组装膜等。

1.1 有机硫化物在金属表面的自组装膜

硫化物在金属表面形成的自组装膜是目前研究得最广泛、最深入的一类。除硫醇类化合物外,硫醚、双硫化合物、苯硫酚、巯基吡啶、原磺酸盐等(图 1-1)均可在金属表面形成自组装膜。适合于含硫化合物形成自组装膜的基底材料除单晶或多晶的金以外,还包括银、铜、铂、汞、铁以及胶体金微粒等[5]。

一般认为硫醇分子在金属表面的结合是 S—H 键断裂并伴随着氢分子的生成:

$$R—S—H + Au_n^0 \rightarrow R—S—Au_n^0 + 1/2 H_2$$

在完全无氧的气相条件下形成分子自组装单层膜[6],是以上机理的证据之一。

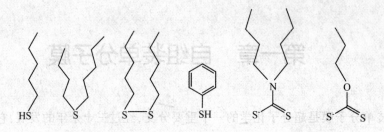

图1-1 几种可以用来分子自组装的硫化物的分子结构式示意图

1.2 有机硅烷在羟基化表面的自组装膜

有机硅烷衍生物在羟基化的表面形成的自组装膜是基于有机硅烷与基底表面的羟基结合,同时伴随着横向的 Si—O—Si 形式的交联(图1-2)。从1980年 Sagiv 报道 $C_{18}H_{37}SiCl_3$(OTS) 在玻璃片上形成自组装膜以来[2],可供利用的基底包括二氧化硅[7]、氧化铝[8]、石英[9]、云母[10]和氧化锗[11]等。相对来说,构造硅烷衍生物自组装膜有一定难度,主要存在三个方面的问题:一是完全无水和含水过多均会引起自组装膜的不完整性[12];二是自组装膜形成时的温度较难控制[13];三是基底的影响[14]。从这个意义上讲,无论从实验的重复性,还是膜的最终结构方面有机硅烷均不如其他自组装膜理想。

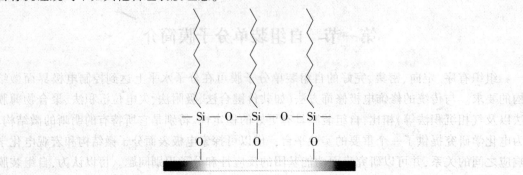

图1-2 有机硅烷衍生物自组装单层膜示意图(基底可以是 SiO_2、Al_2O_3、SnO_2、GeO_2、TiO_2 和 ZnSe 等)

1.3 脂肪羧酸在金属氧化物表面的自组装膜

长链的脂肪羧酸(C_nH_{2n+1}COOH)是典型的 L-B 膜成膜分子,但是它也可以在金属氧化物表面形成自组装膜,其动力来源于羧基阴离子与金属氧化物表面阳离子的成盐过程(图1-3)。Allara 和 Nuzzo[3]研究了直链脂肪酸在氧化铝表面形成的自组装膜;Pemberton 等[15]对上述化合物在氧化银表面上的组装结构进行了考察;Tao 等[16]利用接触角等方法详细考察了双炔酸化合物的自组装膜。

1.4 其他类型的自组装膜

硅表面的烷烃分子自组装单层膜是在硅表面通过形成 C—Si 键而形成的。Linford 和 Chidsey[17]利用 RC(O)O—O(O)CR 与硅表面[H—Si(111)或 H—Si(100)]的多步自由基反应,获得了较为致密的自组装膜。Si—R 与 Si—O(O)CR 的形成增加了成膜性,但其在热水中浸泡,表面分子浓度则有较大的下降,可能是 Si—O(O)CR 被还原脱落所致。用烯烃和[RC(O)O]$_2$ 的混合物进行反应也可以得到较致密、有序的分子单层膜,但其覆盖度只有90%左右,表明其表面仍有较多的缺陷[18]。同时椭圆偏振与红外的数据也证明,烷烃分子亚甲基链在硅表面垂直方向的夹角接近45°,分子轴向的偏转角为53°,可以认为是由于分子间距较大(0.665 nm)而造成的[19]。

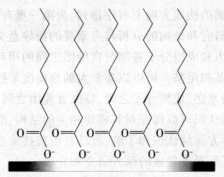

图1-3 羧酸及其衍生物的自组装单层膜示意图(基底可以是 Ag、Al$_2$O$_3$、AgO 和 CuO 等)

1.5 分子自组装单层膜表面的进一步功能化

功能基团化合物可以和自组装膜表面的尾基反应且不影响自组装膜本身的结构,而尾基的反应主要受膜的空间效应及静电作用影响,因此自组装膜上可以进一步连接更多样的功能分子从而有目的地裁剪表面的组成与特性,扩展了自组装膜的研究和应用范围。自组装膜表面上的酯基可以在亲核试剂作用下水解,转化为羧基或羟基[20]。

构成自组装膜的分子间可以发生聚合反应,聚合后的自组装膜会增加膜的稳定性。减小受热脱附,并能抑制自组装膜与溶液中硫醇分子的交换反应。自组装膜的聚合包括末端乙烯基团的交联[21],末端吡咯基团的电聚合[22]等。

通过配对键合反应也可以在自组装膜表面共价连接新功能化合物。最常用的共价转换反应是在偶联试剂碳化二亚胺(EDC、DCC 等)作用下,形成酰胺键和酯键。大多数蛋白质、—COOH 基或—NH$_2$ 基化合物发生配对反应而键合到电极表面[23]或在戊二醛作用下发生胺基缩合反应被固定[24]。利用生物分子的特异性相互作用也可以较稳定地键合,如生物素—亲合素间的识别反应[25],NADH 与其对应的酶分子间的重组[26],自组装膜上的 DNA 单链与其互补链之间的杂交反应[27]等。

2. 自组装膜的制备

自组装膜是分子通过化学键相互作用自发吸附在液—固或气—固界面而形成的热力学稳定和能量最低的有序膜,常用的有机硫化合物多为烷基硫醇和二烷基二硫化物,它们在金表面形成自组装膜的机理相似,都可能是形成了金的一价硫醇盐[28]。RS—H、H—H、RS—Au 的键能分别为 360 J/mol、435 J/mol 和 170 J/mol,用电化学方法可以求得下面过程的反应热为 -23 J/mol,即硫醇分子与基底的结合是一个放热反应,吸附分子自发在基底表面形成紧密堆积的有序薄膜[29]。

$$RS—SR + 2Au_n^0 \longrightarrow 2RSH^-Au^+ \cdot Au_{n-1}^0$$

$$RS—H + Au_n^0 \longrightarrow RS^-Au^+ \cdot Au_{n-1}^0 + 1/2H_2$$

3. 自组装膜的结构

自组装膜属于高度有序的复杂组装体系,这种高度有序表现为有序的多重性。如图1-4

所示,硫醇自组装膜主要包括三部分:分子头基、烷基链和取代尾基。具体而言,化学吸附在基底表面上的头基在二维平面空间具有准晶格结构,为第一重有序;长链结构的自组装分子在轴方向通过烷基链间的范德华力相互作用有序排列,为第二重有序;镶嵌在烷基链内或其末端的特殊官能团在平面的法线方向上有序排列,为第三重有序。自组装膜有序结构的多重性的划分对组装过程的研究和结构的分析具有重要的指导意义,可以使我们更精确地研究膜内分子有序化的驱动力和成膜分子各部分官能团之间的相互作用。

设计成膜分子的头基和尾基,可以以非常大的自由度来控制自组装膜体系所涉及的主要相互作用:成膜分子与基底、成膜分子之间、特殊官能团之间、成膜分子与溶剂之间的相互作用等。这些相互作用的认识可以加深对其密切相关的结构、润湿性、黏结、润滑以及特殊电化学性质、光化学性质等方面的认识和了解,推动自组装技术在半导体、非线性光学材料、生物传感器以及分子、电子学器件领域的深入研究和发展。

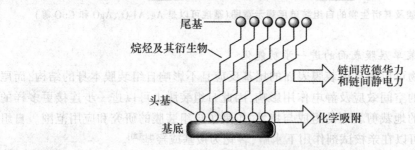

图 1-4　自组装单分子膜结构示意图

有序结构是自组装膜的重要特征,是对自组装膜研究的重要方向之一。现代物理和化学技术的研究均已提示了直链硫醇自组装膜中分子的高度有序排列[30-32]。在 Au(111) 面上,硫原子呈六方堆积,以 $(\sqrt{3}\times\sqrt{3})R30°$ 结构覆盖金表面。其中,相邻硫原子的间距为 0.497 nm(单个硫醇分子的占有面积为 0.214 nm^2),大于硫原子的范德华直径(0.185 nm),也大于最近的烷基间距(0.412 nm)。因此,相邻硫原子之间的相互作用比较小。聚亚甲基链的取向倾斜是为了增大自组装膜中范德华相互作用力。烷基硫醇单分子膜的结构要求每个硫原子的平均投射面积为 21.4 Å^2,而烷基链的所需面积为 18.4 Å^2。密度上的差异要求烷基链相对于金表面倾斜 30°,以使自由空间最小化和使聚亚甲基链间的范德华力最大[33]。

对硫醇分子进行设计,合成末端带有功能基团的硫醇分子,在表面设计领域具有重要的意义。烷基链末端可导入的官能团有—CH$_3$、—CF$_3$、—CH=CH$_2$、—C≡CH、—X、—CN、—OH、—OCH$_3$、—NH$_2$、—N(CH$_3$)$_2$、—SO$_3$H、—S(OCH$_3$)$_3$、—COOH、—COOCH$_3$、—COONH$_2$ 等。另一方面的工作是将特殊活性基团引入自组装膜中进行考察,如含电活性官能团的自组装膜已被广泛用于电子传递的动力学研究[34]。

4. 自组装膜的特点

从自组装膜的形成过程和结构可以看出它具备以下特点:

(1) 简便易得

当金或其他基底暴露在硫醇分子的溶液或气氛中时,自组装膜便可以自发地形成,它不需要特殊的环境和条件,也不需要特殊的仪器。在基底上沉积形成单分子膜的初期过程仅需几秒到几分钟的时间。而且,尽管无有机物的洁净金属是自组装膜的理想成膜表面,但硫对金属基底的强亲和力可以取代许多弱吸附的杂质。金属表面的曲率或可及性也不是影响自组装效率和性质的主要因素,其尺寸和形状可以从宏观到亚微观,在光滑到多孔的范围内变化。

(2) 取向有序

自组装膜是原位自发形成的,其热力学稳定,能量最低。与用分子子束外延生长(MBE)、化学气相沉积(CVD)等方法制备的超薄膜相比,自组装单分子膜具有较高的分子有序性和取向性,并具有高的密度堆积和低的缺陷浓度等优点。

(3) 稳定可靠

基于硫原子对金属的亲和力及所成键的强度,自组装膜可以在空气中长时间暴露而不损坏,可以被几乎所有的化学的或物理的表征方法进行结构和性质分析。

(4) 性质多样

无论在组装分子的合成中还是在自组装的过程中,自组装膜都有很大的灵活性和方便性。选择和修饰组装分子中的官能团的范围很广,它不会破坏自组装过程,也不会使自组装膜不稳定。自组装分子的末端或分子链间可以是多种基团,这有利于我们对其进行性质和应用研究。

(5) 预期结构

通过自组装膜技术的应用,可以达到人为设计分子和表面结构来获得制备预期物理和化学性质界面的目的。如自组装膜在组成上一致并且是密堆积的,单一的官能团暴露在外表面,这一特性可以研究表面组成对表面敏感性质的影响,如润湿、摩擦和吸附现象等。

5. 自组装膜的表征

自组装膜所具有的有序、密集和稳定等特点,有利于用近代物理和化学技术对其进行表征[35,36]。有关研究多集中在金表面,以硫醇类自组装膜体系为代表。物理法包括接触角、椭圆偏振、X 射线光电子能谱(XPS)、石英晶体微天平(QCM)、红外光谱(IR)、表面增强拉曼散射(SERS)、扫描隧道显微镜(STM)、原子力显微镜(AFM)、荧光光谱、表面等离子体共振(SPR)等方法。化学法中主要应用电化学的循环伏安法(CV)和交流阻抗法(EIS)等。应用电化学方法特别便于检测自组装膜的品质,还可以给出膜中缺陷的分布、膜的形成机理和过程动力学、接触基团的氧化还原行为,以及电子转移速率与距离的关系等。

第二节 自组装单分子膜修饰电极的电极过程动力学

单分子层覆盖于基底电极表面,使电极具有某种特定的功能和效应[37,38]。电极表面单分子层的量主要取决于分子的大小及其在电极表面上的取向,对于相对分子质量低的物质,其单层的覆盖量通常为 $10^{-9} \sim 10^{-10}$ mol/cm^2。

单分子层化学修饰电极源于吸附现象,它的理论也是基于吸附模型。Laviron、Hubbard 和 Anson 得到的单分子层化学修饰电极的伏安理论都是从吸附伏安理论中得到的。

在给定温度下,单位电极面积上物质 i 的吸附量 Γ_i 与物质 i 在本体溶液中的活度 a_i^b、体系的电状态 E 或 q^M 之间的关系由吸附等温式给出。它是根据在平衡时本体溶液中的和被吸

附的物质 i 的电化学势相等的条件下得到的,即

$$\bar{\mu}_i^a = \bar{\mu}_i^b \tag{1-1}$$

式中,上标 a 和 b 分别表示吸附和本体,因此,

$$\bar{\mu}_i^{0,a} + RT\ln\bar{\alpha}_i^a = \bar{\mu}_i^{0,b} + RT\ln\bar{\alpha}_i^b \tag{1-2}$$

式中,$\bar{\mu}_i^0$ 是标准电化学势。标准吸附的自由能 $\Delta\bar{G}^0$ 是电极电势的函数,它可定义为

$$\Delta\bar{G}^0 = \bar{\mu}_i^{0,a} - \bar{\mu}_i^{0,b} \tag{1-3}$$

于是

$$\alpha_i^a = \alpha_i^b \exp[-\Delta\bar{G}_i^0/(RT)] = \beta_i \alpha_i^b \tag{1-4}$$

其中

$$\beta_i = \exp[-\Delta\bar{G}_i^0/(RT)] \tag{1-5}$$

方程(1-4)是以 α_i^b 和 β_i 的函数 α_i^a 表示的等温式的一般表示形式。各种特殊的等温式起因于 α_i^a 和 Γ_i 之间关系的不同假设或不同模型。

1. Langmuir 等温吸附

Langmuir 等温式包括以下几点假设:

(a) 电极表面的吸附物质之间无相互作用;

(b) 表面均一;

(c) 本体活度高,电极上被吸附物的饱和覆盖量是 Γ_s(例如形成单分子层)。

于是

$$\Gamma_i/(\Gamma_s - \Gamma_i) = \beta_i \alpha_i^b \tag{1-6}$$

吸附等温式有时写成表面覆盖分数 $\theta = \Gamma_i/\Gamma_s$;这种形式的 Langmuir 等温式是

$$\theta/(1-\theta) = \beta_i \alpha_i^b \tag{1-7}$$

Langmuir 等温式可以把浓度系数包含在 β 项中,用溶液中物质 i 的浓度来描述:

$$\Gamma_i = \Gamma_s \beta_i c_i/(1+\beta_i c_i) \tag{1-8}$$

如果 i 和 j 两种物质竞争吸附,那么相应的 Langmuir 等温式为

$$\Gamma_i = \Gamma_{i,s} \beta_i c_i/(1+\beta_i c_i+\beta_j c_j) \tag{1-9}$$

$$\Gamma_j = \Gamma_{j,s} \beta_j c_j/(1+\beta_i c_i+\beta_j c_j) \tag{1-10}$$

式中,$\Gamma_{j,s}$ 和 $\Gamma_{i,s}$ 分别是 j、i 两种物质的饱和覆盖量。这些方程式可以从假设 θ_i 和 θ_j 是独立的覆盖分数,且每种物质的吸附速度正比于自由面积 $1-\theta_i-\theta_j$ 和溶液的浓度 c_i 和 c_j 的动力学模型导出,并假设各自的吸附速度正比于 θ_i 和 θ_j。

2. Frunkin 等温吸附

Langmuir 等温吸附是一种理想的情况,实际体系中吸附物质之间往往是相互作用的,这常常用 Frunkin 等温吸附理论来描述

$$\beta_i \alpha_i^b = \Gamma_i/(\Gamma_s - \Gamma_i)\exp[2g\Gamma_i/(RT)] \tag{1-11}$$

Frunkin 等温吸附式中包括参数 $g[(J/mol)/(mol/cm^2)]$,它表示提高物质覆盖量时物质 i

的吸附能改变的方式。若g为正,电极表面两物质之间是相互吸引的;若g为负,则电极表面两物质之间是相互排斥的。值得注意的是,当$g\rightarrow 0$时,Frunkin等温吸附式趋近于Langmuir等温式。这种等温式也可以写成下面的形式(β项包含活度系数)

$$\beta c_i=\theta/(1-\theta)\exp(g'\theta) \tag{1-12}$$

式中,$g'=2g\Gamma_s/(RT)$。g'的范围一般为$-2\leqslant g'\leqslant +2$,$g'$也可以是电势的函数。

3. Temkin 等温吸附

Temkin 等温吸附式描述了吸附物质之间的相互作用。Laviron 等人认为 Temkin 等温吸附式可以描述非均匀电极表面的吸附行为。Temkin 对数等温吸附式为

$$\Gamma_i=(RT/2g)\ln(\beta_i a_i^b) \quad (0.2<\theta<0.8) \tag{1-13}$$

这里,参数g具有$(J/mol)/(mol/cm^2)$的因次,它表示提高覆盖量时物质i的吸附能改变的方式。若g为正,表面上i物质之间是相互吸引的;若g为负,表面上i物质之间是相互排斥的,与Frunkin等温式中的g相似。

上述三种吸附等温式是自组装单分子膜化学修饰电极伏安理论的基础。

第三节 自组装成膜过程动力学研究

现代材料科学对异相环境中的各种应用提供了巧妙的解决策略,自组装代表了一种先进的表面功能化,并创造了良好可控制的、可剪裁的以及功能化的纳米结构。实际上,自组装在固体表面和延伸的二维分子层已得到广泛的研究和应用。然而,分子自组装在工程中的应用要求其结构更加先进、易控。最近,研究证明了分子自组装可以获得分子簇、单向性排列以及具有良好定义尺寸的孔穴和周期性的多孔网状表面结构[39-43]。这些研究也促进了另一个应用领域——分子电子学的研究和发展。

早在1965年,莫尔在其著名的莫尔定律中已预言以硅为基础的电路不可避免的终结,组合线路将被限制于几十个纳米,解决这一挑战的关键就是将功能分子结合于硅装置中[44]。今天,合成化学的发展已可以设计并合成预设结构的有机分子,然而,在可裁剪电子结构的单个分子的合成以及与其他器件连接定址的电子线路之间仍然存在差距。除了在微观世界中创造一个界面外,对分子电子器件一个最重要的挑战是分子构筑单元的精确定位,并最终形成功能性的结构。因此,对分子簇、一维分子线,甚至更为复杂的结构提供一种技术路线是分子电子器件研究和应用中最为重要的。相对于每个分子的分别定址,分子自组装对于复杂分子系统的大量制备,已经证实是一种最为先进的策略[45,46]。

为了利用分子自组装技术,需要精确理解自组装的机制,并可以在一个预设方式中制备良好定义的分子结构。由此,分子自组装在最近几十年得到了广泛研究,大多数研究是利用扫描探针显微镜(STM)在超真空(UHV)条件下,以单晶金属或半导体作为基底,考察分子自组装的结构及其影响因素。本节主要论述了有机分子在金属表面及UHV条件下自组装的基本原理,并以硫醇分子在单晶金表面自组装作为模型系统,证明了通过改变基底温度以及表面覆盖可以获得分子簇、准一维排列和二维重叠的分子结构。

1. 自组装成膜过程动力学研究概述

关于自组装成膜过程的研究方法详见表1-1。一般研究认为自组装膜的制备存在两步

表 1-1 自组装成膜过程研究方法

研究者	组装分子	研究方法	研究结论
Bain[47]	1-巯基癸烷	椭圆偏振光测量吸附层厚度,接触角测量湿润角	成膜过程分为两个阶段:第一个阶段是快速吸附过程,膜的厚度快速增长;第二个阶段是膜的重组过程,自组装膜缓慢重组
Dannenberger[48]	$C_5H_{11}SH$, $C_{23}H_{47}SH$	二次谐波技术	硫醇分子的浓度在 1~10 μmol/L 范围内其吸附过程可以被 Langmuir 吸附动力学过程描述
Hong[49]	$CH_3(CH_2)_nSH$ ($n = 4,6,8,10,12$)	循环伏安法	调查了浓度对自组装成膜过程的影响,证实了吸附分子的典型成膜过程包括两个步骤:快速吸附过程和随后的慢速重组过程;吸附过程可以被简单的 Langmuir 吸附动力学过程所描述
Bensebaa[50]	廿二硫醇	傅里叶红外技术	同[49]
De Bono[51]	烷基硫醇	表面等离子体共振	同[49]
Tamada[52]	$CH_3(CH_2)_nSH$ ($n = 8,10,12,16,18$)	原子力形貌图像	链长对吸附过程动力学存在影响,短链吸附分子在金表面的吸附灵活性更高,使其吸附速率常数要略大于长链化合物
Subramanian[53]	$C_{18}H_{37}SH$	电化学交流阻抗技术	考察了链长对吸附过程动力学的影响
Blanchard[54]	$C_8H_{17}SH$	石英晶体微天平	自组装成膜过程存在着吸附/脱附平衡
Bard[55]	$C_{16}H_{33}SH$	扫描电化学显微镜	考察了自组装过程针孔尺寸及分布的变化

动力学过程:第一步,即初始阶段,组装速率非常快,仅需几秒至几分钟的时间,自组装膜的接触角便接近其最大值,膜厚达到最大值时的 80%~90%。这一步可认为是受 Langmuir 扩散控制的吸附过程,组装速率强烈依赖于硫醇化合物的浓度。第二步的组装速率非常慢,在几小时乃至几天后,其接触角和膜厚才达到最大值,形成有序完好的自组装膜,这一步也称为表面结晶过程,它较为复杂,可以认为是表面重组过程。更准确地说,组装过程的第二步是硫醇分子的头基在基底表面上的物理吸附转变为化学吸附的过程,也是成膜分子由无序到有序、由单重有序到多重有序的复杂过程。第一步的动力学过程主要是硫醇分子与表面反应点的结合,其反应活化能可能依赖于吸附硫原子的电荷密度。第二步的动力学过程主要和分子链的无序性(比如非对称缺陷)、分子链之间的作用形式(范德华力相互作用,偶极子与偶极子之间的相互作用等)以及分子在基底表面的流动性等因素有关,最后形成了紧密堆积的全反构象的单分子层。另外,研究还发现,长链硫醇分子(链长 $n>9$)与短链硫醇分子(链长 $n<9$)

形成自组装膜的动力学有显著的区别,这可能与分子间范德华作用力的强弱对第二步动力学速率的影响有关。Ulman已经对自组装成膜的动力学过程进行了评述。

2. 分子自组装的基本原理

分子自组装被定义为分子在热力学平衡下的自发排列形成稳定的非共价聚集,这个定义包括了构筑单元间弱的相互作用以及键发生断裂直到形成稳定的平衡结构[56]。但许多实验研究表明系统不能达到平衡态,而是存在一个动力学限态。自组装被建立在这样一个动力限系统中,需要考虑分子流动和表面扩散的比率,如果流动高、扩散低,分子将不能达到其平衡结构,而是存在于一个扩散限态;如果流动低、扩散高,分子可以在表面自由移动,这将产生热力学平衡结构[45]。

自组装分子结构由分子间及分子—基底相互作用的平衡来控制,分子在表面的扩散可以通过不同的基底温度来控制。增加基底温度,热力学能转移给分子,由此提供了足够的动力学能E_{kin}去克服表面上的扩散能垒E_d,这是形成热力学平衡结构的先决条件。当然,分子的动力学能E_{kin}必须不超过分子在基底上的结合能E_b,否则,分子将从表面上脱附。另外,还要考虑分子间相互作用能E_{inter},这种作用能对于形成有序的结构非常关键,因为它包括了预设构筑单元排列的信息。分子间相互作用一般较弱,这对于研究自组装分子结构非常重要,因为如果分子间相互作用非常强,分子将不可避免地粘住,这将阻止有序平衡结构的形成。一般分子间相互作用能E_{inter}与分子的动力学能E_{kin}具有相同的数量级,稍微比E_{kin}大,完成分子自组装的能量条件为$E_b > E_{inter} \geq E_{kin} > E_d$。

研究表明,分子在表面既存在物理吸附,也存在化学吸附。当分子以化学吸附方式吸附于基底表面上时,通常的扩散能垒便非常大,因为化学吸附比物理吸附要强得多,化学吸附分子在基底表面有序的结构通常并不为分子间相互作用所控制,而是由较强的化学吸附所控制,这也是共价键分子通常并不认为是分子自组装结构,而被认为是"典型"的超结构的原因。然而,化学吸附的分子自组装也可能存在特殊情况,例如,当基底原子的移动足够高时,化学吸附的分子聚集于结合基底,并扩散扮演新的构筑单元,引起了母体基底以及大尺寸表面的重构,这对于分子吸附于金属表面是一个相当普遍的问题[57,58]。

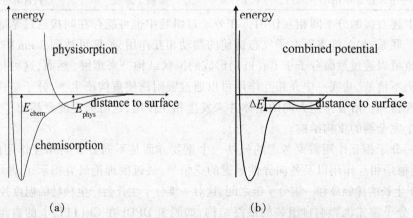

(a)表示距离超过0.3 nm时发生物理吸附,并伴随的弱的范德华作用力,其能量约为0.1 eV,而短距离的离子键或强的共价键发生的是化学吸附;(b)表示发生化学吸附时需要克服的能垒。

图1-5 物理吸附和化学吸附之间的能量图[59]

如图1-5所示,一个活化能垒ΔE存在于弱的物理吸附分子和化学吸附态之间,如果热力学能不足以克服物理吸附和化学吸附之间的能垒,分子将在表面物理吸附,并允许研究物理吸附对分子的自组装结构的影响。随着基底温度升高,热力学能克服了能垒,将发生表面上的化学吸附,从而再一次形成自组装结构[59]。

需要考虑活化能的还有脱氢作用,在最近的研究中,DPDI(4,9-二氨基萘醌-3,10-二酰亚胺)在570 K将发生分子脱氢现象,分子间相互作用(氢键)当脱氢后将处于开关开的状态[60]。

除了上面所讨论的能量因素,分子构建单元也可以裁剪去创造良好定义的分子结构和性质,如相互作用的强度以及方向性等。研究者最感兴趣的分子构建单元是生物分子,如氨基酸和DNA碱基对,后者在分子自组装中得到广泛研究归因于它们能够选择性地形成氢键,并伴随方向的良好控制[61]。

分子自组装的基本分子—分子相互作用类型见表1-2[59]。根据相互作用的强度、键长以及性质可以对不同相互作用类型进行分类。氢键作用在自组装中是非常重要的相互作用,它们既可以提供分子选择性,也可以提供良好的分子定位;偶极—偶极相互作用具有选择性,借此可以很好地控制自组装结构形式。Yokoyama等人[62]已对偶极—偶极相互作用对分子自组装的影响进行了报道。除了这两种键型,金属原子络合的相互作用也可以提供选择性和方向性。其他相互作用,如范德华力和静电相互作用,是非选择性的。尽管范德华相互作用相对较弱,但它们对拥有长的烷基链的有机分子的分子自组装也具有显著的影响。

表1-2 自组装分子间的相互作用类型、强度、键长以及性质

相互作用类型	强度/eV	键长/nm	性质
范德华相互作用	0~0.1	0.5~1	非选择性
氢键作用	0.1~0.5	0.20~0.35	选择性,方向性
静电作用	0.1~3	超过几个纳米	非选择性
偶极—偶极相互作用	0.1~0.5	0.2~0.3	方向性
金属配位键	1~3	0.2~0.3	选择性
基底—中介相互作用	0.1~1	可达到7纳米	摆动性
重构—中介相互作用	1	依赖于系统	共价作用

除了上述直接的分子间相互作用,在分子自组装中也可能存在间接通过基底中介相互作用的分子联系,如二维表面电子气态诱使的摆动相互作用,长度可达到7 nm[40,63]。基底—中介相斥也可以通过蒽醌分子于Cu(111)形成的网状结构[42]来理解。然而,这种排斥的详细机制目前仍不清楚。基底—中介相互作用可以通过吸附诱使重构产生[64],分子结构形式直接形成单向性排列。沿着排列方向的单体并未发生相互作用,这可能是能量超出了重构能,难以形成一个完全新的重构结构。

分子—分子相互作用需要考虑的另外一个重要方面是不同分子间相互作用的类型,如方向性、范德华相互作用以及各向异性氢键的竞争[65]。最近的理论研究揭示了动力学控制的转移存在于生长的接触岛和一维分子链之间,这对一维分子生长会产生独特的温度控制区域[65]。

最后,分子覆盖也影响自组装的最终结构,如脱氢DPDI在Cu(111)上的自组装揭示了三种清晰的依赖于覆盖的不同结构,在低温下,形成的疏状网络仅仅在最大覆盖0.7单层膜时稳定;在中间覆盖区域达0.85时,形成一个三聚单元延伸层的排列;当达到完全的单层膜覆盖时,分子在金属表面形成密集的排列[60]。

最近的研究也包括不同分子的共吸附或是连续吸附,共吸附将形成更加先进的结构,如主—客体分子形成的网状结构[66]。在萘基—四羧基—二酰亚胺(PTCDI)的氢键网状结构的主—客体系统中,以三聚氰胺作为连接体,C_{60}作为客体分子,这种共吸附系统的结构形式已经通过Monte Carlo模拟理论来说明,并依赖于特定相互作用能和分子覆盖,这种模型揭示了两种分子形成的是隔离的或是混合的结构[41,67]。

运用上述策略产生了分子自组装可控及预设的紧密性,最近的研究也证明了自组装结构形式上的灵活性。在分子自组装中,分子间相互作用主导了分子—基底相互作用,尤其是对于大的有机分子,已经识别不对称的或是所谓的准晶体取向附生的薄膜生长比晶体取向附生更有可能,当增加分子尺寸时,可以通过相似吸附位置能量的增加来理解,如萘基—四羧基—二酸酐(PTCDA)在Ag(111)上,可以形成良好有序、对称的重叠膜[68]。

除了基底表面的分子延伸层,通过控制分子间相互作用,如氢键[39,62,69]或是基底—中介相互作用[40,64],可用来观察准一维结构的分子链。在这些研究中,吸附的有机分子在最初发生了聚合反应,在表面上形成了良好定义的一维分子链。其他结构如分子簇、三聚、四聚、十聚、多孔网状以及更复杂的结构[39,41,42,62],包括有金属离子规则性的排列,通过精确地设计分子的构筑单元,在适当的位置保持分子的性质现在也已实现。

当前,由于金属基底能够运用各种技术进行表征,而且单晶金属表面代表了良好定义的基底,包括有机分子的大多数自组装实验已经在单晶金属表面运行,尤其是单晶金表面得到了广泛的应用,这主要是因为它们具有惰性,并且在UHV环境中容易保持清洁。除了Au,Ag[66,68,69]和Cu[40,42,70]在分子自组装研究中也作为标准基底。

3. 硫醇分子在不同单晶Au表面自组装的模型系统

为了进一步考察分子自组装的基本理论,研究者以天然氨基酸cys吸附于两种不同金表面进行了对比[59,71-73]。cys拥有—SH,可以与Au强烈地相互作用,而且cys拥有不对称C原子,具有两个对映体,可以研究其手性效应。上面所提及的不同自组装分子结构,从延伸的、二维重叠的、一维分子排列的、良好定义的分子簇到多孔结构,都可运用此系统进行说明。

研究者以两个不同的金基底,即Au(111)和Au(110)-(1×2),来说明分子—基底相互作用的影响。其中六方紧密堆积的Au(111)表面属惰性基底,呈$22×\sqrt{3}$重构结构,最终以其人字形外观来表征。在UHV条件下,金重构的(110)面进入所谓缺失排列的重构,每一紧密堆积在(110)方向上的排列是缺失的。对于Au(111),其表面原子是九配位的,Au(110)-(1×2)表面具有高的起皱现象,其表面具有7个配位原子,活性比Au(111)更高。当硫醇分子挥发至金的表面上时,很容易反映出表面结构和反应的差异。

3.1 Au(111)

Au(111)表面在分子自组装研究中最为广泛。在室温下,当cys挥发至Au(111)表面时,无序的cys岛将从人字形重构的台阶边缘处生长,增加覆盖,无序岛尺寸增大,并展示了二次对称的有序性。而且,在室温下,分子作用时人字形重构仍然保持,已知化学吸附强烈地影响人字形重构,这个结果表明此过程是非共价分子—基底相互作用。

在高的cys覆盖下,发生分子重叠现象,继续增加cys覆盖,将形成多层膜。当基底在380 K退火时,发生了分子岛的扰动,cys将从物理吸附转变为化学吸附,在脱氢—SH和表面之间形成了共价S—Au键。而且,在此温度下,高覆盖区域产生$(\sqrt{3}×\sqrt{3})R30°$重叠,结构并不优美,表明 cys 吸附在 Au(111)表面与原型系统的烷基硫醇相比,存在明显的差异。

3.2 Au(110)-(1×2)

相比较于 Au(111),cys 沉积于 Au(110)-(1×2)表面的分子重叠揭示了更多分子结构的信息,而且分子结构的变化也依赖于分子吸附态对诱使的表面重构和分子覆盖。cys 吸附于 Au(110)-(1×2)变化的过程和结构见图 1-6。

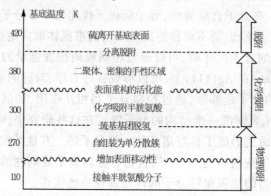

图 1-6 依赖于基底温度的 cys 在 Au(110)-(1×2)表面的吸附结构

由图 1-6 可知,在低温下,分子物理吸附于 Au(110)-(1×2)表面,并没有获得足够的移动去形成有序结构,当加热到 270 K 时,分子拥有足够的动力学能扩散形成单分散性的 cys 分子簇,进一步加热至室温,分子将由物理吸附向化学吸附转变,这可由分子簇的消失来证明。因为在化学吸附中分子强烈地与 Au 表面键合,有序结构的自组装本质上将被禁止,当加热至 340~380 K 时,分子形成了有序结构,并发生基底诱使的表面重构。加热提供了足够的能量使表面重构克服了能量壁垒,这对于 cys 分子化学吸附自组装为有序结构是必需的,进一步提高温度将导致分子脱附。cys 在 Au(110)-(1×2)表面相对简单的结构系统可以说明分子自组装对于加工可剪裁分子结构的潜在应用,因为其不同的分子结构可以通过温度和分子覆盖两个参数的变化来进行调整。

自组装是创造易控、可剪裁性质及表面功能化纳米结构的一种先进技术,理解分子自组装的基本原理,对于灵活控制分子合成以及自组装分子结构和性质的延伸具有重要意义。

分子自组装进一步的发展趋势包括组装的反应分子在自组装之前反应,这对于较大分子尤其重要,因为大分子一般很难加热挥发;其次,系统的设计也是非常重要的,它应用于自组装反应之后,因为自组装可以精确地定位分子形成一个预先设定的结构,随后的反应可以继续加工此结构获得更加稳定的、可以对抗苛刻环境的结构;最后,分子自组装在非导体基底上的延伸也是非常重要的,因为许多反应都局限于非导体表面,尤其是对于分子电子学研究,非导体基底是目前最佳的选择。

第四节 自组装膜长程电子转移研究

1. 自组装膜的电子转移理论

异相间的长程电子转移机制在化学和生物过程中具有重要意义,它是许多生命活动的基础步骤。研究者们一直努力寻求理想模型来研究电子转移的过程与机理。自从自组装膜出现和发展以后，这类结构上高度有序的单分子体系便很快被应用于电极表面电子转移方面的研究。从化学修饰电极方面考虑，利用高度有序、结构可控的自组装膜对电极表面进行修饰或直接设计具有电活性中心的自组装膜体系，为我们从新的角度研究界面电子转移现象及其机制提供了强有力的工具。与电极表面有序分子组装体系的电化学动力学理论的发展相适应，一系列关于电极表面的自组装膜中长程电子转移动力学的研究使我们对界面电子转移的一些基本问题有了更明晰的认识。而利用电化学活性的自组装膜进行这方面的研究有其独特的优点：它可以使我们在分子水平上设计和改变电子转移距离以及电活性中心所处的微环境。因此，关于自组装膜长程电子转移近年来已成为自组装膜领域一个非常重要的研究方向。自组装膜为考察界面电化学的一些基本问题，如电子转移速率与距离的关系、界面结构、微环境的影响等提供了较为理想的模型。

有关电化学电极表面长程电子转移的理论可以追溯到20世纪初。其发展过程详见表1-3。

表1-3 长程电子转移的理论发展简表

研究者	时间	相关理论或贡献
Randles[74,75]		Butler–Volmer 理论 *
Marcus[76-78]		Marcus 理论 **
Hubbard[79]	1973	分别从经典的电子转移理论出发，以 Langmuir 吸附等温式为基础，推导了通过峰分离得到表观电子转移速率的关系式
Anson[80]	1977	
Laviron[81]	1979	
Schmickler[82]	1986	以 Marcus 理论为基础提出了电子隧穿理论 ***
Ueyama[83,84]	1990	推导了单层 L-B 膜及多层 L-B 膜中的电子转移速率的理论求解过程
White[85]	1993	给出了自组装膜中界面电势分布和可逆循环伏安的模拟方法
Creager[86]	1993	对自组装膜中界面电势分布以及通过电势阶跃法求取电子转移速率的理论有所贡献
Bowden[87]	1994	以细胞色素 c 固定的硫醇自组装膜表面的体系为模板，推导了不可逆电活性中心在自组装膜中的电子转移理论处理
Creager[88]	1994	从发展 Laviron 的理论出发研究了自组装膜中电子转移速率与过电位的关系，给出了循环伏安的模拟公式
Murry[89]	1994	立足于数字积分给出了通过循环伏安分析得到自组装膜电子转移速率常数的工作曲线与经验方法

*Butler–Volmer 理论：对于发生在金属电极和溶液中氧化还原中心之间简单、可逆的电子转移反应，

$$O_x + e^- \underset{k_\eta}{\overset{k_c}{\rightleftarrows}} Red$$

Butler–Volmer 电极反应动力学的速率常数可以表达为：

$$K_{c,\eta}=k^{\theta}\exp[\alpha\eta/(k_BT)]$$
$$K_{a,\eta}=k^{\theta}\exp[-(1-\alpha)\eta/(k_BT)]$$

式中,k^{θ}为标准速率常数,α为电子转移系数,η为过电位,k_B为Boltzmann常数,T为绝对温度。根据公式,速率常数随过电位的增加而呈指数增加,因此Butler-Volmer理论不能解释在高过电位下实际观察到的速率常数的减小,也不能准确地显示出速率常数与温度的依赖关系。

**Marcus理论:Marcus由于对电子转移动力学理论作出的贡献,1992年获诺贝尔化学奖。下面方程给出了Marcus理论的简化速率公式:

$$K_{c,\eta}=k^{\theta}\exp[\alpha\eta/(k_BT)-\eta^2/(4\lambda k_BT)]$$
$$K_{a,\eta}=k^{\theta}\exp[-(1-\alpha)\eta/(k_BT)-\eta^2/(4\lambda k_BT)]$$

式中,λ为重组能,η为过电位,k_B为Boltzmann常数。根据公式,当η/λ远小于1时,Marcus关系等同于Butler-Volmer关系;当η接近λ时,速率常数并不随过电位的增加连续地呈指数增加,而在$\eta=\pm\lambda$时达到最大值;在较大η时,速率常数事实上是减小的,即所谓的经典Marcus翻转区。

***Schmickler认为氧化还原中心固定在电极表面上时,对电子转移影响的主要因素有:与氧化还原电对有关的电子给体与受体的能态密度;金属被占能级与未占能级的能态密度;隔开电极与氧化还原中心的能垒的高度、形状与厚度等。电子传递反应速率常数可由下式计算得到:

$$k_s=k\int\rho(E)n(E)Dox(E)p(E)d(E)$$

式中,E相对于金属费米能级,$\rho(E)$是金属电极上的电子态密度函数,$n(E)$是金属被占能级的费米分布函数,$Dox(E)$为电子受体的能态密度函数,$p(E)$是电子穿越能垒的隧穿几率,$d(E)$是电子隧穿能垒厚度函数。在弹性隧穿的条件下,可近似表示为:

$$p(E)=(E_B-E+e\eta/2)\exp(-\beta d)$$
$$\beta=(2(2m)^{1/2}/h)(E_B-E+e\eta/2)^{1/2}$$

其中,E_B是过电位为零时的隧穿势垒的平均高度,d为势垒厚度,β为隧穿系数。

2. 电子传递距离对电子转移速率的影响

利用自组装膜的分子设计和结构可控等特点,在自组装的过程中,使用一系列带有不同长度烷基链的硫醇分子,可以精确控制电极与活性基团之间的电子传输距离,进而考察电子转移速率对距离的依赖性,这可以使我们从实验层次上深入了解电子传输的机制。1984年Weaver[90]利用5种硫醚分子在金属金和汞表面进行组装,基于线性电位扫描方法求出表面吸附的Co(Ⅲ)在-300 mV条件下的电子转移的速率常数,首次在实验中报道了电子转移速率随电子转移距离呈线性下降关系,并得到了电子隧穿系数β=14.5 nm^{-1}。但由于他的体系在结构上没有得到明确的证实,电子转移的正确距离也便无法准确确定。因此,Weaver结果的定量意义便很难估计。在此之后,围绕自组装膜的界面电子传递与距离关系等问题的研究工作主要集中在两个方面:一是溶液中氧化还原活性中心与电极之间的跨膜电子传递;另一方面是将电化学活性氧化还原中心固定在自组装膜上,考察其与电极之间的跨膜电子转移。表1-4列出了部分氧化还原体系长程电子转移隧穿系数。

3. 膜表面分子的设计和状态对长程电子转移的影响

通过自组装膜表面分子基团的设计,考察影响异相电子转移的速率,从而进一步研究电子转移机制变化方面的工作也是自组装膜的重要研究领域。这方面不仅能直接体现界面结构和功能性的关系,同时也是自组装膜在分子识别、传感控制等方面得以应用的基础。较早的工作是Uosaki[103]报道的二茂铁自组装膜的电荷中介作用:自组装膜表面的二茂铁基团是溶液中Fe(Ⅲ)EDTA电子转移反应的良好中介。末端基团分别为—OH、—COOH和—NH$_2$的自组装膜对溶液中铁氰化钾与六氨合钌的氧化还原特性有直接的影响,并对带不同电荷氧化还原活性中心的法拉第过程表现出选择性影响[104]。对于溶液中取代二茂铁衍生物的异相电子转移过程也有同样的作用[105]。Bowden[106]用交流阻抗法考察了铁氰化钾在硫醇自组装膜表面的电化学过程,并以一个同时考虑跨膜隧穿与缺陷反应的模型进行了解释。另外的有意义的工作是通过改变溶液的 pH,调节自组装膜表面基团的状态,进一步考察影响异相电子转移过程的研究。如 Crooks[107]在 1990 年研究了巯基苯胺自组装膜在不同 pH 下所显示的对溶液中电活性分子的特征吸附行为;Dong[108]发现了双硫吡啶自组装膜的表面状态对于溶液中亚铁氰根离子氧化还原可逆性的直接影响。Nakashima 等[109]则报道了羧基末端的自组装膜对溶液中 pH 值相关电活性物质所起的"界面质子库"作用等,从本质上通过控制膜的表面状态达到了控制整个分子组装体系电化学行为的目的。

表 1–4　几种氧化还原体系长程电子转移隧穿系数

研究者	研究体系	隧穿系数(β)	主要贡献
Miller[91-94]	$Fe(CN)_6^{3-/4-}$,$Ru(NH_3)_6^{3+/2+}$/$HO(CH_2)_nSH$/SAMs ($n=2\sim16$)	10.8 ± 2 nm^{-1}	基于隧穿效应和 Marcus 理论测算了重组能和频率因子,为 Marcus 理论预见翻转区的存在提供了新的证据,并且 SAMs 的电子隧穿势垒与电势基本无关
Miller[95]	$Fe(CN)_6^{3-/4-}$,$Ru(NH_3)_6^{3+/2+}$/$HO(CH_2)_nX(CH_2)_mSH$/SAMs	10.8 ± 2 nm^{-1}	从理论上考察了 X 基团的改变对于跨膜电子耦合的影响
Finklea[96]	联吡啶钌/硫醇衍生物/SAMs	10.6 ± 0.4 nm^{-1}	认为 SAMs 上电子转移是通过键耦合隧穿(through bond tunnelling)机制进行的
Willner[97]	紫精/SAMs 体系	0.06 nm^{-1}	膜上不同氧化还原活性基团的电子传递特征研究
Chidsey[98]	二茂铁/SAMs 体系	12.1 ± 0.5 nm^{-1}	
Takehara[99]	萘醌/SAMs 体系	4.3 nm^{-1}	
Mallouk[100]	二茂铁/多层双磷酸锆/SAMs 体系	4.3 nm^{-1}	膜层数的不同对电子传递的影响
Mclendon[101]	二茂铁单层 L–B 膜	9.6 nm^{-1}	L–B 膜长程电子传递研究
刘忠范[102]	偶氮苯/SAMs 体系	1.35 ± 0.2/CH$_2$	氧化还原活性基团在膜中不同位置及膜结构的不同对电子传递的影响

除了可以精确控制电活性中心与电极之间的距离之外,自组装膜赋予我们最大的功能是可以通过膜结构的设计和体系环境的改变来控制氧化还原中心所处的微环境,这对于深入探讨微环境因素与自组装膜功能的关系有直接的帮助。对膜结构的设计一般可以通过两种手段来实现:一是混合组装;二是后置换方法,这种方法在理论上可以用来考察电子转移速率与距离、介质和空间阻碍结构的关系。

在自组装膜用于研究简单氧化还原活性中心的电子转移方面比较成熟以后,研究重点逐渐转移到自组装膜对复杂分子以及生物蛋白分子的电子传递方面。由于蛋白等生物分子结构本身的复杂性,目前阶段的研究主要集中在一些结构比较简单和明确的分子如细胞色素 C、紫精及醌类衍生物等上。Niki 小组[110-112]利用多种方法研究了固定在分子末端为羧酸基团的硫醇自组装膜上细胞色素 C 与电极间的电子转移机理和混合自组装膜上细胞色素 C 的电化学行为。实验得到的电子转移速率常数比 Marcus 理论预计的要小,这归结于电子转移是通过键耦合机制进行的,其电子隧穿系数$\beta=8.21$ nm^{-1}。董绍俊研究小组在研究紫精自组装膜的双电子转移过程方面也进行了卓有成效的工作[113]。并进一步用紫精作为桥梁连接辣根过氧化物酶或 DNA 等蛋白分子来研究了酶的直接电子转移行为[113~115]。

从上述自组装膜在电化学电子转移方面的一些研究成果可以看出,自组装膜在研究简单氧化还原活性中心的电子转移方面有不可比拟的优势,研究相对比较透彻。但其在研究生物分子在生命过程中的电子转移方面还有许多问题需要解决,并且受制于其他学科关键技术和理论的突破。目前,该领域一方面在拓展新的研究体系,如从二维平面扩展到零维、一维和三维空间,或从简单分子扩展到复杂生物分子,在这方面它将与分子电子学、生物传感器等紧密联系,这将会对生命科学中的最基本问题的认识产生非常大的帮助;另一方面它将与纳米科技等领域进行交叉,构筑有机、无机复合体系,更进一步研究异相电子转移机理,并渴望在催化等领域有所作为。我们相信,随着自组装膜研究的更加完善和深入,尤其是自组装技术在不同领域的应用成果的涌现,自组装膜在电子转移方面的研究将会出现新的高峰。

4. 硫醇自组装膜末端基团对其电荷输运特性的影响

硫醇类分子在金属表面形成稳定且致密有序的自组装单层膜,这种膜的制备方法简单可控,因此被广泛应用于电化学以及分子器件等领域的研究中[116~119]。近来,硫醇分子自组装膜的电输运特性也受到很多研究工作者的关注,并开展了很多相关研究。许多研究已经表明:电子以隧穿的方式通过硫醇分子膜[120~124],且电荷在硫醇分子膜内存在两种输运机理——链内隧穿与链间隧穿[125~128]。在分子与外界接触相同的条件下,分子膜的导电能力受到组装分子构型与分子尺度等因素的影响,其中分子的化学构成决定了分子隧穿势垒的高度和隧穿衰减因子β的大小[129],分子链的长度则决定隧穿势垒的宽度[125,127,128,130]。此外,分子的电荷输运性能还受到分子与外界电极接触的影响[131,132],显然,不同性质的分子末端基团也将改变分子膜的导电能力。例如,Pflaum 等人利用扫描隧道显微镜研究了末端基团分别为—CF 和—CH 的烷烃硫醇分子,发现前者在低偏压电压小于 2 V 下其电导值基本不随偏压变化,高偏压下电导则随电压升高迅速上升,而后者却没有这种特性[133]。但是,由于 STM 针尖无法同分子膜表面的末端基团形成良好的接触,所以上述结果忽略了针尖与分子末端基团之间真空结的影响。因此,需要用更好的方法研究末端基团对分子输运特性的影响。

最近,研究者采用导电原子力显微镜(CAFM)实现导电探针与分子膜表面的直接接触,从而更为可靠地研究不同末端基团对分子自组装膜电输运特性的调控作用。研究者利用 CAFM 对三种具有不同末端基团(—CH、—OH、—COOH)的硫醇分子自组装单层膜进行了电输运的测量,发现不同末端基团自组装膜的电输运能力相差很大。在相同电压和链长的条件下,具有—CH$_3$末端的分子膜的电导最大,而以—COOH 为末端的分子膜的电导最小。此外,虽然末端基团对分子链间的隧穿过程影响不大,但却影响电流—电压(I—V)曲线的对称性。借助 X 射线光电子能谱分析发现:末端基团碳原子的结合能越高,其对应的电子局域化程度越高,使得电子通过分子末端进入分子链的势垒增大,从而减弱了分子膜的导电能力。

另外,研究也发现不同末端基团的硫醇分子膜有不同的表面电势,导致分子膜电流—电压特性曲线的零点产生偏离。结果表明可以通过改变分子的末端基团实现对分子自组装膜电输运特性的调控,该调控方法将有望在设计和构筑分子器件中得到应用。

第五节 自组装单分子膜的性质和电化学行为

1. 双电层结构和电容

自组装膜/电解液界面与常见的金属/电解液界面不太一样。通常描述金属/电解液界面的双层模型为 GCS(gouy-chapman-stern)模型[134],在长链烷基自组装膜修饰的电极/电解液界面(此时的自组装膜尾基是不可解离或不带电荷的)上,对 GCS 理论加以修正就能很好地用于该界面。如 GCS 模型的紧密层可看成一层介电常数小的、排列紧密的碳氢链层,我们就可以将自组装膜电极/电解液界面看成是两个串联的电容器。总电容(C_T)表达式为

$$1/C_T = 1/C_m + 1/C_d \tag{1-14}$$

式中,C_d 是扩散层电容;C_m 是自组装膜电容。上列方程中的 C_d 和 C_m 可以分别计算,C_m 符合 Helmholtz 模型:

$$C_m = \varepsilon \varepsilon_0 A / d \tag{1-15}$$

式中,ε 是自组装膜的净介电常数;ε_0 是真空介电常数;A 是电极面积;d 是膜厚度。而扩散层电容值可以直接从 Gouy-Chapman 理论公式求得。对于烷基硫醇 $HS(CH_2)_nCH_3(n>8)$ 自组装膜,当 $n>10$ 时,其界面电容倒数(C_T^{-1})与链长(n)呈线性关系[135]。并由该斜率求算的有效介电常数为 2.3(假设烷链倾角为 0°)或 2.6(假设烷链倾角为 30°),该值与纯脂肪烃的介电常数(2.0)相一致,表明长链硫醇自组装膜对水和离子的渗透性极低。

尾端为羟基的 $HS(CH_2)_nOH(n=2\sim16)$ 自组膜修饰金电极也呈现出线性的 C_T^{-1}—n 关系曲线。在链构象和倾角的通常假定下,由斜率求得的表观介电常数为 2.6,所得的介电常数同这类硫醇自组装膜本身具有较好的阻塞效应相一致。各类硫醇,如 $HS(CH_2)_9CH_3$,$HS(CH_2)_{10}OH$,$HS(CH_2)_{10}COOH$,$HS(CH_2)_{10}CN$ 和 $HS(CH_2)_7CF_3$ 自组装膜修饰电极的电容在 0.1 mol/L NaF 溶液中呈现出如下的变化次序[136]:—CF_3<—CH_3<—OH<—CN<—$COOH$。

为深入了解穿越分子单层电子转移动力学,必须知道在分子单层内和接触电解质内的电位差。通常,电位 φ 是相对于电解质内的某个远点定义的;φ_m 是金属表面电位,φ_s 是自组装膜外表面电位。由于 $C_T \approx C_m$ 且存在下述关系

$$(\varphi_m - \varphi_s)/\varphi_m = C_T/C_m \tag{1-16}$$

所以简单分析表明几乎所有电位降都发生在分子单层内,更复杂的分析需要扩散层内电位降的存在,该分析需要知道对应于零表面电位(E_{zsp})的施加电位。一般地,E_{zsp} 由扩散层电容对电极电位曲线的最小电容来决定。在没有特性阴离子吸附的空白电极上,$E_{zsp}=E_{pzc}$(即零电荷电位);烷基硫醇自组装膜的最小电容应该较空白金的值移向更负的施加电位;带负电性基团的自组装膜的最小电容应该较 $HS(CH_2)_nCH_3$ 的值移向更正的施加电位[137]。若自组装膜含有可通过氧化还原或酸碱反应改变的功能尾基,那么 φ_s 将对电极电位和电解液 pH 的变化十分敏感。White[138]和 Fawcett[139]等已对该情况进行了处理。

在实际的烷基硫醇自组装膜内存在着各种类型的缺陷,如针孔、畴边界等。通常认为膜的电容由两部分构成:(1)未被吸附分子占据的裸金面积,此处的微分电容值较大;(2)被自组装膜占据的那部分电极面积,此处的微分电容值较小。假设其覆盖度为 θ,则

$$C_T = C_{bare}(1-\theta) + C_{SAM}\theta \tag{1-17}$$

结合下面要提到的,用多种物理、化学方法对自组装膜缺陷结构的表征,可以计算膜电容的准确值。

2. 自组装膜的稳定性和致密度

自组装膜具有很高的稳定性,可以经受住很高的电场强度,故在较大电位范围内作线性电位扫描而不被破坏,它对于本体溶液的组成也表现出很高的稳定性,改变溶液的酸碱性,烷基硫醇自组装膜的结构几乎没有变化。但是在强碱性的溶液(0.5 mol/L KOH 溶液)中进行电位扫描(向负方向扫至-1.5 V vs Ag/AgCl)时硫基覆盖层呈定量的脱落[140]。

硫基的还原峰出现在$-0.7\sim-1.4$ V[141]。峰出现的位置与烷基链的长度有关并且峰面积为(90 ± 7) $\mu C/cm^2$。硫基在电极表面按照下列机理进行还原:

$$Au|S(CH_2)_nX + ne^- \Leftrightarrow Au^0 |+ ^-S(CH_2)_nX \tag{1-18}$$

但硫基的还原峰面积并不依赖于碳链长度。通过还原峰面积可以求出自组装膜的覆盖度。由该峰面积转换为单位面积上的摩尔分数为7.8×10^{-10}[Au(Ⅲ)]、7.7×10^{-10}[Ag(Ⅲ)]和9.8×10^{-10}(Hg)。该数值与硫醇在$(\sqrt{3}\times\sqrt{3})$R30°晶格上的理论吸附量[7.7×10^{-10}Au(Ⅲ),7.6×10^{-10}Ag(Ⅲ)]很一致。Buttery等[142]认为还原峰含有较大的充电电流成分,因此充电电流的基线应该校正。尽管所得的吸附量只是近似值,但该方法用于求算硫醇的表观覆盖度还是可行的。

还原溶出峰电位可以用来考察 Au—S 键的强度、键间相互作用、中间体的生成或弱吸附物的存在等[140]。溶出峰电位主要与烷基链长度有关,同时还与自组装膜沉积的基底有关。例如在多晶金电极上自组装膜的还原溶出峰电位要比 Au(Ⅲ)自组装膜的更负[143]。电化学研究表明基底的性质会影响自组装膜的电化学稳定性。另外溶液 pH、电位、扫描速度等都会影响自组装膜的还原脱附和氧化再吸附。当电极电位正向扫描时,由于已脱附的硫醇处在电极表面附近,所以会很容易地再吸附到电极表面。

3. 自组装膜的微结构和覆盖度

即使是在单晶电极表面,自组装膜的结构也不是完全理想的。单晶电极的本身结构缺陷(如台阶、空位、异质原子、晶界)能导致自组装膜排列不致密或出现针孔型缺陷。在该缺陷处,在电极表面存在三种对自组装膜电极过程影响较大的结构:

(1) 针孔型缺陷,在该缺陷处,在电极/电解液面上溶液一侧的溶剂分子、电解质离子及电活性探针分子能够自由扩散到达电极表面;

(2) 非针孔型缺陷,在此缺陷处溶液中的分子、离子等能够一定程度地较为接近电极表面,在较大的过电位下电子转移反应也能由隧穿效应发生;

(3) 排列完好的区域,在此区域处溶液中的离子、分子均不能通过,电子转移动力学基本上可以完全忽略不计。

4. 自组装膜的针孔型缺陷

紧密排列的自组装膜能阻碍多种电极过程的发生,如金属电极的氧化、金属离子的欠电位沉积、溶液中电活性物种的电子得失等。上述几种过程都需要溶液中的物质扩散到电极表面才能发生。针孔型缺陷存在与否可从自组装膜修饰电极的电化学行为是否与理想的阻化行为有偏离,或是通过能在该缺陷处发生的特殊电化学反应来判断。

判断自组装膜的针孔型缺陷,主要有四种电化学方法:

第一种方法是根据金的氧化物的还原。当长链烷基硫醇自组装膜覆盖于金电极表面时,金的氧化或还原都被强烈地抑制。通常在循环伏安实验中观察不到氧化峰电流,但是氧化电流随着扫描圈数的增加会缓慢变大;而由金的氧化物还原所产生的还原峰电流却能很容易地被观察到。很显然,由长链烷基硫醇紧密排成的自组装膜不但能抑制膜基底金的氧化而且能抑制硫原子的氧化。这种抑制作用可能来自自组装膜的憎水烷基链,它能有效地阻止水分子到达电极表面。可以认为,只有与水分子相接触的金电极表面才会发生金的氧化和还原。从金的还原峰的面积,可以估计出自组装膜的覆盖度,即针孔型缺陷所占的比率。

第二种方法是金属离子的欠电位沉积。金属的沉积包括体相沉积和欠电位沉积两种模式。体相沉积发生在该金属离子的热力学电位处,而欠电位沉积发生的电位较其热力学电位为正。研究者应用循环伏安法对铜的两种沉积模式进行了研究。铜离子($Cu^{2+/0}$)的标准电位约在 0.15 V(vs.AgCl),铜的欠电位沉积和溶出所出现的电位范围为 0.2~0.3 V(vs.Ag/AgCl)。铜的欠电位沉积和溶出峰的形状和位置与金电极的晶面取向有很大的关系。和金的氧化物还原一样,铜的欠电位沉积和溶出峰的面积,也与自组装膜上针孔型缺陷的面积成正比。

尽管铜的欠电位沉积能定量地给出有关自组装膜缺陷方面的信息,但这类实验通常需要结合 STM 表征来进行。Crooks[144]等率先开展了这方面的工作,结果表明铜在自组装膜上的欠电位沉积与在金属基底上的欠电位沉积并不完全一致,铜在自组装膜上的沉积呈簇状,且高度并不是预计的原子单层厚度(2.5 Å)而是 6 Å,簇状沉积物的直径约为 20 Å。这些簇状物所处的位置说明,针孔型缺陷可能是自组装膜的岛状生长过程中膜的边界不完整所致。研究者发现了较高浓度 Cu^{2+} 溶液中在金电极上铜的欠电位沉积的稳定行为,其沉积峰和溶出峰的电量与电极表面积有严格的定量关系[145],还可用于分子自组装膜缺陷的表征。比较自组装前后的欠电位沉积电量便可了解到针孔型缺陷所占比例:

$$1-\theta=Q_{SAM}^{UPD}/Q_{bare}^{UPD} \tag{1-19}$$

但是当缺陷面积很小时所产生的法拉第电流与充电电流相当条件下,这种方法不再有效。

第三种方法是氧化还原探针。通过检测自组装膜上的伏安行为,也能定性和定量地了解自组装膜表面的针孔型缺陷。这是由于在没有缺陷的自组装膜区域内,由碳氢原子构成的憎水烷基链能有效地阻止溶液中的水分子、离子等的渗透。溶液中存在的电活性物种只有当其到达电极表面时才能发生电子转移反应,但有些憎水分子在特定的重要条件下能有效地透过自组装膜。如烷基硫醇自组装膜能强烈地阻碍水溶液中的 Fe^{2+}/Fe^{3+} 离子的电子得失,然而乙腈溶液中的二茂铁探针分子却能轻易地透过自组装膜,到达电极表面而发生氧化还原反应[146]。总的来说,自组装膜电极上的法拉第电流响应包括针孔型缺陷部分、非针孔型缺陷部分和探针分子在膜外侧的隧道电流三个部分。若在伏安实验中能观察到比较明显的法拉第电流,且这些电流出现在所选探针的式电位±200 mV 范围内,可以初步判断该自组装膜结构中可能存在着针孔型缺陷。如果在这个电位范围内电流出现峰形或平台便可以肯定该膜结构中存在着针孔型缺陷。

第四种方法是交流阻抗法。它是目前研究自组装膜中针孔型缺陷最为有效和准确的方法。为了简便地分析所得的阻抗数据,通常的阻抗测试都在相同浓度的氧化还原探针的电解液中进行。所设置的工作电极电位是开路电位或探针分子式电位,用一个小幅的交流电压对体系进行扰动,所测量的信号是交流电流信号。通过改变电压的频率,可以在较大的频率范围内得到体系的阻抗谱，即全谱。由于所设扰动的过电位很小，整个体系基本上处于稳态，而且还可认为在如此小的过电位之下，氧化还原探针分子不会在非针孔缺陷处以较大的速度得失电子，克服了采用伏安法中根据金的氧化物还原法判断的不足[147]。自组装膜界面的阻抗可由一些相联的电路元件来表示，最常见的是 Randles 模型[147]，如图 1-7 所示。

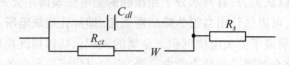

图 1-7 Randles 电路模型

自组装膜电极上所有的针孔型缺陷可看成是一个微电极阵列。当微电极阵列的表观活性面积占整个电极的 5% 以下时，微电极阵列上的扩散问题可作为一个偶联了前置和后继化学反应的电子转移反应[148]。将微电极阵列的相关参数代入耦合化学步骤的电子转移反应的阻抗表达式中，可以得到微阵列电极上的法拉第阻抗相关表达式。

在上面提到的金氧化物还原、铜欠电位沉积、伏安法和阻抗法四种方法中，伏安法和阻抗法相结合能够快速、准确地得到有关自组装膜针孔型缺陷参数的信息。通过研究电子转移速率较大的、具有外球氧化还原机理的探针分子或离子在自组装膜电极上的伏安行为，可以很快知道该自组装膜电极上是否存在针孔型缺陷。若有，则可用交流阻抗法进一步将缺陷加以量化，求得缺陷大小、间距和分布等参数。

5. 自组装膜表面的酸碱性

自组装膜的研究发展已使人们通过分子设计来控制二维体系的结构成为可能。研究者进一步通过对自组装膜存在环境的控制来调制膜的表面结构和性质。在这里，自组装膜表面的 pK_a 是一个关键因素，它在利用静电作用吸附蛋白质、DNA、聚电解质、金属离子及其他物种等方面具有重要作用。如果选择其末端可解离的分子制备自组装膜，则可以方便地通过调节溶液的 pH 来控制自组装膜表面的荷电情况。精确测定膜的表面解离常数（表面 pK_a）来控制自组装膜表面的荷电状况，在理论研究和实际中均占重要地位。目前已有不少方法被用于表面酸碱性的研究。Bain 和 Whitesides 利用接触角滴定研究直链羧基末端自组装膜表面酸碱性[149]，刘忠范等也曾利用接触角滴定研究巯基乙胺自组装膜表面的 pK_a 值[150]。Kaifer 等利用紫精—聚电解质体系作为探针研究了羧基末端自组装膜表面的 pK_a[151]。Brajter-Toth 小组[152]发展了电化学研究硫辛酸自组装膜体系，Bard[153]和刘忠范小组[154]分别提出了 AFM 的力曲线滴定法，研究了巯基丙酸和巯基十一酸自组装膜的表面 pK_a。最近，Marsh 等[155]采用化学力显微镜（CFM）研究了自组装膜所带磷酸基团离子化状态随 pH 及溶液中离子强度的变化，从而测定了自组装膜表面的 pK_a。该法在 CFM 的探头上化学修饰的功能基团，使探头与基底之间产生共价吸附，利用这种黏滞力研究基底纳米区内的相互作用，从而得到纳米区内的解离状态。此外，石英晶体微天平法、光谱方法等亦曾被尝试过测量表面 pK_a 值。一般认为组装在电极表面的分子表面的 pK_a（或 pK_b）值与其在溶液体系中的游离态的 pK_a（或 pK_b）值有一定的差别，而这种差别一般归结于表面分子氢键的形成。Aoki 通过计算模拟，也进一步证实了氢键对电极表面的 pK_a（或 pK_b）值的影响[156]。

用电化学方法（如伏安法、电化学阻抗法[157,158]）测定表面 pK_a 值，因其简单、方便的特点而具有更大的实用价值。基于 Brajter-Toth 工作原理,研究者发展了电化学测定表面 pK_a 的方法,包括循环伏安法[157,159]、扫描电化学显微镜法(SECM)[160]等。采用具有良好的电化学可逆性的 $Fe(CN)_6^{3-}$ 作为探针分子,可以观察到双硫吡啶自组装膜的表面状态对溶液中探针离子氧化还原可逆性的直接影响。

按照可解离物在溶液中 pK_a 的定义,自组装膜的表面 pK_a 可由末端基团的解离平衡定义。

$$AH \rightleftharpoons A^- + H^+$$

$$K_a = [A^-][H^+]/[AH] \tag{1-20}$$

式中,$[A^-]$ 和 $[AH]$ 分别是解离和非解离态部分的表面浓度(自组膜表面总浓度定义为1)。可以从式(1-21)得到

$$pK_a = pH + \lg[A^-]/[AH] \tag{1-21}$$

假设探针分子在自组膜表面总的表观电流密度(i)可以分解为两个独立部分的贡献和:即一部分贡献来自探针分子在解离的自组膜表面 $[A^-]$ 的电子转递 i_{A^-};另一部分贡献来自探针分子在非解离的自组膜表面 $[AH]$ 的电子转递 i_{AH},则总的电流密度 i 可以描述为

$$i = i_{A^-}[A^-] + i_{AH}[AH] \tag{1-22}$$

利用 $[A^-]+[AH]=1$ 和式(1-21)、式(1-22),可以得到以下表达式:

$$pK_a = pH + \lg[(i_{AH}-i_{A^-})/(i-i_{A^-})-1] \tag{1-23}$$

研究者以 $Fe(CN)_6^{3-}$ 为探针分子测定巯基丙酸单层膜的 pK_a,i_{A^-} 和 i_{AH} 可以由高于 6 和 pH 低于 4 时电流的平均结果得到。以相关的微分曲线的顶点可以确定表面 pK_a 值为 5.2±0.1。对于自组装膜结构较疏松的体系,自组装膜尾基的表面 pK_a 值大于其在溶液体系中的游离态的 pK_a,而且有较长分子链的巯基十一酸自组装膜,其表面 pK_a 值明显小于在溶液体系的 pK_a 值[159],并且随覆盖度的增加而降低。

研究者亦采用 SECM 法研究了 pH 对 4-氨基苯甲基酸修饰电极表面的羧基解离状态的影响,通过对该修饰膜电极表面在 $K_3Fe(CN)_6$ 溶液中的电流—距离($i-d$)曲线进行拟合,准确得到 $K_3Fe(CN)_6$ 在修饰电极表面的电子转移反应速率常数,从而得出反应速率常数与 pH 的依赖关系。测得 pK_a 值为 2.9,这与用其他方法测定的值相符合。

参考文献

[1] Bigelow W C, Pickett D L, Zisman W A. Oleophobic monolayers. I. Films adosrbed from solution in non-polar liquids. J. Colloid. Interface. Sci., 1946, 1: 513-538.

[2] Sagiv J. Formation and structure of oleophobic mixed monolayers on solid surfaces. J. Am. Chem. Soc., 1980, 102: 92-98.

[3] Nuzzo R G, Allara D L. Adsorption of bifunctional organic disulfides on gold surfaces. J. Am. Chem. Soc., 1983, 105: 4481-4483.

[4] Ulman A. Formation and structure of self-assembled monolayers. Chem. Rev., 1996, 96: 1533-1554.

[5] 董绍俊,车广礼,谢远武. 化学修饰电极. 北京:科学出版社, 2003.

[6] Poirier G E, Pylant E D. The self-assembly mechanism of alkanethiols on Au(111). Science, 1996, 272: 1145-1148.

[7] Wasserman S R, Tao Y T, Whitesides G M. Structure and reactivity of alkylsiloxane

monolayers formed by reaction of alkyltrichlorosilanes on silicon substrates. Langmuir, 1989, 5: 1074–1087.

[8] Gun J, Sagiv J. On the formation and structure of self-assembling monolayers: III. Time of formation, solvent retention, and release. J. Colloid Interface Sci., 1986, 112: 457–472.

[9] Brandriss S, Margel S. Synthesis and characterization of self-assembled hydrophobic monolayer coatings on silica colloids. Langmuir, 1993, 9: 1232–1240.

[10] Kessel C R, Granick S. Formation and characterization of a highly ordered and well-anchored alkylsilane monolayer on mica by self-assembly. Langmuir, 1991, 7: 532–538.

[11] Allara D L, Parikh A N, Rondelez F. Evidence for a unique chain organization in long chain silane monolayers deposited on two widely different solid substrates. Langmuir, 1995, 11: 2357–2360.

[12] Tripp C P, Hair M L. Reaction of methylsilanols with hydrated silica surfaces: the hydrolysis of trichloro-, dichloro-, and monochloromethylsilanes and the effects of curing. Langmuir, 1995, 11: 149–155.

[13] Gao W, Reven L. Solid-state NMR studies of self-assembled monolayers. Langmuir, 1995, 11:1860–1863.

[14] Ulman A. An introduction to ultrathin organic films. From L-B to SA. San Diego CA: Academic Press Inc., 1991.

[15] Thompson W R, Pemberton J E. Characterization of octadecylsilane and stearic acid layers on Al_2O_3 surfaces by raman spectroscopy. Langmuir, 1995, 11: 1720–1725.

[16] Huang D Y, Tao Y T. Self-assembled monolayer: behavior of diacetylenic amphiphiles. Bull. Inst. Chem. Acad. Sin., 1986, 33: 73–80.

[17] Linford M R, Chidsey C E D. Alkyl monolayers covalently bonded to silicon surfaces. J. Am. Chem. Soc., 1993, 115: 12631–12632.

[18] Yoneyama M, Fujii A, Maeda S. Oscillating luminescence in a ruthenium complex $R_u(bpy)_3^{2+}$ Langmuir monolayer on the Belousov-Zhabotinskii reactor. J. Am. Chem. Soc., 1993, 115: 11630–11631.

[19] Ulman A, Eilers J E, Tillman N. Packing and molecular orientation of alkanethiol monolayers on gold surfaces. Langmuir, 1989, 5: 1147–1152.

[20] Van Ryswyk H, Turtle E D, Watson-Clark R, et al. Reactivity of ester linkages and pentaammineruthenium (iii) at the monolayer assembly/solution interface. Langmuir, 1996, 12: 6143–6150.

[21] Peanasky J S, McCarley R L. Surface-confined monomers on electrode surfaces. 4. Electrochemical and spectroscopic characterization of undec-10-ene-1-hiol self-assembled monolayers on Au. Langmuir, 1998, 14: 113–123.

[22] Willicut R J, McCarley R L. Electrochemical polymerization of pyrrole-containing self-assembled alkanethiol monolayers on Au. J. Am. Chem. Soc., 1994, 116: 10823–10824.

[23] Ruzgas T, Gaigalas A, Gorton L. Diffusionless electron transfer of microperoxidase-11 on gold electrodes. J. Electroanal. Chem., 1999, 469: 123–131.

[24] Lixin Cao, Peisheng Yan, Kening Sun, et al. Gold 3D brush nanoelectrode ensembles with enlarged active area for the direct voltammetry of daunorubicin. Electroanalysis, 2009, 21

(10): 1183-1188

[25] Spinke J, Liley M, Schmitt F J, et al. Supra-biomolecular architectures at functionalized surfaces. J. Phys. Chem., 1993, 99: 7012-7019.

[26] Katz E, Heleg-Shabtai V, Willner B, et al. Electrical contact of redox enzymes with electrodes: Novel approaches for amperometric biosensors. Bioelectro. Bioenerg., 1997, 42: 95.

[27] Peterlinz K A, Georgiadis R M, Herne T M, et al. Observation of hybridization and dehybridization of thiol-tethered DNA using two-color surface plasmon resonance spectroscopy. J. Am. Chem. Soc., 1997, 119: 3401-3402.

[28] Rryce B M, Petty M C. Electrically conductive Langmuir-Blodgett films of charge-transfer materials. Nature, 1995, 374: 771-776.

[29] Schlenof J B, Li M, Ly H. Stability and self-exchange in alkanethiol monolayers. J. Am. Chem. Soc., 1995, 117: 12528-12536.

[30] Chidsey C E D, Liu G F, Rowntree Y P, et al. Molecular order at the surface of an organic monolayer studied by low energy helium diffraction. J. Chem. Phys., 1989, 91: 4421-4423.

[31] Alves C A, Smith E L, Poter M D. Atomic scale Imaging of alkanethiolate monolayers at gold surfaces with atomic force microscopy. J. Am. Chem. Soc., 1992, 114: 1222-1227.

[32] Dubois L H, Zegarski B R, Nuzzo R G. Molecular ordering of organosulfur compounds on Au(111) and Au(100)—Adsorption from solution and in ultrahigh-vacuum. J. Chem. Phys., 1993, 98: 678-688.

[33] Laibinis P E, Whitesides G M, Allara D L, et al. Comparison of the Structures and Wetting Properties of Self-assembled Monolayers of n-Alkanethiols on the Coinage Metal Surfaces, Cu, Ag, Au. J. Am. Chem. Soc., 1991, 113: 7152-7167.

[34] 李景虹. 自组装膜电化学. 北京: 高等教育出版社, 2003.

[35] Poirier G E. Characterization of organosulfur molecular monolayers on Au(111) using scanning tunneling microscopy. Chem. Rev., 1997, 97: 1117-1128.

[36] Kurth D G, Bein T. Thin films of (3-aminopropyl)triethoxysilane on aluminum oxide and gold substrates. Langmuir, 1995, 11 (8): 3061-3067.

[37] Lane R F, Hubbard A T. Electrochemistry of chemisorbed molecules. I. Reactants connected to electrodes through olefinic substituents. J. Phys. Chem., 1973, 77: 1401-1410.

[38] Mirkin C A, Ruther M A. Molecular electronics. Annu. Rev. Phys. Chem., 1992, 43: 719-754.

[39] B hringer M, Morgenstern K, Schneider W D, et al. Two-dimensional self-asselmbly of supramolecular clusters and chains. Phys Rev Lett., 1999, 83: 324-327.

[40] Lukas S, Witte G, Wöll C. Novel mechanism for molecular self-assembly on metal substrates: unidirectional rows of pentacene on Cu(110) produced by a substrate mediated repulsion. Phys Rev Lett., 2002, 88: 028301.

[41] Theobald J A, Oxtoby N S, Phillips M A, et al. Controlling molecular deposition and layer structure with supramolecular surface assemblies. Nature, 2003, 424: 1029-1031.

[42] Pawin G, Wong K L, Kwon K Y, et al. A homomolecular porous network at a Cu(111) surface. Science, 2006, 313: 961-962.

[43] Stepanow S, Lingenfelder M, Dmitriev A, et al. Steering molecular organization and

[44] Schulz M. The end of the road for silicon? Nature, 1999, 399: 729-730.

[45] Barth J V, Costantini G, Kern K. Engineering atomic and molecular nanostructures at surfaces. Nature, 2005, 437: 671-679.

[46] Barth J V. Molecular architectonic on metal surfaces. Annu Rev Phys Chem., 2007, 58: 375-407.

[47] Bain C D, Troughton E B, Tao Y T, et al. Formation of monolayer films by the spontaneous assembly of organic thiols from solution onto gold. J. Am. Chem. Soc., 1989, 111: 321-335.

[48] Dannenberger O, Buck M, Grunze M. Self-assembly of n-alkanethiols: A kinetic study by second harmonic generation. J. Phys. Chem. B., 1999, 103: 2202-2213.

[49] Hong H G, Park W. Study of adsorption kinetics and thermodynanics of ω-mercaptoalkylhydroquinone self-assembled monolayer on a gold electrode. Electrochimica Acta., 2005, 51: 579-587.

[50] Bensebaa F, Voicu R, Huron L, et al. Kinetics of formation of long-chain n-alka-ethiolate monolayers on polycrystalline gold. Langmuir, 1997, 13: 5335-5340.

[51] De Beno R F, Loucks G D, Manna D D, et al. Self-assembly of short and long-chain n-alkyl thiols onto gold surfaces: A real-time study using surface plasmon resonance techniques. Can. J. Chem., 1996, 74: 677-688.

[52] Tamada K, Hara M, Sasabe H, et al. Surface phase behavior of n-alkanethiol self-assembled monolayers adsorbed on Au (111): An atomic force microscope study. Langmuir, 1997, 13: 1558-1566.

[53] Subramanian R, Lakshminarayanan V. A study of kinetics of adsorption of alkanethiols on gold using electrochemical impedance spectroscopy. Electrochim. Acta., 2000, 45: 4501-4509.

[54] Karpovich D S, Blanchard G J. Direct measurement of the adsorption kinetics of alkanethiolate self-assembled monolayers on a microcrystalline gold surface. Langmuir, 1994, 10: 3315-3322.

[55] Forouzan F, Bard A J, Mirkin M V. Water molecules in short-and long-distance proton transfer steps of bacteriorhodopsin proton pumping. Israel Journal of Chemistry, 1997, 37: 155-161.

[56] Whitesides G M, Mathias J P, Seto C T. Molecular self-assembly and nanochemistry: a chemical strategy for the synthesis of nanostructures. Science, 1991, 254: 1312-1319.

[57] Maksymovych P, Sorescu D C, Yates J, et al. Gold adatom-mediated bonding in self-assembled short-chain alkanethiolate species on the Au (111) surface. Phys. Rev. Lett., 2006, 97: 146103-146104.

[58] Chen Q, Richardson N V. Surface facetting induced by adsorbates. Prog. Surf. Sci., 2003, 73: 59-77.

[59] Kühnle A. Self-assembly of organic molecules at metal surfaces. Current Opinionin Colloid & Interface Science, 2009, 14: 157-168.

[60] St hr M, Wahl M, Galka C H, et al. Controlling molecuar assembly in two dimensions:

concentration dependence of thermally induced 2D aggregation of molecules on a metal surface. Angew. Chem., 2005, 117: 7560-7564.

[61] Otero R, Schöck M, Molina L M, et al. Guanine quartet networks stabilized by cooperative hydrogen bonds. Angew. Chem. Int. Ed., 2005, 44: 2270-2275.

[62] Yokoyama T, Yokoyama S, Kamikado T, et al. Selective assembly on a surface of supramolecular aggregates with controlled size and shape. Nature, 2001, 413: 619-621.

[63] Repp J, Moresco F, Meyer G, et al. Substrate mediated long-range oscillatory interaction between adatoms: Cu/Cu(111). Phys. Rev. Lett., 2000, 85: 2981-2984.

[64] Kühnle A, Molina L M, Linderoth T R, et al. Growth of unidirectional molecular rows of cysteine on Au (110)-(1×2) driven by adsorbate-induced surface rearrangements. Phys. Rev. Lett., 2004, 93: 086101.

[65] Taylor J B, Beton P H. Kinetic instabilities in the growth of one dimensional moleeular networks. Phys. Rev. Lett., 2006, 97: 236102.

[66] De Wild M, Berner S, Suzuki H, et al. A novel route to molecular self-assembly. Self-intermixed monolayer phases. Chem. Phys. Chem., 2002, 3: 881-885.

[67] Zhdanov V P. Ostwald ripening of close-packed and honey-comb islands during coadsorption. Phys. Rev. B., 2007, 76: 033406.

[68] Eremtchenko M, Schaefer J A, Tautz F S. Unders tanding and tuning the epitaxy of large aromatic adsorbates by molecular design. Nature, 2003, 425: 602-605.

[69] Barth J V, Weckesser J, Cai C, et al. Building supramolecular nanostructures at surfaces by hydrogen bonding. Angew. Chem. Int. Ed., 2000, 39: 1230-1234.

[70] Chen Q, Frankel D J, Richardson N V. Self-assembly of adenine on Cu(110) surfaces. Langmuir, 2002, 18: 3219-3225.

[71] Kühnle A, Linderoth T R, Hammer B, et al. Chiral reeognition in dimerization of adsorbed cysteine observed by seanning tunneling microscopy. Nature, 2002, 415: 891-893.

[72] Kühnle A, Linderoth T R, Besenbacher F. Self-assembly of monodispersed, chiral nanoclusters of cysteine on the Au (110)-(1×2) surface. J. Am. Chem. Soc., 2003, 125: 14680-14681.

[73] Kühnle A, Linderoth T R, Besenbacher F. Enantiospecific adsorption of cysteine at chiral kink sites on Au(110)-(1×2). J. Am. Chem. Soc., 2006, 128: 1076-1077.

[74] Randles J E B. Kinetics of rapid electrode reactions. Part 2. Rate constants and activation energies of electrode reactions. Trans. Faraday Soc., 1952, 48: 828-832.

[75] Conway B E. Theory and Principles of Electrode Processes. New York: Ronald, 1965.

[76] Marcus R A. On the theory of oxidation-reduction reactions involving electron transfer. I.J. Chem. Phys., 1956, 24: 966-978.

[77] Marcus R A. On the theory of electron-transfer reactions. vi. Unified treatment of homogeneous and electrode reactions. J. Chem. Phys., 1965, 43: 679-701.

[78] Marcus R A. Electron transfer at electrodes and in solution: Comparison of theory and experiment. Electeochim. Acta., 1968, 13: 995-1004.

[79] Lane R F, Hubbard A T. Electrochemistry of chemisorbed molecules. i. reactants connected to electrodes through olefinic substituents. J. Phys. Chem., 1973, 77: 1401-1410.

[80] Brown A P, Anson F C. Cyclic and differential pulse voltammetric behaviour of reactants

confined to the electrode surface. Anal. Chem., 1977, 49: 1589-1595.

[81] Laviron E. General expression of the linear potential sweep voltammogram in the case of diffusionless electrochemical systems. J. Electroanal. Chem., 1979, 101: 19-28.

[82] Schmickler W. A theory of adiabatic electron-transfer reactions. J. Electroanal. Chem., 1986, 204: 31-43.

[83] Daifuku H, Aoki K, Tokuda K, et al. Electrode kinetics of surfactant polypyridine osmium and ruthenium complexes confined to tin oxide electrodes in a monomolecular layer by the Langmuir-Blodgett method. J. Electroanal. Chem., 1985, 183: 1-26.

[84] Ueyama S, Isoda S, Maeda M. Interlayer electron transfer in porphyrin Langmuir-Blodgett multilayer-modified gold electrodes. J. Electroanal. Chem., 1990, 293: 111-123.

[85] Creager S E, Weber K. On the interplay between interfacial potential distribution and electron-transfer kinetics in organized monolayers on electrodes. Langmuir, 1993, 9: 844-850.

[86] Smith C P, White H S. Voltammetry of molecular films containing acid/ base groups. Langmuir, 1993, 9: 1-3.

[87] Nahir T M, Clark R A, Bowden E F. Linear-sweep voltammetry of irreversible electron transfer in surface-confined species using the marcus theory. Anal. Chem., 1994, 66: 2595-2598.

[88] Weber K, Creager S E. Voltammetry of redox-active groups irreversibly adsorbed onto electrodes. treatment using the marcus relation between rate and overpotential. Anal. Chem., 1994, 66: 3164-3172.

[89] Tender L, Carter M T, Murry R W. Cyclic voltammetric analysis of ferrocene alkanethiol monolayer electrode kinetics based on marcus theory. Anal. Chem., 1994, 66(19): 3173-3181.

[90] Li T T T, Weaver M J. Intramolecular electron transfer at metal surfaces. 4. Dependence of tunneling probability upon donor-acceptor separation distance. J. Am. Chem. Soc., 1984, 106: 6107-6108.

[91] Miller C, Cuendet P, Gratzel M. Adsorbed omega-hydroxy thiol monolayers on gold electrodes: evidence for electron tunneling to redox species in solution. J. Phys. Chem., 1991, 95: 877-886.

[92] Miller C, Gratzel M. Electrochemistry at omega-hydroxythiol coated electrodes. 2. Measurement of the density of electronic states distributions for several outer-sphere redox couples. J. Phys. Chem., 1991, 95: 5225-5233.

[93] Becka A M, Miller C. Electrochemistry at omega-hydroxy thiol coated electrodes. 3. Voltage independence of the electron tunneling barrier and measurements of redox kinetics at large overpotentials. J. Phys. Chem., 1992, 96: 2657-2668.

[94] Becka A M, Miller C. Electrochemistry at omega-hydroxy thiol coated electrodes. 4. Comparison of the double layer at omega-hydroxy thiol and alkanethiol monolayer coated Au electrodes. J. Phys. Chem., 1993, 97: 6233-6239.

[95] Cheng J, Saghi-Szabo G, Tossell J A, et al. Modulation of electronic coupling through self-assembled monolayers via internal chemical modification. J. Am. Chem. Soc., 1996, 118: 680-684.

[96] Finklea H O, Hanshew D D. Electron-transfer kinetics in organized thiol monolayers with attached pentaamine (pyridine)ruthenium redox centers. J. Am. Chem. Soc., 1992, 114:

3173-3181.

[97] Katz E, Itzhak N, Willner I. Electron transfer in self-assembled monolayers of N-methyl-N´-carboxy-alkyl-4,4´-bipyridinium linked to gold electrodes. Langmuir, 1993, 9: 1392-1396.

[98] Smalley J F, Feldberg S W, Chidsey C E D, et al. The kinetics of electron transfer through ferrocene-terminated alkanethiol monolayers on gold. J. Phys. Chem., 1995, 99: 13141-13149.

[99] Mukae F, Takemura H, Takehara K. Electrochemical behavior of the naphtoquinone anchored onto a gold electrode through the self-assembled monolayers of aminoalkanethiol. Bull. Chem. Soc. Jpn., 1996, 69: 2461-2464.

[100] Hong H G, Mallouk T E. Electrochemical measurements of electron transfer rates through zirconium 1,2-ethanediylbis (phosphonate) multilayer films on gold electrodes. Langmuir, 1991, 7: 2362-2369.

[101] Guo L H, Facci J S, McLendon G. Distance dependence of electron transfer rates in bilayers of a ferrocene Langmuir-Blodgett monolayer and a self-assembled monolayer on gold. J. Phys. Chem., 1995, 99: 8458-8461.

[102] Yu H Z, Shao H B, Luo Y, et al. Evaluation of the tunneling constant for long range electron transfer in azobenzene self-assembled monolayers on gold. Langmuir, 1997, 13: 5774-5776.

[103] Uosaki K, Sato Y, Kita H. A self-assembled monolayer of ferrocenylalkane thiols on gold as an electron mediator for the reduction of Fe(III)-EDTA in solution. Electrochimica Acta, 1991, 36: 1799-1801.

[104] Takehara K, Takemura H, Ide Y. Electrochemical studies of the terminally substituted alkanethiol monolayers formed on a gold electrode: Effects of the terminal group on the redox responses of $Fe(CN)_6^{3-}$, $Ru(NH_3)_6^{3+}$ and ferrocenyldimethanol. Electrochimica. Acta., 1994, 39: 817-822.

[105] Takehara K, Takemura H. Electrochemical behaviors of ferrocene derivatives at an electrode modified with terminally substituted alkanethiol monolayer assembly. Bull. Chem. Soc. Jpn., 1995, 68: 1289-1296.

[106] Nahir T M, Bowden E F. Impedance spectroscopy of electroinactive thiolate films adsorbed on gold. Electrochimica. Acta., 1994, 39: 2347-2352.

[107] Sun L, Johnson B, Wade T, et al. Selective electrostatic binding of ions by monolayers of mercaptan derivatives adsorbed to gold substrates. J. Phys. Chem., 1990, 94: 8869-8871.

[108] Xie Y W, Dong S J. Electrochemical behaviour of hexacyanoferrate (II) on a gold electrode modified with bis (4-pyridyl) disulfide. J. Chem. Soc. Faraday Trans., 1992, 88: 2697-2700.

[109] Kunitake M, Deguchi Y, Kawatana K, et al. Interfacial effect of self-assembled monolayers of a carboxylic acid terminated alkanethiol on a gold electrode. J. Chem. Soc. Chem. Comm., 1994, 563-564.

[110] Arnold S, Feng Z Q, Kakiuchi T, et al. Investigation of the electrode reaction of cytochrome c through mixed self-assembled monolayers of alkanethiols on gold(111) surfaces. J.

Electroanal. Chem., 1997, 438: 91-97.

[111] Imabayashi S, Mita T, Feng Z Q, et al. Redox reactions of cytochrome c on self-assembled monolayers of 4-mercaptobenzoic acid and 4-mercaptohydrocinnamic acid adsorbed Au(111) electrodes. Denki Kagaku, 1997, 65: 467-470.

[112] Feng Z Q, Imabayashi S, Kakiuchi T, et al. Electroreflectance spectroscopic study of the electron transfer rate of cytochrome c electrostatically immobilized on the ω-carboxyl alkanethiol monolayer modified gold electrode. J. Electroanal. Chem., 1995, 394: 149-154.

[113] Li J H, Cheng G J, Dong S J. Electrochemical study of the interfacial characteristics of redox-active viologen thiol self-assembled monolayers. Thin Solid Films, 1997, 293: 200-205.

[114] Li J H, Cheng G J, Dong S J. Electrochemical study of interactions between DNA and viologen-thiol self-assembled monolayers. Electroanalysis, 1997, 9: 834-837.

[115] Li J H, Yan J C, Deng Q, et al. Viologen-thiol self-assembled monolayers for immobilized horseradish-peroxidase at gold electrode surface. Electrochimica. Acta., 1997, 42: 961-967.

[116] Chen J, Reed M A, Rawlett A M, et al. Large on-off ratios and negative differential resistance in a molecular electronic device. Science, 1999, 286, 1550-1552.

[117] Sortino S, Petralia S, Conoci S, et al. Novel self-assembled monolayers of dipolar ruthenium (Ⅲ/Ⅱ) pentaammine (4,4-bipyridinium) complexes on ultrathin platinum films as redox molecular switches. J. Am. Chem. Soc., 2003, 125, 1122-1123.

[118] Li Z L, Wang C K, Li Y, et al. Influence of external voltage on electronic transport properties of molecular junctions: the nonlinear transport behavior. Chin. Phys., 2005, 14, 1036-1040.

[119] 王伟, 黄岚, 张宇, 等. 用分子自组装技术制备的单电子器件的Monte Carlo模拟. 物理学报, 2002, 51, 63-67.

[120] Slowinski K, Chamberlain R V, et al. Evidence of inefficient chain-to-chain coupling in electron tunneling through liquid alkanethiol monolayer films on mercury. J. Am. Chem. Soc., 1996, 118, 4709-4710.

[121] 李红海, 李英德, 王传奎. 分子和金表面相互作用的第一性原理研究. 物理学报, 2002, 51, 1239-1243.

[122] 陈永军, 赵汝光, 杨威生. 长链烷烃和醇在石墨表面吸附的扫描隧道显微镜研究. 物理学报, 2005, 54, 284-290.

[123] Cui X D, Primak A, Zarate X, et al. Changes in the electronic properties of a molecule when it is wired into a circuit. J. Phys. Chem. B, 2002, 106, 8609-8614.

[124] Wold D J, Frisbie C D. Formation of metal-molecule-metal tunnel junctions: Microcontacts to alkane thiol monolayers with a conducting AFM tip. J. Am. Chem. Soc., 2000, 122, 2970-2071.

[125] Cui X D, Zarate X, Tomfohr J, et al. Making electrical contacts to molecular monolayers. Nanotechnology, 2002, 13, 5-14.

[126] Slowinski K, Chamberlain R V, Miller C J, et al. Through-bond and chain-to-chain coupling. Two pathways in electron tunneling through liquid alkanethiol monolayers on mercury electrodes. J. Am. Chem. Soc., 1997, 119, 11910-11919.

[127] 胡海龙，张琨，王振兴，等. 自组装硫醇分子膜电输运特性的导电原子力显微镜研究. 物理学报, 2006, 55, 1430-1434.

[128] Wold D J, Frisbie C D. Fabrication and characterization of metal-molecule-metal junctions by conducting probe atomic force microscopy. J. Am. Chem. Soc., 2001, 123, 5549-5556.

[129] Wold D J, Haag R, Rampi M A, et al. Distance dependence of electron tunneling through self-assembled monolayers measured by conducting probe atomic force microscopy: unsaturated versus saturated molecular junctions. J. Phys. Chem. B, 2002, 106, 2813-2816.

[130] Holmlin R E, Haag R, Chabinyc M L, et al. Electron transport through thin organic films in metal-insulator-metal junctions based on self-assembled monolayers. J. Am. Chem. Soc., 2001, 123, 5075-5085.

[131] Beebe J M, Engelkes V B, Miller L L, et al. Contact resistance in metal-molecule-metal junctions based on aliphatic SAMs: Effects of surface linker and metal work function. J. Am. Chem. Soc., 2002, 124, 11268-11269.

[132] Engelkes V B, Beebe J M, Frisbie C D. Length-dependent transport in molecular junctions based on SAMs of alkanethiols and alkanedithiols: Effect of metal work function and applied bias on tunneling efficiency and contact resistance. J. Am. Chem. Soc., 2004, 126, 14287-14296.

[133] Pflaum J, Bracco G, Schreiber F, et al. Structure and electronic properties of CH_3- and CF_3-terminated alkanethiol monolayers on Au(111): a scanning tunneling microscopy, surface X-ray and helium scattering study. Surf. Sci., 2002, 498, 89-104.

[134] Bard A J, Faulkner L R. Electrochemical methods: fuandamentals and applications. New York: Wiley, 1980.

[135] Plant A L. Self-assembled phospholipid/alkanethiol biomimetic bilayers on gold. Langmuir, 1993, 9: 2764-2767.

[136] Chidsey C E D, Loiacono D N. Chemical functionality in self-assembled monolayers. Structural and electrochemical properties. Langmuir, 1990, 6: 682-691.

[137] Evans S D, Ulman A. Surface potential studies of alkyl-thiol monolayers adsorbed on gold. Chem. Phys. Lett., 1990, 170: 462-466.

[138] Smith C P, White H S. Voltammetry of molecular films containing a acid/base groups. Langmuir, 1993, 9: 1-3.

[139] Andreu R, Fawcett W R. Discreteness-of-charge effects at molecular films containing acid/base groups. J. Phys. Chem., 1994, 98: 12753-12758.

[140] Hatchett D W, Uibel R H, Stevenson K J, et al. Electrochemical measurement of the free energy of adsorption of n-alkanethiolates at Ag (111). J. Am. Chem. Soc., 1998, 120: 1062-1069.

[141] Widrig C A, Chung C, Porter M D. The electrochemical desorption of n-alkanethiol monolayers from polycrystalline Au and Ag electrodes. J. Electroanal. Chem., 1991, 310: 335-359.

[142] Schneider T W, Buttry D A. Electrochemical quartz crystal microbalance studies of adsorption and desorption of self-assembled monolayers of alkyl thiols on gold. J. Am. Chem. Soc., 1993, 115: 12391-12397.

[143] Schoenfisch M H, Pemberton J E. Air stability of alkanethiol self-assembled monolayers

on Ag and Au surfaces. J. Am. Chem. Soc., 1998, 120: 4502–4513.

[144] Sun L, Crooks R M. The first quantitative report of defect formation in self-assembled monolayers (using a combined underpotential deposition/STM counting approach). J. Electrochem. Soc., 1991, 138: L23–L25.

[145] Zhao J, Niu L, Dong S. Technique for determination of gold electrode. Area and its application of characterization in self-assembling process. Mol. Crys. Liq. Cryst., 1999, 337: 265–268.

[146] Finklea H O, Avery S, Lynch M, et al. Blocking oriented monolayers of alkyl mercaptans on gold electrodes. Langmuir, 1987, 3: 409–413.

[147] Finklea H O, Snider D A, Fedyk J, et al. Characterization of octadecanethiol-coated gold electrodes as microarray electrodes by cyclic voltammetry and ac impedance spectroscopy. Langmuir, 1993, 9: 3660–3667.

[148] Amotore C, Saveant J M, Tessier D. Charge transfer at partially blocked surfaces a model for the case of microscopic active and inactive sites. J. Electroanal. Chem., 1983, 147: 39–51.

[149] Bain C D, Whitesides G M. A study by contact angle of the acid-base behavior of monolayers containing omega-mercaptocarboxylic acids adsorbed on gold: an example of reactive spreading. Langmuir, 1989, 5: 1370–1378.

[150] 赵建伟, 于化忠, 王永强, 等. 巯基乙胺自组装膜表面润湿性的变化. 高等学校化学学报, 1998, 19: 464–469.

[151] Goldnez L A, Castro R, Kaifer A E. Adsorption of viologen-based polyelectrolytes on carboxylate-terminated self-assembled monolayers. Langmuir, 1996, 12: 5087–5092.

[152] Cheng Q, Brajter-Toth A. Permselectivity, sensitivity and amperometric pH sensing at thioctic acid monolayer microelectrodes. Anal. Chem., 1996, 68: 4180–4185.

[153] Hu K, Bard A J. Use of Atomic force microscopy for the study of surface acid?base properties of carboxylic acid-terminated self-assembled monolayers. Langmuir, 1997, 13: 5114–5119.

[154] He H X, Li C Z, Xu X J, et al. Force titration of self-assembled monolayers using chemical force micrroscopy. Acta. Phys. Chim., 1997, 13: 293–295.

[155] Marsh A, Wong M. Determination of the ionization state of 11-thioundecyl-1-phoshonic acid in self-assembled monolayers by chemical force microscopy. Anal. Chem., 2000, 72: 1973–1978.

[156] Aoki K, Kakiuchi T. pKa of an ω-carboxylalkanethiol self-assembled monolayer by interaction model. J. Electroanal. Chem., 1999, 478: 101–107.

[157] Xie Y, Dong S. Effect of pH on the electron transfer of cytochrome c on a gold electrode modified with bis(4-pyridyl)disulphide. Biosens. Bioenerg., 1992, 29: 71–79.

[158] Molinero V, Calvo E J. Electrostatic interactions at self-assembled molecular films of charged thiols on gold. J. Electroanal. Chem., 1998, 445: 17–25.

[159] 罗立强, 赵建伟, 杨秀荣, 等. 电流滴定法对自组装膜表面酸碱性的研究. 高等学校化学学报, 2000, 21: 380–382.

[160] Ye J, Liu J, Zhang Z, et al. Investigation of the model compounds at 4-aminobenzoic acid modified modified glassy carbon electrode by scanning electrochemical microscopy. J. Electroanal. Chem., 2001, 508: 123–128.

第二章 自组装膜研究方法

传统的超薄有机膜通常很柔软,一经暴露于真空中便倾向于升华,而且它们的热不稳定性及对辐射损伤的敏感性,都使得它们不适于用电子或离子散射技术进行结构表征。自组装膜是利用分子的活性基团与基底之间发生了化学或物理作用而形成的,其稳定性非常好,能禁受住高真空环境和光子及离子的"撞击"而不会引起结构上大的变化,所以几乎所有的物理方法和化学方法均可应用于自组装膜的表征。同时由于自组装膜高度有序和完美,它反过来又为各种表征技术提供了一个理想模型。

第一节 自组装膜电化学研究方法

电化学方法是以界面上电荷传递及相关的过程和现象为主要研究对象,是对带电相界面(尤其是电子导体—离子导体界面)的结构进行表征与性质研究的方法。电化学的各种方法已被广泛地应用于自组装膜的研究,可以给出自组装膜的界面结构和性质的直接信息。如电容法和阻抗法可以给出自组装膜的双电层结构,循环伏安法可获得自组装膜的表面覆盖情况。此外,研究自组装膜对溶液中电活性物质传递的阻碍作用及其对金属的防腐性,并通过研究含电活性基团自组装膜的电化学性质还可以完善和发展电子转移理论。

虽然自组装膜几乎可以完全覆盖基底电极的表面,但是一些针孔型缺陷总是存在,这些针孔可以引起溶液中的氧化还原活性分子与电极表面的直接接触。电化学方法研究自组装膜的另一个主要优点是可以现场给出自组装膜中缺陷大小及形态分布的最直接证据。因此,电化学方法对于监测单分子膜的性质就显得十分重要。除此之外,电化学方法还可以给出自组装膜的其他信息,如吸附组分的氧化还原性质、动力学特性和单分子膜的形成机理等。

1. 循环伏安法

循环伏安法是获得电化学反应定性信息应用最为广泛的方法。其功能在于能够快速提供氧化还原反应大量的热力学信息、异相电子转移反应的动力学信息以及偶联反应和吸附过程的相关信息。该方法能够快速确定电活性组分的氧化还原电位,方便评价介质对氧化还原过程的影响。

循环伏安法由静止电极上(在未搅动溶液中)等腰三角形波的线性电位扫描实现。根据获取信息的不同,可采用单循环或多循环。在电位扫描过程中,恒电位仪测定了由施加电位产生的电流,获得的电流—电位曲线图,即循环伏安图。循环伏安图是大量的物理和化学参数与时间关系的复杂函数。

对于完全可逆体系的理论循环伏安曲线。假定初始时溶液中仅有氧化性组分存在,初始半圈扫描选择负电位方向,初始电位选择为还原反应未发生的电位处,这样,当电位扫描到接近氧化还原电对的特征电位E^{θ}时阴极电流开始增加,一直到达到峰值。当电位在还原区内扫描到一定范围时,电位扫描方向反转。在反向扫描过程中,还原形R(在正扫时产生的并累积在电极表面附近)被再次氧化为O,并产生阳极峰。

循环伏安图中的特征峰是由于电极表面附近扩散层的形成引起的。此结果可以通过分析电位扫描过程中的浓度—距离分布曲线来理解。表面浓度的连续变化伴随着扩散层厚度的扩展(在与静止溶液中预测的情况一样),产生的电流峰恰好反映了浓度梯度随时间的连续变化。这样,峰电流的增加对应于扩散控制过程。而电流的降低表现为与$t^{1/2}$相关(与施加的电位无关)。由于此种原因,反向电流具有与正向电流相同的形状。当使用微电极时,传质过程受径向扩散(而不是线性扩散)控制,将获得S形循环伏安曲线。

1.1 循环伏安数据的解释

循环伏安图以若干重要的参数为特征。其中由 Nicholson 和 Shain[1]提出的两个峰电流和两个峰电位这四个参数是分析响应的基础。

1.1.1 可逆体系

25 ℃时,可逆电对的峰电流可由 Randles-Sevcik 方程描述,

$$i_p = (2.69 \times 10^5) n^{3/2} A c D^{1/2} v^{1/2} \tag{2-1}$$

式中,n 为电子转移数;A 为电极表面积,单位为 cm²;c 为浓度,单位为 mol/L;D 为扩散系数,单位为 cm²/s;v 为电位扫描速度,单位为 V/s。由上式可见,电流与浓度成正比,并随扫描速率的平方根的增加而增加。这种扫速关系表明电极反应是受物质传输控制的(半无限线性扩散)。对于简单的可逆电对,逆向与正向峰电流之比为1。峰电流之比受到氧化还原过程中伴随的化学反应的强烈影响。电流峰通常是由峰前的基线电流外推法进行测量的。

峰在电位轴上的位置(E_p)与氧化还原过程的形式电位相关。对于可逆体系,形式电位位于 E_{pa} 和 E_{pc} 两者之间:

$$E^{\theta} = (E_{pa} + E_{pc})/2 \tag{2-2}$$

峰电位之差(对于可逆体系)为,

$$\Delta E_p = E_{pa} - E_{pc} = (0.059/n) \text{ V} \tag{2-3}$$

这样,峰电位之差可用于确定电子转移数,并可作为判断能斯特行为的标准之一。由此可知,快速的单电子过程其 ΔE_p 为 59 mV 左右。阳极峰电位和阴极峰电位与扫描速率无关。可将半峰电位 $E_{p/2}$ 相关联,

$$E_{p/2} = E_{1/2} \pm (0.028/n) \text{ V} \tag{2-4}$$

+表示为还原过程,−表示为氧化过程。

对于多电子转移过程(可逆),如果每步电子转移的 E^{θ} 值是连续升高且彻底分开的,循环伏安图则由几个清晰的峰组成。例如,富勒烯 C_{60} 和 C_{70} 就符合此机理,可观察到6个相继的还原峰[2],如图 2-1 所示。

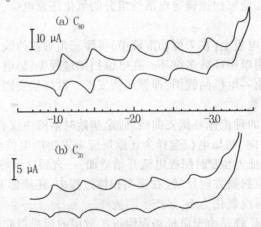

图 2-1 C_{60} 和 C_{70} 在乙腈—甲苯溶液中的循环伏安图[2]

如果氧化还原反应较慢或伴随有偶联的化学反应,情况会有较大的不同,这种情况称为"非理想"过程。这些非理想过程通常具有更重要的化学意义,也是最能展示循环伏安强大功能之处。这些机理信息通常是将实验循环伏安图与理论模拟的循环伏安图比较而获得的。适当补偿欧姆降对循环伏安的表征应用也是非常重要的准则之一。

1.1.2 不可逆及准可逆体系

对于不可逆过程(即电子转移过程是缓慢的),峰高度减小,峰—峰分离较宽。完全不可逆体系以峰电位随扫描速度变化而移动为特征。

$$E_p = E^\theta - (RT/\alpha n_a F)[0.78 - \ln(k^0/D^{1/2}) + \ln(\alpha n_a Fv/RT)^{1/2}] \tag{2-5}$$

式中,α 为传递系数;n_a 为电荷转移步骤中转移的电子数。当 E_p 出现在比 E^θ 高的电位处时,就有了与 k^0 和 α 相关的过电位。此峰位移与 k^0 值无关,可由适当地改变扫描速度进行适当补偿。此时的峰电位和半峰电位(25 ℃)之间相差 $48/(\alpha n)$ mV。此时的循环伏安图随 αn 值的减小而变得拉长,其峰电流可表示为:

$$i_p = (2.99 \times 10^5) n(\alpha n_a)^{1/2} A c D^{1/2} v^{1/2} \tag{2-6}$$

峰电流仍然正比于本体浓度,但是峰高降低(与 α 值有关)。假定 α 为 0.5,可逆与不可逆峰电流之比为 1.27(即不可逆过程的峰电流为可逆过程的 80%)。

对于准可逆体系(10^{-1} cm/s$>k^0>10^{-5}$ cm/s),峰电流受电荷传递和传质两者控制。循环伏安图的形状是 $k^0/(\pi\alpha D)^{1/2}$($\alpha = nFv/RT$)的函数。随着 $k^0/(\pi\alpha D)^{1/2}$ 的增加,趋近于可逆情况。而较小的 $k^0/(\pi\alpha D)^{1/2}$(如较快的扫速下),体系表现为不可逆体系的循环伏安图更易被拉长,且表现出比可逆体系更大的峰—峰电位差。

1.2 反应机理研究

循环伏安法最为重要的应用是定性表征伴随氧化还原反应的前行或后行化学反应,通常使用 E 和 C 分别表示氧化还原反应和前行或后行化学反应步骤。前行或后行化学反应的发生直接影响了电活性组分的表面浓度,并由于与电化学的反应物及产物之间形成竞争的化学反应,而引起循环伏安图形状的变化,对于获取反应历程和反应中间产物的化学信息非常重要。

例如,当一个氧化还原体系受到后行化学反应的微扰,即 EC 机理,

$$O + ne^- \rightleftharpoons R \rightarrow Z \tag{2-7}$$

循环伏安法给出一个小的逆向峰(由于产物 R 被后行化学反应从表面去除),峰电流比要小于单位值,峰电流比的精确值可用于估计化学反应的速率常数。在极端情况下,化学反应足够快,使电化学步骤中产生的 R 全部转化为 Z,而观察不到反向峰。

此外,有关偶联化学反应速率的一些信息也可通过改变扫描速度获得(即调整实验的时间窗坐标)。特别是扫描速度控制了转向电位与峰电位之间的时间间隔(为化学反应发生的时间段)。因此,正是化学反应的速率常数(化学步骤)与扫描速率之比控制了峰电流之比。当化学反应时间落入电化学实验的时间窗内时,就能获得重要的信息。对于扫速在 0.02~200 V/s 范围内(在常规电极上),可获得的时间窗为 0.1~1000 ms。超微电极可提供更快的扫描速率,由循环伏安法连续可测的速率常数的上限进一步提高[3]。例如,由电子转移产生的寿命期为 25 ns 的高活性组分,可通过 10^6 V/s 扫描速率下的循环伏安法测定。在一个较大范围内的快速反应(包括异构化和二聚合反应)都能够由循环伏安法检测。获取这样的信息通常需要扣除由于超快扫描速率引起的大的背景充电电流。

例如在抗坏血酸存在下多巴胺的氧化[4]。在氧化还原步骤中产生的多巴胺醌被抗坏血酸离子重新还原成多巴胺。在该催化反应中，峰电流之比为单位值。

其他类型的反应机理可通过相似方式进行解释。例如对CE机理而言，一个较慢的纵深反应发生在电子转移步骤之前(前行化学反应)，峰电流之比通常大于1并随着扫速的减小而趋近于1。反扫峰电流通常不受伴随的化学反应的影响，同时，正向峰电流不再与扫速的平方根成正比。

ECE过程是一个化学反应插入两个电化学反应之间的过程，由于可观察到两对峰各自分开，所以也可很容易地通过循环伏安法进行研究：

$$O+ne^- \rightleftharpoons R_1 \rightarrow O_2+ne^- \rightarrow R_2 \qquad (2-8)$$

化学反应的速率常数可由两个循环伏安峰相对大小进行估计。

许多阳极氧化反应都属于ECE机理。例如，神经传递子肾上腺素可以通过电子转移氧化为肾上腺素醌，肾上腺素醌经环化生成无色的肾上腺素红，无色的肾上腺素红经历一个快速电子转移反应生成甲肾上腺素红[5]。苯胺的电化学氧化是另一个经典的ECE机理的例子[6]。苯胺氧化生成的阳离子自由基经历一个快速的二聚化反应，生成易于氧化的对氨基联苯胺。另一个例子是活化的烯烃还原性偶联生成自由基阴离子，进一步与母体烯烃反应获得一还原性二聚体[7]。如果化学反应步骤非常快(与电子转移步骤相比)，体系的行为与EE机理相似(两步相继的电子转移过程)。表2-1总结了常见的化学偶联反应的电化学机理[8]。事实上，已经开发出用于揭示各种类型机理的功能强大的循环伏安计算机模拟软件[9]。这些模拟的循环伏安图能够与实验曲线进行比较与拟和。

表2-1 包含化学偶联反应的电化学机理

机理	反应式
1.可逆电子转移，无并行化学反应	$O+ne^- \rightleftharpoons R$
2.可逆电子转移伴随可逆化学反应——ErCr机理	$O+ne^- \rightleftharpoons R \quad R \underset{k_{-1}}{\overset{k_1}{\rightleftharpoons}} Z$
3.可逆电子转移伴随不可逆化学反应——ErCi机理	$O+ne^- \rightleftharpoons R \quad R \xrightarrow{k} Z$
4.可逆化学反应前期伴随可逆电子转移——CrEr机理	$Z \underset{k_{-1}}{\overset{k_1}{\rightleftharpoons}} O \quad O+ne^- \rightleftharpoons R$
5.可逆化学反应前期伴随不可逆电子转移——CrEi机理	$Z \xrightarrow{k_1} O \quad O+ne^- \rightleftharpoons R$
6.可逆电子转移伴随着初始物质的不可逆再生——催化机理	$O+ne^- \rightleftharpoons R \quad R+Z \xrightarrow{k} O$
7.不可逆电子转移过程伴随着初始物质的不可逆再生	$O+ne^- \rightarrow R \quad R+Z \xrightarrow{k} O$
8.化学反应中存在多电子转移过程——ECE机理	$O+n_1e^- \rightleftharpoons R \quad R \rightleftharpoons Y \quad Y+n_2e^- \xrightarrow{k} O$

1.3 吸附过程研究

循环伏安法也能够用于评价电活性化合物的界面行为。反应物和产物都可能包含吸附—解吸过程。在许多有机化合物和金属络合物的研究中(如果配体为特性吸附),经常会出现这样的界面行为。限定在表面上的非反应性组分和理想的能斯特行为给出的为一个对称的循环伏安峰形曲线($\Delta E=0$)。半峰宽为 90.6/nmV。峰电流正比于表面覆盖度(Γ)和电位扫描速率:

$$i_p=n^2F^2\Gamma Av/(4RT) \tag{2-9}$$

扩散控制组分的能斯特响应的峰电流与扫速为$v^{1/2}$的关系。实际上,对于没有分子间相互作用的电子转移很快的吸附层,在扫描速率很慢时接近理想状态。在饱和吸附状态下的峰面积(即在还原或吸附过程中消耗的电量)可用于计算表面覆盖度:

$$Q=nFA\Gamma \tag{2-10}$$

此值也可用于计算吸附分子所占据的电极表面积,预测分子在电极上的定向。表面覆盖度通常与本体浓度相关,并由等温吸附方程描述,其中最常见的是 Langmuir 等温吸附式:

$$\Gamma=\Gamma_m BC/(1+BC) \tag{2-11}$$

式中,Γ_m为表面浓度相应的单层覆盖度,单位为 mol/cm^2;B为吸附常数。当吸附物质浓度较低时,获得线性吸附等温曲线$\Gamma=\Gamma_m BC(1\gg BC)$。Langmuir 等温吸附曲线可应用于吸附质间无相互作用的情况。而其他的吸附曲线(如 Frumkin 或 Temkin 吸附曲线)将考虑相互作用。事实上,Langmuir 等温吸附曲线是 Frumkin 吸附曲线之间无相互作用条件下的一个特例。当反应物或产物之一被吸附时(但两者不同时发生),能够分别观察到前置峰或后置峰(在扩散控制的峰电位之前或之后)。

非理想情况包括准可逆和不可逆的电活性组分的吸附以及不同强度的反应物和产物吸附[10,11]。较快速吸附过程的速率常数可由超微电极上的高速循环伏安法测定[12]。

吸附动力学通常可由两个常规模型描述:一个是传质控制的快速吸附;另一个为吸附动力学控制的体系。在后者(Langmuir)条件下,被吸附物质在t时刻的表面覆盖度Γ_t可表示为:

$$\Gamma_t=\Gamma_e[1-\exp(-k'C_t t)] \tag{2-12}$$

式中,Γ_e为平衡表面覆盖度;k'为吸附速率常数。

定域在化学修饰电极表面上的氧化还原修饰物及导电聚合物的电化学行为与吸附过程相似,可由循环伏安法表征,并由循环伏安响应的变化常数来检测绝缘膜的封闭/阻碍行为(如自组装单层膜)。

1.4 定量应用

基于峰电流的测定,循环伏安法也应用于定量分析,但需要适当的方法确定其基线。对于相邻峰(混合物的),第二个峰的基线可由第一个峰的衰减电流(相应于$t^{-1/2}$)外推获得。背景响应电流主要与双电层充电电流和氧化还原的表面过程相关,使检测下限限制在 1×10^{-5} mol/L 水平上。扣除背景的循环伏安法可用于测定较低浓度的物质。特别是在碳纤维的微电极上,快扫(500~1000 V/s)扣除背景的循环伏安法越来越多地应用于人脑中活体神经传递子(如多巴胺、复合胺)的检测[13]。将数值扣除背景和快速循环伏安法的结合提供了一个具有毫秒级时间分辨的检测方法,用于监测脑的细胞间液中毫摩尔水平的物质动态浓度变化。这种具有较好时间分辨和化学物质识别的循环伏安法增进了人们对脑化学的了解。这些连续扫描循环伏安法获得了大量的数据,最好使用三维(电位、电流、时间)色阶图像来描述[14]。例如,电刺激伴随着多巴胺的瞬间释放就是从峰电位附近色阶的快速增加中获得证据的。超快扫描也能消除吸附过程以及与儿茶酚神经传递素氧化反应偶联的化学反应的干扰[15]。

对于循环伏安理论的详细信息以及循环伏安曲线的解释,可参考相关文献[1,7,16,17]。

2. 交流阻抗法

阻抗谱是一个表征化学修饰电极特性和了解化学反应速率的有效技术[18]。当电流流经一个由电阻、电容和电感组成的电路时,阻抗表现为一个复数电阻。在电极—溶液界面上发生的电化学反应可用与实验阻抗谱相应的等效电路元件模型描述。描述界面现象使用Randles 和 Ershler 等效电路模型(图 2-2)。该模型是由双电层电容C_d、电解质溶液电阻R_s、电子传递R_p及离子从本体溶液向电极表面扩散引起的 Warburg 阻抗W组成。根据欧姆定律,界面阻抗可由两部分组成,即一个实数部分Z'和一个虚数部分Z''。

$$Z(\omega)=R_s+R_p/(1+\omega^2 R_p^2 C_d^2)-j\omega R_p^2 C_d/(1+\omega^2 R_p^2 C_d^2)=Z'+jZ'' \qquad (2-13)$$

式中,$j=(-1)^{1/2}$。交流阻抗谱是将一小幅度的正弦电压信号(频率为ω)加到电解池上,测量电流的效应。获得的法拉第交流阻抗谱称为 Nyquist 图,相应于阻抗的虚数部分与实数部分的相互关系(图 2-2),其中包含丰富的荷电界面和电子转移反应的信息。Nyquist 图通常是由一个与坐标轴相交的半圆和其后伴随的直线部分组成。半圆部分(在高频区)相应于电子传递限制过程,而直线部分(低频区)代表了扩散限制过程。这样的谱图能够用于获得电子转移动力学和扩散特征参数。在较快的电子转移过程中,交流阻抗谱仅包含直线部分,而较慢的电子转移过程则以一个大的半圆区为特征。半圆的直径等于电子转移电阻。半圆在实轴(Z_{re})上的截距相应于溶液电阻(R_s)。交流阻抗谱除了应用于基础的电化学研究外,对生物亲和反应的研究也是非常有用的,如现代电化学免疫传感器及 DNA 生物传感器[19]。这种生物亲和性信号转换成阻抗是以含有铁氰化钾的氧化还原体系溶液为参考,由于大的生物分子的键合(如抗原的捕获将阻碍电子转移),电极表面绝缘性将增加。

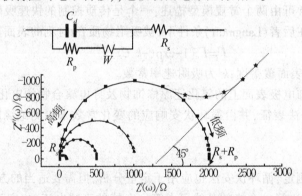

图 2-2　法拉第交流阻抗谱的 Nyquist 图以及荷电界面的等效电路[18]

3. 光谱电化学方法

光谱与电化学方法相结合,称为光谱电化学法,自 20 世纪 80 年代初就已经广泛地应用于无机、有机及生物的氧化还原体系中[20,21]。这种电化学微扰与分子的特征光谱检测相结合成功地解决了与电流响应相对应的分子结构信息有限的问题,并且对于解释反应机理和描述动力学及热力学参数是非常重要的。

3.1 实验装置

光透电极(OTEs)允许光通过其界面以及相邻的溶液层,这是光谱电化学实验的关键。

其中一类光透电极是由一种金属微网(金、银、镍等)组成,中间含有 10~30 μm 的孔,这种电极具有很好的的透光率(>50%)和很好的导电性。微网通常与两个微光透片制成"夹心"式的薄层池,微腔内盛装含有电活性组分的溶液,并与一个插有参比电极和对电极的较大的容器相接触。OTEs 放置在分光光度的腔中,让光束垂直通过光透电极和溶液。这种薄层池的工作体积只有 30~50 μL,溶质的完全电解只需 30~60 s。另外,OTEs 也可由透明材料(石英片或玻璃片)镀有薄金属膜(如金、铂)(厚度约为 100~5000Å)或镀有半导体材料(如二氧化锡)的物质组成。其电导率和透光率常常取决于薄膜的厚度。

研究者也改进了电解池的设计,如使用光学纤维引入光源和进行电极表面附近光的收集[22]。长光程 OTEs 的制作也可通过在固体导电材料中钻一小孔,常用于弱的光吸收系数组分的灵敏度检测[23],也常常将具有网眼的开口的多孔材料(特别是网状玻璃碳材料)[24]用于弱的光吸收系数组分的灵敏度检测[23]。

3.2 原理与应用

光谱电化学的主要优势是电化学和光谱同时获得的信息可以相互认证。如氧化还原过程生成或消耗组分引起的光吸收的变化,光吸收值的变化与浓度及光程长度有关。分析电生成或消耗的光化学活性组分的吸光度的瞬间响应($A—t$曲线),能够获得有关反应机理和动力学的非常重要的信息。当反应物或产物的光谱差别很大时,这类实验特别有效。

例如,考虑一般性氧化还原反应过程:

$$O + ne^- \rightleftharpoons R \tag{2-14}$$

当施加在 OTEs 上的一个电位跃,使式(2-14)过程以扩散控制速率发生时,R 的吸光度随时间的变化可表示为:

$$A = 2C_O \varepsilon_R D_O^{1/2} t^{1/2} / \pi^{1/2} \tag{2-15}$$

式中,ε_R 为 R 的摩尔吸光系数;D_O 和 C_O 分别为 O 的扩散系数和浓度。这样,A 随时间的平方根($t^{1/2}$)线性增加,反映了 R 的增加取决于 O 向电极表面上扩散的速度。如果电生组分是稳定的,式(2-15)成立。而当 R 为短寿命的组分时(如 ES 机理),吸光度的响应将小于式(2-15)理论值。分解反应的速率常数可由吸光度的降低进行计算。许多其他类型的反应机理都可以相似的方式,通过分析对 $A—t$ 曲线[由方程式(2-15)计算]的偏离进行研究。这种电位跃实验称为计时吸收光谱法。

薄层光谱电化学方法对于测定氧化还原形式电位(E^θ)和电子转移数(n)特别有用。这种方法是通过测定在每一外加电位下氧化组分和还原组分的浓度比 ([O]/[R])(可从适当波长下吸光度之比获得)来完成。在薄层条件下,由于本体电解可在几秒内完成,整体溶液能够很快与外加电位达到平衡(根据能斯特方程)。

除电位跃实验外,在 OTEs 上也可使用线性电位扫描微扰[25]。这种伏安—吸收实验获得了一个与伏安实验相似的光学吸收结果。其 $dA/dE—E$ 曲线(通过反应产物的吸收对施加电位的微分)获得的吸收—电位微分曲线的形状与氧化还原过程的伏安曲线形状是一样的。根据被检测组分的摩尔吸收系数,微分光谱响应可能是一个比伏安法更灵敏的方法。此信号没有类似用于充电电流的背景电流干扰,更加有利于化学耦合反应的机理分析和动力学表征。

光谱电化学实验能够用于监测各种吸附—脱附过程。实际上,此过程中吸光度的变化可由电极表面积与溶液体积比较大的长光程 OTEs 进行检测[26]。此外,拉曼光谱电化学实验也能获得此类信息。

除透射实验外,OTEs 也可使用更灵敏的反射光谱方法。特别是内反射光谱(IRS)法。在此方法中,光束以一定角度入射到电极表面上,记录从固—液界面上反射光束的光谱。棱镜常用来控制辐射的入射和反射。该方法除了高灵敏度外,IRS 很少受到溶液阻抗的影响。

红外光谱电化学方法,特别是基于傅里叶变换(FTIR)的红外光谱电化学方法能够提供一些紫外—可见光谱不能提供的结构信息。FTIR 光谱电化学方法在表征电极表面发生的反应方面越来越引起研究者的关注。但是该技术需要使用很薄的电解池以克服溶液的吸收问题。

光谱电化学除了广泛地应用于研究氧化还原过程机理外,对于以分析为目的的研究也是非常重要的。在实践中,同时获得体系的光谱和电化学性质,增加了不同方法的总的灵敏度[27]和检测的总选择性[28]。通过巧妙地选择应用电位和波长,可以方便地实现两种方法选择性的结合。

4. 扫描电化学显微镜

4.1 实验装置

电化学研究中特别引人注目的是扫描电化学显微镜(SECM)[29,30]。在 SECM 实验中,电极浸在含有电活性物质的溶液中,超微电极的针尖在靠近工作电极基底表面上移动时(由压电控制),记录针尖上的法拉第电流。针尖电流是电导率、基底的化学性质以及针尖—基底间距的函数。获得的图像为深入了解基底的电化学和化学活性的微观分布以及形貌提供了重要信息(图 2-3)。SECM 已经广泛地应用于不同领域的电化学体系中。

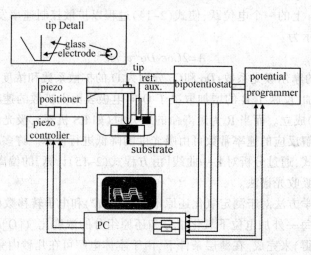

图 2-3　扫描电化学显微镜(SECM)实验基本装置[29]

4.2 原理与应用

最常见的 SECM 模式是反馈模式,它包括电活性物质在针尖和基底表面间的循环(图 2-4)。当微电极距离基底表面几倍于电极直径时,在针尖上获得一稳态电流 $i_{T,\infty}$[图 2-4(a)]。当针尖接近导电性基底时(施加有足够正的电位),针尖上生成的 R 被再次氧化成 O,针尖的电流将大于 $i_{T,\infty}$[图 2-4(b)]。相反地,当针尖扫过绝缘性基底区域时,向针尖方向的扩散被阻止,反馈电流减小[图 2-4(c)]。SECM 图像能够描绘出基底电极表面上电化学活性的分布,图像的精细程度由针尖的尺度和形状控制,并可通过图像的数字化处理进一步改善。然而,与STM或AFM不同,SECM不能获得原子水平上的分辨率。

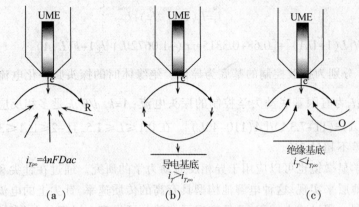

(a)针尖远离基底表面,O的扩散导致稳态电流;
(b)针尖接近导电基底,带有O的正反馈;(c)针尖接近绝缘基底表面,O的扩散受阻
图2-4 SECM反馈模式原理图[29]

SECM具有较为成熟的定量分析理论基础,这些理论因对不同的异相或均相过程以及探头和基底形貌的不同而有所不同。在不同过程中,Fick定律会有不同的边界条件,从而使探头电流有所改变。对于常见的反馈模式,且符合如下假设:溶液中氧化还原分子探针的反应是受扩散控制的;氧化还原分子探针的氧化型和还原型扩散系数相等且基底无限大。对于导体基底,一般以超微圆盘电极作为探针。在稳态条件下,探头的归一化电流 I_T ($I_T=i_T/i_{T,\infty}$, $i_{T,\infty}=4nFDac$)与归一化距离 L ($L=d/a$, d 为探头到基底的距离, a 为探头的半径)之间有如下的表达式[31,32]:

$$I_T^c(L)=i_{T(L)}/i_{T,\infty}=0.68+0.78377/L+0.3315\exp(-1.0672/L) \tag{2-16}$$

L 在 0.05 到 20 之间时, I_T 与 L 曲线的偏差小于 0.7%(见图 2-5a)。

对于绝缘基底,扩散控制的异相反应则符合公式(2-17)。

$$I_T^{ins}(L)=i_{T(L)}/i_{T,\infty}=1/[0.15+1.5385/L+0.58\exp(-1.14/L)+0.0908\exp(L-6.3)/1.017L] \tag{2-17}$$

L 值在 0.05 到 20 之间时, I_T 与 L 曲线的偏差小于 0.5%(见图 2-5b)。

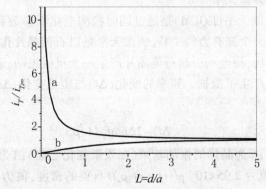

图2-5 SECM反馈模式的典型曲线[31]

当基底电极上发生的不可逆异相反应是慢反应时,其速率可以通过公式(2-18)和公式

(2-19),进行拟合求算。

$$I_T^k = I_S^k(1 - I_T^{ins}/I_T^c) + I_T^{ins} \tag{2-18}$$

$$I_S^k = 0.78377/[L(1+1/\Lambda)] + [0.68+0.3315\exp(-1.0672L)]/[1+F(L,\Lambda)] \tag{2-19}$$

式中,I_T^c、I_T^{ins}分别为扩散控制的基底为导体、绝缘体时的探头归一化电流;I_S^k为动力学控制的基底电流;I_T^k为有限基底动力学控制的探头电流;$\Lambda=k_f d/D$,k_f是多相反应的表观速率常数(cm/s);$F(L,\Lambda)=(11+7.3\Lambda)/[\Lambda(110-40L)]$。在$0.1 \leqslant L \leqslant 1.5$,且$-2 \leqslant \lg\Lambda \leqslant 3$,$\Lambda=k_f a/D$时,上述表达式的误差不超过2%。

扫描电化学显微镜也可以应用于异相反应动力学的研究。通过在针尖和导电性基底间形成的双电极薄层来实现。这种电解池构型具有高的传质速率,针尖上的电流只受固有的电子转移速率限制。通过减小体积,已经实现了在单分子尺度上进行的电化学研究,从而揭示出在大量分子存在时不可能出现的新效应[33]。为此,要对针尖进行绝缘处理,以使针尖能够在小的空腔中捕获单一分子。

扫描电化学显微镜也可以应用于局部生物活性研究,例如,现场表征生物传感器[34,35]。为此,针尖用来检测生物反应产生或消耗的电活性组分,如酶的表面反应产物。该技术还有希望应用于各种化学刺激对细胞活性影响的研究。SECM也已经用于局域膜传输研究,如监测通过皮肤的离子流量[36],通过监测氧在敏感探针上的传感效应测定单个活细胞的呼吸成像[37],或DNA微区列阵的双成像[38]。

利用电位测量的(pH是选择性的)针尖也能进行pH分布成像,包括那些由酶反应引起的(尿酶)pH分布[39]。这类电位测量针尖[40]有望测定伏安探针无法检测的反应,如非电活性物质的局部浓度的检测。与其他伏安技术不同,电位测量针尖纯属被动型传感器。当针尖在基底表面进行移动时,多种电化学过程(如电镀、蚀刻、腐蚀),都能以高分辨率表征出来。SECM除了广泛应用于表面表征外,将针尖作为一个电化学"笔"或"橡皮",还可将SECM作为一种微加工工具[41]。SECM在功能和应用范围上的进一步发展是将其与扫描探针或与光学成像技术[43](如AFM[42]或单分子荧光光谱技术)相结合。SECM-AFM结合技术将力学传感与电极结合起来,使形貌与电化学同时成像。

5. 电化学石英晶体微天平

电化学石英晶体微天平(EQCM)是通过同时检测电化学参数和电极表面质量的变化来解释电极界面反应的一个强有力的工具。该微天平是以石英晶片作为基础材料,夹在两个电极中间以产生感应电场,而电场引起石英晶片产生一个机械振动。表面反应引起的微小质量变化对晶体振荡频率产生了微扰。频率的变化(Δf)与质量变化(Δm)之间遵从Sauerbrey方程:

$$\Delta f = -2\Delta m n f_0^2 /A(\mu\rho)^{1/2} \tag{2-20}$$

式中,n为谐波数;f_0为晶体的振动基频(在质量变化之前);A为面积,单位为cm^2;μ为石英晶片的剪切模数,值为2.95×10^{11} g/(cm·s);ρ为石英的密度,值为2.65 g/cm^3。负号表示随着频率的增加质量减小,反之亦然。Sauerbrey方程确定了EQCM具有优良的质量灵敏度。能够现场检测出1 ng/cm^{2}的质量变化。EQCM能够有效地监测均匀发生在表面上的过程,包括表面层的沉积与溶解过程,各种吸收过程(如导电聚合物的掺杂/去掺杂,或在聚合膜上发生的离子交换反应)。监测过程由各种控制电位或控制电流实验来完成。在实验过程中,一个电极(在基材上的电极)与溶液接触作为电解池中的工作电极,并能够实现频率与电流的同时测定。Sauerbrey 方程用于研究溶液中的聚合物膜时需要附加一个钢性近似膜(即为弹性的无溶剂薄层膜)。当聚合物的厚度小于晶体的厚度时,此近似是有效的,与外涂层的晶体的基频相比,共振频率变小,通常当晶体质量的变化≤0.05%时,符合上述近似。由于不存在分子的特异性,EQCM 不能用于分子水平的表面表征。电化学石英晶体微天平也具有基于亲和关系方面的化学传感器特征,它允许进行质量和电流的同时测定。EQCM 的原理和应用已有综述性报道[44,45]。其与扫描电化学显微镜结合,研究各种薄膜的溶解与蚀刻的工作也有报道[46]。基于共振子列阵的多通道石英晶体微天平的发展必将进一步增加其功能和扩大其应用领域[47]。

6. 电化学发光技术

电化学发光(ECL)是另一种研究电生自由基寿命和发光性质的有用方法。电化学生成的反应物,经历了高效、快速的高能量电子转移反应,生成光致激发态物质[48-51]。许多有机和无机物质,例如多环芳烃、硝基化合物、鲁米诺、联吡啶钌络合物 $Ru(bpy)_3^{3+}$ 能够在电极上发生电子转移,产生 ECL,生成自由基。这种电生自由基是一种很强的氧化还原剂,它们之间的反应或与其他物质的反应具有足够的能量,可以跃迁至激发态。ECL 实验通常是在除氧的非水溶剂(例如高纯乙腈或 DMF)中记录发射光谱实现的。因为 $Ru(bpy)_3^{2+}$ 标记物能够在这类介质中发生 ECL,所以实验通常是在含有钌的标记物的非水介质中进行的。

ECL 的分析应用是基于 ECL 光强度与反应物浓度之间的线性关系实现的[52]。由于 ECL 能够检测很低的光强度(例如单光子计数方法),已证明具有非常低的检测限,ECL 惊人的灵敏度,$Ru(bpy)_3^{2+}$ 标记物和三丙胺(TPA)发光体系被广泛用于免疫传感器或 DNA 生物分析的研究中[53]。为了发光,$Ru(bpy)_3^{2+}$ 和 TPA 在电极表面上发生氧化反应,分别生成强氧化剂 $Ru(bpy)_3^{3+}$ 和一个阳离子自由基 TPA^{+}。后者失去一个质子并与 $Ru(bpy)_3^{3+}$ 反应形成一个激发态 $Ru(bpy)_3^{2+}$,该组分发生弛豫,在 620 nm 处发光,并释放一个光子。ECL 作为一种检测方法,已经被用于液相色谱和微芯片装置中[54,55]。

第二节 自组装膜其他研究方法

分子自组装体系是一个高度组织、高度有序、结构化、功能化和信息化的复杂系统。通过分子自组装获得的具有新奇光、电、催化等功能和特性的材料,在非线性光学器件、化学生物传感、分子信息存储以及生物大分子合成等方面都有广泛的应用前景,而且,分子自组装体系的特性与生命活动密切相关,已成为研究生命现象和发展仿生技术极有价值的体系,受到研究者广泛的重视和研究。

1. 自组装膜计算机模拟研究方法

已经有大量的实验方法用于分子自组装体系的结构分析,如静态和动态光散射、小角光散射、小角中子散射等散射技术、傅里叶变换红外、荧光光谱技术、核磁共振等光谱技术以及各种电化学方法。除射线技术可以得到聚集体直接的结构信息外,其他实验研究不能直接提供自组装体系的微观结构和动态性质,均需借助于数据的解释和处理,因此许多研究者利用热力学和动力学方法对分子自组装体系的特性进行计算机模拟研究,力求从微观上揭示其自组装过程的作用机理及规律[56]。

1.1 胶束化热力学模拟

表面活性剂胶束化热力学实验研究的重点在于测定并探究焓、熵效应等热力学参数对胶束聚集体的结构及其相互作用等因素的影响,而对其进行的热力学模拟则更注重于通过胶束化过程的自由能来表征胶束的特征。最简单的胶束化热力学模型是相分离模型,它把胶束作为分开的一相,从单体与胶团间平衡的角度获得胶束化的自由能变:

$$\Delta G^\theta = RT\ln X_{CMC} \qquad (2-21)$$

式中,X_{CMC}是以摩尔分数表示的临界胶团浓度;ΔG^θ对温度拟合直线的斜率和截距分别对应胶团化的熵变和焓变,这样胶团化的热力学参数都与体系的临界胶团浓度关联起来,显然,它只能计算胶团化的热力学函数而无法对相行为作出预测[57]。

胶束化的热力学模型只能初步解释和预测某些体系的热力学性质及相行为,对体系弱相互作用的调控机理反映并不充分,需要建立形式简单、物理意义明确的热力学模型使各种弱相互作用的贡献量化。表面活性剂胶团化的热力学模拟只能静态反映胶团结构和表面活性剂分子的平衡,并不能再现自组装体系的动态过程和微观结构,因而对分子自组装进行动力学研究就显得尤为重要。

1.2 分子自组装动力学模拟

近年来发展的运用计算机模拟分子动力学的方法可以弥补实验研究的不足,它可以在分子水平上加深我们对胶束结构和聚集过程的理解。计算机模拟分子动力学的基本思想就是运用统计力学的方法来分析系统中分子的运动轨迹,即以很大的自由度来模拟真实粒子的聚集行为[58]。通过计算机对自组装的模拟,可以获得一些基本物理现象的深层次信息,如无规则界面构象、粗糙表面的结构和动力学以及膜中的相转变过程等。分子动力学模拟应用最多的有自洽场理论和粗粒分子动力学模型。

自洽场理论是在平均场层次上研究平衡态最准确、系统的理论方法,可以方便地考虑共聚物的结构并给出聚合物链的构象信息,预测分子共聚物的相图及其相结构。由于自洽场理论比较完善和全面,自建立以来其理论表述基本没有变化,发展和变化的主要是各种计算方法。如 Drolet 等[59]提出实空间计算法(Real-Space Method),在周期性的计算盒子中数值解自洽场方程,不需要预先知道结构的对称性,就可以预测和发现复杂嵌段共聚物的微相分离形态。但是,要消除盒子的有限尺寸效应对模拟结果的影响就要增大盒子,这又导致计算量的增加。研究者[60]发展了柔性高分子—小分子液晶混合物连续自洽场理论。将小分子液晶模型化为取向与位置无关的单体分子,小分子液晶间存在各向异性的 Maier-Saupe 相互作用。该理论可还原成高分子和各向同性小分子组成的 Flory-Huggins 溶液理论和纯液晶的 Maier-Saupe 液晶理论。通过数值解自洽场方程组,发现用于研究柔性高分子—小分子液晶混合物相分离界面性质得到的结果与用 Helfand 格子界面理论 Monte Carlo(MC)模拟的结果一致。Bohbot-Raviv 等[61]提出的周期单胞法(Periodic Unit Cell)继承了实空间方法不需要预先假设相结构对称性的优点,通过对盒子尺寸扫描优化解决了有限尺寸限制和计算量之间的矛盾。

粗粒分子动力学模型常把特定的原子基团看做是一个计算单元,对溶液体系进行分子动力学模拟,而且在模拟中常常忽略静电势能和二面角势能。晶格动力学模型在模拟时使用了一个立方晶格,其中的模拟点和周围晶格点之间的相互作用完全等效。这些模拟点既可以是亲水的也可以是亲油的,而两亲分子由一个亲油点和一个亲水点构成。在这种模型基础上对表面活性剂聚集体结构用 MC 方法进行动力学模拟十分有效。Laradji 等[62]对在液—液界面上自组装单分子层的黏弹性的 MD 模拟发现,界面张力会随着表面活性剂覆盖度的增大或表面活性剂碳链的增长而降低。而在覆盖度较小的时候,表面活性剂覆盖度的增大会导致弯曲度下降,但覆盖度进一步增大则会导致弯曲度上升。这些结论对生物膜体系的研究颇有帮助。Shelley 等[63]在粗粒尺度上对双肉豆蔻磷脂酰胆碱的自组装双分子层膜使用系统进行了动力学模拟。在他们的模拟中,对不同组分使用了不同的模型。他们的模拟结果对于了解磷脂层状相的溶胀、层状相到六角相状的转变以及磷脂双分子层的横向分离等具有重要意义。研究者[64]利用正则系统分子动力学模拟方法研究了锚定单链聚乙烯与无限大基底表面的相互作用,监测了高分子链的有序化行为,计算了能量参数和结构参数随模拟时间的演化,讨论了聚集成核的微观过程。发现锚定聚乙烯在 300 K 时只在锚定点附近形成一个局部有序结构,然后以此为核进行生长。在模拟得到的有序程度不同的聚集体结构中,锚定点都位于聚集体链轴的端面。Mattice 等[65,66]模拟了三嵌段共聚物在选择性溶剂中胶团、凝胶的形成过程,发现三嵌段共聚物链有丰富的拓扑构型——末端摇摆、环状、交联,其中不同链间的交联使三嵌段共聚物体系凝胶的弹性模量和弹性持续时间都要大于两嵌段共聚物体系凝胶。难溶嵌段与溶剂的相互作用越大、难溶嵌段比例越大、链间的交联程度越大、弹性模量越大、弹性持续时间越长。研究者还考察了小分子的加入对三嵌段共聚物在选择性溶剂中胶团化的影响和增溶机理[67]。另外,Akinchina 等[68]研究了水溶液中带相反电荷的双嵌段共聚物接枝到球形颗粒上形成分子刷的结构。Houdayer 等[69]模拟了随机嵌段共聚物熔融体的相分离过程。

除此之外,发展分子自组装动力学模拟的还有介观动力学模拟。介观层次上的计算机模拟方法比较成熟的方法有 Fraaije 等[70]提出的介观动力学(MesoDyn,平均场密度泛函理论,基于时间相关的 Ginzburg-Landau 模型)和 Groot 等[71]提出的耗散颗粒动力学(dissipative particle dynamics,DPD)。与分子模拟相比,介观模型使用比原子尺度大的基本单元,通过动力学模拟来确定体系的结构、性质和动力学演变过程,在较大的空间和时间尺度上描述介观体系。

分子自组装作为化学、物理、生命科学和材料科学的交叉学科,它将在光电材料、人体组织材料、高性能高效率分离材料以及纳米材料中发挥应有的作用。研究探索分子自组装特性特别是运用计算机进行模拟,将会对材料科学、生命科学等领域中许多物理、化学的性质研究产生重要意义。这些模拟研究都会加强我们对分子自组装原理的认识,也会为其更好地应用于工业生产和生活中积累理论基础。在以分子自组装体为模板制备纳米材料、膜材料以及其他应用领域,自组装体的结构和状态至关重要,目前对自组装体结构的控制还处于实验探索阶段。这就需要研究者对表面活性剂自组装基础理论进行深入研究,特别是对生物大分子自组装领域的模拟研究。实验研究和计算机模拟研究相互辅助,将会为其应用开辟更广阔的领域[72]。

2. 自组装膜非电化学研究方法

目前,用于自组装膜的表面表征技术除了各种电化学方法之外,主要还有:湿润接触角、椭圆偏振光、表面等离子体共振谱法(SPR)、表面红外光谱(IR)、X 射线光电能谱(XPS)、俄歇电子能谱(AES)、表面增强拉曼光谱(SERS)、扫描电子显微镜(SEM)、扫描隧道显微镜

(STM)以及原子力显微镜(AFM)等[73,74]。

红外光谱在诠释自组装膜结构方面始终是一个强有力的工具。有关硫醇自组装膜的第一篇文献就是使用红外技术研究金基底上二硫化物自组装膜的性质[75],从此开辟了一个新纪元。目前,研究者已经很好地解释了在金、银和铜上烷基硫醇自组装膜的红外谱图和分子结构与取向。

拉曼光谱是探测烷基硫醇自组装膜中的反式构象和gauche构象非常有用的技术。拉曼数据说明在金上的烷基硫醇自组装膜都显示了低密度的gauche构象,平均链倾角接近于30°,C—S键几乎平行于金表面[76]。

单分子层最可利用的性质之一是它的湿润性。Whitesides等研究小组已对均一单分子层和混合单分子层的湿润接触角进行了深入研究,并用于解释自组装膜体系的微观结构[77,78]。

椭偏法可以测得自组装膜的厚度,并原位跟踪自组装膜的组装过程。只要烷链硫醇中亚甲基单元大于5,椭圆厚度就随着亚甲基单元数呈线性增加[79,80]。不同端基的硫醇自组装膜$HS(CH_2)_{15}X(X=—CH_3,—CH_2OH,—COOH,—CONH_2$和$—COOCH_3)$均由椭偏法得出厚度为2.2 nm,表明自组装膜是与端基无关的均匀分子的定向和堆积密度单分子层[81]。若混合硫醇自组装膜中的两组分硫醇的链长相差较大,椭偏法能提供自组装膜组成的粗略估计值[82]。椭偏法在确证双层[83]和多层自组装膜的逐步沉积以及检测蛋白质在各种自组装膜表面吸附[84]的表征中也是广泛应用的技术。

表面等离子体共振谱(SPR)是一种灵敏的光谱分析技术。由于它响应于两相界面膜的折射率n、吸收系数k和膜厚度d的变化,因此,SPR能够用于检测与膜的性质及厚度变化有关的反应和过程。目前主要应用的领域包括有机薄膜性质的表征、分子浓度和质量的测定、蛋白质吸附以及组装过程的研究。SPR特别适合于研究生物分子识别及其相互作用,Whitesides研究小组结合自组装膜技术和SPR技术研究了大量生物分子识别及分子相互作用。SPR适合监控自组装膜接触水溶液时的表面反应和组装过程动力学,已成功地应用于类脂二硫化物自组装厚度的测定[85]、聚电解质巯基十一酸在自组装膜上的吸附[86]、抗生蛋白链菌素与附着在自组装膜上生物素的键合过程[87]以及类脂和神经节苷脂在十六烷基硫醇自组装膜上的吸附[88]。

Hubbard等[89]结合X射线衍射方法研究了不同硫醇在Pt(111)和Ag(111)上的吸附。各种电子和原子衍射方法(如HREELS、AES和LEED)和氦衍射实验结果揭示了烷基硫醇在Au(111)形成了$(\sqrt{3}\times\sqrt{3})R30°$的叠层结构。结构谱提供了单分子层形成的可能步骤:球状烷基硫醇的快速吸附;随着更多烷基硫醇的吸附,烷基链段伸展为自由体积。

程序升温脱附用来考察含硫化合物与金的键合方式以及溶剂分子同自组装膜的各种端基间的相互作用[90]。激光脱附质谱或表面解离质谱[91]特别适合于考察自组装膜的组成,并提供了二硫化物自组装膜中二硫键断裂的证据。

扫描隧道显微镜(STM)和原子力显微镜(AFM)的初期工作主要是集中在超真空(UHV)条件下探测洁净的裸露表面。最近,这方面的研究正在从超真空条件扩展到大气状态下、极性和非极性溶液中以及在电化学池中去研究自组装膜体系,并成为自组装膜研究中的常规技术之一,其突出优点在于:可以分辨出单个原子;实时地得到在实空间中的三维图像,适合于自组装膜中的扩散等动态过程的研究;可观察到自组装膜中单个原子层的局部表面结构,而不是整个表面或体相的平均性质,因而可直接研究自组装膜的表面缺陷、表面重构和表面吸附体的形态和位置以及吸附体引起的表面重构等；可在大气、常温等不同环境下工作，不需要特别的制样技术，而且检测过程对样品无损伤，因此可获得自组装膜本身的原位信息。另外，还可以应用 STM 和 AFM 技术对自组装膜进行纳米级加工和分子识别研究。

若不在高电流密度下长时间扫描，就可避免 STM 对自组装膜和金基底的破坏[92]。若隧道电流小于 10 nA，在 Au(111) 基底上的不同链长烷基硫醇自组装膜产生了六角形对称亮点影像，其间隔同衍射实验[93,94]所得到的 $(\sqrt{3}\times\sqrt{3})R30°$ 结构一致。由于影像和链长无关，所以它们应用于硫醇盐头基。

在大约 0.1~1 nA 的隧道电流范围内，各种烷基硫醇自组装膜影像显示了随机凹陷的图形[95]，该图形包含新出现的宽度为 2.0~5.0 nm 和深度为 0.25±0.03 nm 的"坑"。目前认为"坑"的产生是自组装过程中表面金原子解离和扩散协同作用的结果。

当隧道电流减小到几个 pA 时，出现高分辨影像。在这些条件下，金属尖端可能定位于自组装膜的外部。因此，STM 图像呈现出自组装膜中的单分子影像。在仔细退火 Au(111) 上的十一烷基硫醇自组装膜的 STM 图像中可清楚地观察到对应于 $(\sqrt{3}\times\sqrt{3})R30°$ 晶格的六角形排列亮点和几个 c(4×2) 形超晶格图形，这些图形可由烷烃链扭曲角从分子到分子的变化来解释。

Stranick 等[96]通过双分子烷基硫醇混合自组装膜的影像，给出了混合自组装膜中不同的硫醇倾向于相析的证据。使用异常低的 STM 漂移速率检测到了自组装膜在 Au(111) 基底上平台边界的移动。

STM 可用来刻蚀烷基硫醇自组装膜。方法是在 STM 探针的尖端和基底上加电压脉冲[97]，可在自组装膜的表面上产生完备尺寸的针孔(25.0 nm×25.0 nm)。

在同一个烷基硫醇自组装膜上，用 STM 可观察到"坑"的存在，而用 AFM 则观察不到，这反映了 STM 较 AFM 有内在的高分辨率。当 AFM 尖端在高力作用下推进烷基硫醇自组装膜时[98]，观察到一个有趣的现象，就是力—距离曲线的阻滞。该结果表明链—链相互作用倾向于自组装膜产生完好的定向，经历微扰后自组装膜能快速地恢复它的定向。

另外，除紫外—可见光吸收检测外，其他的光谱技术也能够用于检测电化学反应的动力学过程或考察电生组分的寿命。特别是将电化学与电子自旋共振、核磁共振、质谱等技术相结合，能够得到更为丰富的信息。已有研究者构建了各种特殊设计的电解池，并有若干综述性文章发表[99-103]。

参考文献

[1] Nicholson R S, Shain I. Theory of stationary electrode polarography: single scan and cyclic methods applied to reversible, irreversible, and kinetic systems. Anal. Chem., 1964, 36: 706-723.

[2] Echegoyen L, Echegoyen L E. Cheminform abstract: Electrochemistry of fullerenes and their derivatives. Acc. Chem. Res., 1998, 31: 593-601.

[3] Andrieux C P, Hapiot P, Saveant J M. Ultramicroelectrodes for fast electrochemical kinetics. Elctroanalysis, 1990, 2: 183-193.

[4] Gelbert M B, Curran D J. Alternating current voltammetry of dopamine and ascorbic acid at carbon paste and stearic acid modified carbon paste electrodes. Anal. Chem., 1986, 58:

1028-1032.

[5] Hawley M D, Tatawawdi S V, Piekarski S, et al. Electrochemical studies of the oxidation pathways of catecholamines. J. Am. Chem. Soc., 1967, 89: 447-450.

[6] Mohilner D M, Adams R N, Argersinger W R. Investigation of the kinetics and mechanism of the anodic oxidation of aniline in aqueous sulfuric acid solution at a platinum electrode1,2. J. Am. Chem. Soc., 1962, 84: 3618-3622.

[7] Johnson J S, Evans D A. Chiral bis(oxazoline) copper (ii) complexes: Versatile catalysts for enantioselective cycloaddition, adol, michael and carbonyl ene reactions. Acc. Chem. Res., 2000, 33: 325-335.

[8] Mabbott G. A. An introducio to cylic voltammet. J. Chem. Educ., 1983, 60: 697-702.

[9] Rudolph M, Reddy D, Feldberg S W. A simulator for cyclic voltammetric responses. Anal. Chem., 1994, 66: 589A-600A.

[10] Pearce P J, Bard A J. Polymer films on electrodesPart II. Film structure and mechanism of electron transfer with electrodeposited poly(vinylferrocene. J. Electroanal. Chem., 1980, 112: 97-115.

[11] Brown A P, Anson F C. Cyclic and differential pulse voltammetric behaviour of reactants confined to the electrode surface. Anal. Chem., 1977, 49: 1589-1595.

[12] Stamford J, Hurst P, Kuhr W, et al. Characterization of states of adsorption with fast-scan cyclic voltammetry. J. Electroanal. Chem., 1989, 265: 291-296.

[13] Venton B, Wightman R M. Psychoanalytical electrochemistry: dopamine and behavior. Anal. Chem., 2003, 75: 414A-421A.

[14] Michael D, Travis E, Wightman R M. Color images for fast-scan CV measurements in biological systems. Anal. Chem., 1998, 70: 586A-592A.

[15] Jackson B, Dietz S, Wightman R M. Fast-scan cyclic voltammetry of 5-hydroxytryptamine. Anal. Chem., 1995, 67: 1115-1121.

[16] Heinze J. Cyclic voltammetry — "electrochemical spectroscopy". New analytical methods (25). Angew. Chem. Int. Ed. Engl., 1984, 23: 831-847.

[17] Baldwin R P, Ravichandran K, Johnson R K. A cyclic voltammetry experiment for the instrumental analysis laboratory. J. Chem. Ed., 1984, 61: 820-829.

[18] Park S M, Yoo J S. Electrochemical impedance spectroscopy for better electrochemical Measurements. Anal. Chem., 2003, 75: 455A-461A.

[19] Katz E, Willner I. Probing biomolecular interactions at conductive and semiconductive surfaces by impedance spectroscopy: Routes to impedimetric immunosensors, DNA-sensors, and enzyme biosensors. Electroanalysis, 2003, 15: 913-947.

[20] Kuwana T, Heineman W. Study of electrogenerated reactants using optically transparent electrodes. Acc. Chem. Res., 1976, 9: 241-248.

[21] Heineman W, Hawkridge F, Blount H. Spectroelectrochemistry at optically transparent electrodes. New York: Marcel Dekker, 1986.

[22] Van Dyke D A, Cheng H Y. Fabrication and characterization of a fiber-optic-based spectroelectrochemical probe. Anal. Chem., 1988, 60: 1256-1260.

[23] Brewster J, Anderson J L. Fiber optic thin-layer spectroelectrochemistry with long optical path. Anal. Chem., 1982, 54: 2560-2566.

[24] Sorrels J W, Dewald H D. Spectroelectrochemical characteristics of the reticulated vitreous carbon electrode. Anal. Chem., 1990, 62 (15): 1640-1643.

[25] Heineman W R. Spectroelectrochemistry: The combination of optical and electrochemical techniques. J. Chem. Educ., 1983, 60: 305-308.

[26] Gui Y P, Porter M, Kuwana T. Long optical path length thin-layer spectroelectrochemistry: Quantitation of adsorbed aromatic molecules at platinum. Anal. Chem., 1985, 57: 1474-1476.

[27] Shi Y, Slaterbeck, A F, Seliskar C J, et al. Spectroelectrochemical sensing based on multimode selectivity simultaneously achievable in a single device. 1. Demonstration of concept with ferricyanide. Anal. Chem., 1997, 69: 3679-3686.

[28] Dewald H D, Wang J. Spectroelectrochemical detector for liquid chromatography and flow injection systems. Anal. Chim. Acta., 1984, 166: 163-170.

[29] Bard A J, Denuault G, Lee C, et al. Scanning electrochemical microscopy — a new technique for the characterization and modification of surfaces. Acc. Chem. Res., 1990, 23: 357-363.

[30] Mirkin M V. Recent advances in scanning electrochemical microscopy. Anal Chem., 1996, 68: 177A-182A.

[31] Bard A J, Fan F F, Kwak J, et al. Scanning electrochemical microscopy (SECM). Anal. Chem., 1989, 61: 132.

[32] Liu B, Bard A J, Mirkin M V, et al. Electron transfer at self-assembled monolayers measured by scanning electrochemical microscopy. J. Am. Chem. Soc., 2004, 126: 1485-1492.

[33] Fan F, Kwak J, Bard A J. Single molecule electrochemistry. J. Am. Chem. Soc., 1996, 118: 9669.

[34] Wang J, Wu L, Li R. Scanning elecrochemical microscopic monitoring of biological processes. J. Electroanal. Chem., 1989, 272: 285.

[35] Pierce D, Unwin P, Bard A J. Electrochemical microscopy. xvii. Studies of enzyme-mediator kinetics for membrane and surface-immobilised glucose oxidase. Anal. Chem., 1992, 64: 1795.

[36] Bath B D, White H S, Scott E R. Imaging Molecular Transport Across Membranes New York: John Wiley, 2001.

[37] Yasukawa T, Kaya T, Matsue T. Characterization and imaging of single cells with scanning electrochemical microscopy. Electroanalysis, 2000, 12: 653-659.

[38] Yamashita K, Takagi M, Uchida K, et al. Visualization of DNA microarrays by scanning electrochemical microscopy. Analyst, 2001, 126: 1210-1211.

[39] Horrocks B, Mirkin M, Pierce D, et al. Permeability of glucose and other neutral species through recast perfluorosulfonated ionomer films. Anal. Chem., 1993, 65: 1304-1311.

[40] Wei C, Bard A J, Nagy G, et al. Scanning electrochemical microscopy. 28. ion-selective neutral carrier-based microelectrode potentiometry. Anal. Chem., 1995, 67: 1346-1356.

[41] Nowall W, Wipf D, Kuhr W G. Localized avidin/biotin derivatization of glassy carbon electrodes using SECM. Anal. Chem., 1998, 70: 2601-2606.

[42] Kranze C, Friedbacher G, Mizaikoff B, et al. Integrating an ultramicroelectrode in an AFM cantilever: combined technology for enhanced information. Anal. Chem., 2001, 73: 2491-

2500.

[43] Boldt F, Heinze J, Diez M, et al. Real-time pH microscopy down to the molecular level by combined scanning electrochemical microscopy/single-molecule fluorescence spectroscopy. Anal. Chem., 2004, 76: 3473-3481.

[44] Deakin M, Buttry D. Electrochemical applications of the quartz crystal microbalance. Anal. Chem., 1989, 61: 1147A-1154A.

[45] Ward M D, Buttry D. In situ interfacial mass detection with piezoelectric transducers. Science, 1990, 249: 1000-1007.

[46] Cliffel D, Bard A J. Scanning electrochemical microscopy. 36. A combined scanning electrochemical microscope-quartz crystal microbalance instrument for studying thin films. Anal. Chem., 1998, 70: 1993-1998.

[47] Tatsuma T, Watanabe Y, Oyama N, et al. Multichannel quartz crystal microbalance. Anal. Chem., 1999, 71: 3632-3636.

[48] Faulkner L, Bard A J. Electrochemiluminescence // Electroanalytical Chemistry. New York: Marcel Dekker, 1977.

[49] Xu G B, Dong S J. Electrochemiluminescent determination of peroxydisulfate with $Cr(bpy)_3^{3+}$ in purely aqueous solution. Analytica. Chimica. Acta., 2000, 412(1-2): 235-340.

[50] White H S, Bard A J. Electrogenerated chemiluminescence and chemiluminescence of the $Cr(bpy)_3^{2-}$ peroxydisulfate system in acetonitrile-water solutions. J. Am. Chem. Soc., 1982, 104: 6891-6895.

[51] Richter M M. Electrochemiluminescence (ECL). Chem. Rev., 2004, 104: 3003-3036.

[52] Knight A W, Greenway G. Occurrence, mechanisms and analytical applications of electrogenerated chemiluminescence. Analyst, 1994, 119: 879-890.

[53] Kulmala S, Suomi J. Current status of modern analytical luminescence methods. Anal. Chim. Acta., 2003, 500: 21-69.

[54] Forry S P, Wightman R M. Electrogenerated chemiluminescence detection in reversed-phase liquid chromatography. Anal. Chem., 2002, 74: 528-532.

[55] Zhan W, Alvarez J, Crooks R M. Electrochemical sensing in microfluidic systems using electrogenerated chemiluminescence as a photonic reporter of redox reactions. J. Am. Chem. Soc., 2002, 124: 13265-13270.

[56] 李华, 韩永才, 张金利. 大分子自组装的计算机模拟进展. 化学进展, 2004, 16(3): 431-437.

[57] 陈澍, 郭晨, 刘会洲. 两亲性共聚物自组装的计算机模拟进展. 化学通报, 2005, (9): 681-689.

[58] Alexander P, James R K, Pavan K, et al. Importance of micellar kinetics in relation to technological processes. J. Colloid Interface Sci., 2002, 245(1): 1-15.

[59] Drolet F, Fredrickson G H. Combinatorial screening of complex block copolymer assembly with self-consistent field theory. Phys. Rev. Lett., 1999, 83(21): 4317-4320.

[60] 王家芳, 张红东, 邱枫, 等. 柔性高分子、小分子液晶混合物的自洽场理论. 化学学报, 2003, 61(8): 1180-1185.

[61] Bohhot R Y, Wang G Z. Discovering new ordered phases of block copolymers. Phys.

Rev. Lett., 2000, 85(16): 3428-3431.

[62] Laradji M, Mouritsen O G. Elastic properties of surfactant monolayers at liquid-liquid interfaces: a molecular dynamics study. J. Chem. Phys., 2000, 112(19): 8621-8630.

[63] Shelley J C, Shelley M Y, Reedr R C, et al. A coarse grain model for phospholipid simulations. J Phys Chem B, 2001, 105(17): 4464-4470.

[64] 韩铭, 李霆, 杨小霞. 表明上锚定聚乙烯链聚集的分子动力学模拟. 高等学校化学学报, 2005, 26(5): 960-963.

[65] Nguyen M M, Mattice W L. Dynamics of end-associated triblock copolymer networks. Macromolecules, 1995, 28(20): 6976-6985.

[66] Nguyen M M, Mattice W L. Micellization and gelation of symmetric triblock copolymers with insoluble end blocks. Macromolecules, 1995, 28(5): 1444-1457.

[67] Xing L, Mattice W L. Solubilization of small molecules by triblock copolymer micelles in selective solvents. Macromolecules, 1997, 30(6): 1711-1717.

[68] Akinchina A, Shusharina N P, Linse P. Diblock polyampholytes grafted onto sphercal particles: Monte Carlo simulation and lattice mean-field theory. Langmuir, 2004, 20(23): 10351-10360.

[69] Houdayer J, Muller M. Phase diagram of random copolymer melrs: a compute simulation study macromolecules. Macromolecules, 2004, 37(11): 4283-4295.

[70] Fraaije J G E M. Dynamic density functional theory for microphase seperation kinetics of block copolymer melts. J Chem Phys, 1993, 99: 9202-9212.

[71] Groot RD, Madden T J. Dynamic simulation of diblock copolymer microphase separation. J Chem phys, 1998, 108: 8713-8724.

[72] 惠路华, 洪坤, 符意德, 等. 分子自组装的应用与计算机模拟研究进展. 计算机与应用化学, 2007, 24(5): 653-658.

[73] Joseph W. 分析电化学. 朱永春, 张玲, 译. 北京: 化学工业出版社, 2009.

[74] 董绍俊, 车广礼, 谢远武. 化学修饰电极. 北京: 科学出版社, 2003.

[75] Nuzzo R G, Allara D L. Adsorption of bifunc-tional organic disulfides on gold surfaces. J. Am. Chem. Soc., 1983, 105: 4481-4483.

[76] Pemberton J E, Bryant M A, Sobocinski R L, et al. A simple method for determination of orientation of adsorbed organics of low symmetry using surface-enhanced Raman scattering. J. Phys. Chem., 1992, 96: 3776-3782.

[77] Bain C D, Whitesides G M. Modeling organic surfaces with self-assembled monolayers. Angew. Chem. Int. Ed. Engl., 1989, 28: 506-512.

[78] Kumar A, Whitesides G M. Features of gold having micrometer to centimeter dimensions can be formed through a combination of stamping with an elastomeric stamp and an alkanethiol "ink" followed by chemical etching. Appl. Phys. Lett., 1993, 63: 2002-2004.

[79] Miller C J, Cuender P, Gratzel M. Adsorbed. omega -hydroxy thiol monolayers on gold electrodes: evidence for electron tunneling to redox species in solution. J. Phys. Chem., 1991, 95: 877-886.

[80] Evans S D, Ulman A. Surface potential studies of alkyl-thiol monolayers adsorbed on gold. Chem. Phys. Lett., 1990, 170: 462-466.

[81] Nuzzo R G, Dubois L H, Allara D L. Fundamental studies of microscopic wetting on

organic surfaces. 1. Formation and structural characterization of a self-consistent series of polyfunctional organic monolayers. J. Am. Chem. Soc., 1990, 112: 558–569.

[82] Offord D A, John C M, Griffin J H. Contact angle goniometry, ellipsometry, XPS, and TOF-SIMS analysis of gold-supported, mixed self-assembled monolayers formed from mixed dialkyl disulfides. Langmuir, 1994, 10: 761–766.

[83] Sun L, Crooks R M, Ricco A J. Molecular interactions between organized, surface-confined monolayers and vapor-phase probe molecules. 5. Acid-base interactions. Langmuir, 1993, 9: 1775–1780.

[84] Prime K L, Whitesides G M. Adsorption of proteins onto surfaces containing end-attached oligo (ethylene oxide): a model system using self-assembled monolayers. J. Am. Chem. Soc., 1993, 115: 10714–10721.

[85] Lang H, Duschl C, Gratzel M, et al. Self assembly of thiolipid molecular layers on gold surfaces. Optical and electrochemical characterisation. Thin Solid Films, 1992, 210/211: 818–821.

[86] Jordan C E, Frey B L, Kornguth S, et al. Characterization of poly-l-lysine adsorption onto alkanethiol-modified gold surfaces with polarization-modulation fourier transform infrared spectroscopy and surface plasmon resonance measurements. Langmuir, 1994, 10: 3642–3648.

[87] Spinke J, Liley M, Schmitt F-J, et al. Supra-biomolecular architectures at functionalized surfaces. J. Phys. Chem., 1993, 99: 7012–7019.

[88] Terrettaz S, Stora T, Duschl C, et al. Protein binding to supported lipid membranes: investigation of the cholera toxin-ganglioside interaction by simultaneous impedance spectroscopy and surface plasmon resonance. Langmuir, 1993, 9: 1361–1369.

[89] Gui J H, Stern D A, Frank D G, et al. Adsorption and surface structural chemistry of thiophenol, benzyl mercaptan, and alkyl mercaptans. Comparative studies at silver (111) and platinum (111) electrodes by means of Auger spectroscopy, electron energy loss spectroscopy, low energy electron diffraction and electrochemistry. Langmuir, 1991, 7: 955–963.

[90] Nuzzo R G, Zegarski B R. Fundamental studies of the chemisorption of organosulfur compounds on gold (111). Implications for molecular self-assembly on gold surfaces. J. Am. Chem. Soc., 1987, 109: 733–740.

[91] Li Y, Huang J, Melver R G, et al. Characterization of thiol self-assembled films by laser desorption Fourier transform mass spectrometry. J. Am. Chem. Soc., 1991, 114: 2828–2832.

[92] Kim Y-T, Bard A J. Imaging and etching of self-assembled n-octadecanethiol layers on gold with the scanning tunneling microscope. Langmuir, 1992, 8: 1096–1102.

[93] Widrig C A, Alves C A, Porter M D. Characterization of thiol self-assembled films by laser desorption Fourier transform mass spectrometry. J. Am. Chem. Soc., 1992, 114: 2428–2432.

[94] Kim Y-T, McCarley R L, Bard A J. Scanning tunneling microscopy studies of Au (111) derivatized with organothiols. J. Phys. Chem., 1992, 96: 7416–7421.

[95] Kim Y-T, McCarley R L, Bard A J. Observation of n-octadecanethiol multilayer formation from solution onto gold. Langmuir, 1993, 9: 1941–1944.

[96] Stranick S J, Parikh A N, Tao Y-T, et al. Phase separation of mixed-composition

self-assembled monolayers into nanometer scale molecular domains. J. Phys. Chem., 1994, 98: 7636-7646.

[97] Ross C B, Sun L, Crooks R M. Scanning probe lithography. 1. Scanning tunneling microscope induced lithography of self-assembled n-alkanethiol monolayer resists. Langmuir, 1993, 9: 632-636.

[98] Thomas R C, Houston J E, Michalske T A, et al. The mechanical response of gold substrates passivated by self-assembling monolayer films. Science, 1993, 259: 1883-1885.

[99] Bagchi R N, Bond A M, Scholz F. ESR-electrochemical cells and their performance in studies of redox processes. Electroanalysis, 1989, 1: 1-11.

[100] Richards J A, Evans D H. Flow cell for electrolysis within the probe of a nuclear magnetic resonance. spectrometer. Anal. Chem., 1975, 47: 964-966.

[101] Chang H, Johnson D C, Houk R S. Trands. In situ coupling between electrochemistry and mass spectrometry-A literature review. Anal. Chem., 1989, 8: 328-333.

[102] Regino M C, Brajter-Toth A. An electrochemical cell for on-line electrochemistry / mass spectrometry. Anal. Chem., 1997, 69: 5067-5072.

[103] Tong Y, Oldfield E, Wieckowski A. Exploring electrochemical interfaces with solid-state NMR. Anal. Chem., 1998, 70: 518A-527A.

第三章 氧化还原自组装膜界面电子转移研究

电子传递在生命过程、传感、人工光合成以及分子电子器件中获得了广泛研究。Marcus 理论假设电子给体(A)和电子受体(D)之间的电子传递速率依赖于 ΔG、重组能 λ 以及 D 和 A 之间的电子偶合能 H_{AB}[1]。在电极表面通过分子链接氧化还原基团描述的电子传递中,一项重要的工作是关于系统长程电子转移研究,包括人工光合成中心的分子系统、高共轭分子导线以及混合价态分子系统等。对电子传递过程动力学的研究极大地促进了分子电子学的发展(分子整流器、分子结、分子开关、分子二极管以及传感器等)[2-9]。

自组装膜(SAMs)为电子传递动力学过程研究提供了良好的平台,可以有效地通过电化学去研究电子传递过程[10-15]。氧化还原 SAM 被系统地设计研究 SAM 构成的 ΔG、λ 和 H_{AB} 的相关性,每一参数(ΔG、λ、T、H_{AB})的变化均可以通过实验控制,例如,SAM 可以控制双电层效应,并且可以消除质量扩散的问题等(图 3-1)。

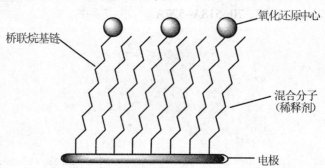

图 3-1 包含氧化还原中心的氧化还原活性 SAM

本章描述了几种常用的电化学方法,包括 CV 法、ACV 法、CA 法、EIS 法等,使用氧化还原修饰的 SAM 系统去测量 K_{ET}、ΔG、λ、H_{AB} 等参数。另外,也从 SAM 应用的角度对氧化还原分子的基本性质、电极和氧化还原中心的连接基团、SAM 结构、各种电极材料的性质以及 SAM 模型的计算途径等进行了探讨。

第一节 氧化还原自组装膜电子传递研究的电化学方法

应用各种电化学技术可以探索影响电子传递的因素,如链接基团可以控制氧化还原中心和电极之间的距离及耦合性,各种溶剂及不同结构 SAM 的氧化还原中心分子环境可以探索重组能 λ 的变化等。

SAM 作为理想的系统可以运用电化学方法去研究发生在电极上的长程电子转移,然而,电化学技术的多样性以及复杂的数据处理,限制了该领域的发展和应用。本节主要描述了几种常用的电化学方法,包括 CV 法、ACV 法、CA 法、EIS 法等,使用氧化还原修饰的 SAM 系统去测量 K_{ET}、ΔG、λ、H_{AB} 等参数。

1. 循环伏安法

循环伏安(CV)法是一种电位扫描方法。CV 法测试仪器相对简单、价廉,因此得到了广泛的应用。但与其他的电化学技术相比,CV 法对异相电子传递动力学缺乏灵敏性[16-18]。异相电子传递动力学归因于由 SAM 缺陷引起的氧化还原中心分子环境的变化。使用以下相关的数学处理,CV 法可以用于 K_{ET} 检测。

从 CV 图可以获得大量的信息,可用于单层膜表面结合电活性物质电子传递的测量,背景电流、峰电流以及峰电位在检测电子传递反应速率常数中是非常重要的。另外,单层膜的完整、有序性也非常重要,因为 SAM 的无序结构将可能引起速率常数的偏离。对于 K_{ET} 的测量,氧化还原中心在均一的分子环境中应当理想地彼此隔离。图 3-2 表明了可以从 CV 图获得的相关参数。

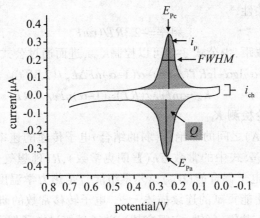

图 3-2 表面结合氧化还原物质 CV 的相关参数,参数包括:E_{Pc},E_{Pa},i_{ch},Q,$FWHM$

背景电流(电荷或电容电流)与 SAM 的厚度密切相关,双电层电容 C_{DL} 被经常用于表面积 A_{SUR} 的拟合(公式 3-1)。

$$i_{ch}/A_{SUR}=C_{DL}v/A_{SUR} \tag{3-1}$$

法拉第电流 i_P 与扫描速率成正比(公式 3-2),表面覆盖 Γ 可以从 i_P vs v 曲线的斜率求得,这个值经常与吸附分子表面积的理论最大值进行比较。

$$i_P = n^2F^2vA_{SUR}\Gamma/(4RT) \tag{3-2}$$

峰的形状(对称性)表明单层膜是同质性的,通过公式 3-3 可以评价峰的半宽度 $FWHM$ 值,与理论值相比稍大或稍小,是邻近电荷物质之间的静电效应所致[16,19,20]。

$$FWHM=3.53RT/(nF)=90.6/n \tag{3-3}$$

氧化还原电势 E^θ,由阴极和阳极峰电位 E_{Pa}、E_{Pc} 以及峰分离 $\Delta E_P(E_{Pa}-E_{Pc})$ 的平均值计算,扫速增加,ΔE 增加。在低扫速下,峰分离 ΔE_P 为 0。

电荷 Q 的量可由背景—基底峰的组合来检测,如图 3-2 阴影部分的面积。根据公式 3-4,过电位 η 是峰电位 E_P 和氧化还原活性物质电极电位 E^θ 的差值。对于不同的扫速,电子传递速率在特定过电位下的速率常数 k_s 可由公式 3-5 给出,对 η vs lg $k_s(\eta)$ 作图,获得的 Tafel 曲线,使用 Marcus 密度态模型,可用于 K_{ET} 的检测。

$$\eta=E_P-E^\theta \tag{3-4}$$

$$k_s=i_P/Q \tag{3-5}$$

总之,当氧化还原物质强烈吸附于电极上时,理想可逆的电化学响应将产生还原和氧化形式,对于完全可逆系统,CV图表明:峰对称;峰电流和扫速呈线性关系;在低扫速下ΔE_P值为0。例如,对于单电子转移过程的峰形,如以$FWHM$描述,在25℃时为90.6 mV[16]。

2. 检测K_{ET}的Laviron方法

1979年,Laviron应用线性扫描伏安法,并应用相关数学处理对吸附在电极上的物质的K_{ET}进行检测[21]。这种方法以B-V法为基础,唯一应用的实验数据是过电势η。

Laviron法广泛地应用于K_{ET}的检测,然而,由于条件的限制,其应用受到了影响。首先,这种方法依赖于转移系数α,α是氧化还原反应能量势垒的对称性表示,对于所有的η,理想的$\alpha=0.5$。然而,在更多的情况下,α偏离0.5,因此,检测α对于K_{ET}是非常关键的。为了检测α,以峰电位E_P vs $\lg v$做图,在较高的扫速下,当$v > 100$ mV时,E_{Pa}和E_{Pc}分别得到两条曲线,依据公式3-6,斜率与α具有相关性[21]。

$$斜率 = -2.3RT/(\alpha nF) \quad (3\text{-}6)$$

通过应用$\eta=0$时的截距,由公式3-7可以控制K_{APP},进而通过公式3-8可以检测K_{ET}。

$$\lg k_{APP} = \alpha \lg(1-\alpha) + (1-\alpha)\lg\alpha - \lg RT/(nFv) - \alpha(1-\alpha)nF\Delta E_p/(2.3RT) \quad (3\text{-}7)$$

$$K_{ET} = \alpha nFv_C/(RT) = (1-\alpha)nFv_C/(RT) \quad (3\text{-}8)$$

3. Marcus密度态理论检测K_{ET}

在溶液中A和B(D—A)之间的非绝热(弱的结合)电子传递的速率常数可以通过Marcus半经验公式3-9来描述,公式中的常数有h(普朗克常数),R(理想气体状态常数),K_B(波尔兹曼常数),A和B之间的电子偶合能(H_{AB}),λ是重组能,T是热力学温度。Marcus理论是基于溶剂对电子传递反应活化能贡献的连续描述[22-25]。电子转移常数的研究是通过改变D和A之间链的长度和结构(共价、共轭、氢键、空间穿越),并通过探讨电子偶合能H_{AB}来进行的[26,27]。

$$K_{ET} = 4\pi^2 H_{AB}^2 \exp[-(\Delta G + \lambda)^2/4\lambda K_B T]/[h(4\pi\lambda K_B T)^{1/2}] \quad (3\text{-}9)$$

重组能λ适用于所有原子(从平衡态到反应产物),其包括两个部分:$\lambda = \lambda_i + \lambda_o$。内部$\lambda_i$与键长变化的能量,如旋转态有关;外部$\lambda_o$与溶剂的能量相关。公式3-10应用了简单的几何假设:D和A是球体,并且溶剂被处理为介电连续体,r_A和r_B是氧化还原中心A和B的半径,d是氧化还原中心之间的距离,ε_{OP}和ε_S分别是溶剂光学和静态介电常数。在$-\Delta G = \lambda$条件下,应用公式3-9可检测Marcus翻转区,其电子转移常数将达到最大值,这是重组能重要的实验研究[28-32]。

$$\lambda_o = N_A e^2 (1/2r_A + 1/2r_B - 1/d)(1/\varepsilon_{OP} - 1/\varepsilon_S) \quad (3\text{-}10)$$

对于发生在电极上的氧化还原过程,电势依赖于电化学速率的参数分别由公式3-11和3-12表示,在公式中,λ、e_0、A、K_B、η和T分别表示重组能、电子电荷、指前因子、波尔兹曼常数、过电位及温度。在经典的B-V方程中,氧化反应和还原反应的活化能假定是η的函数,并且传递系数α假定是0.5[16,33]。

$$k_{ox} = A\exp[-(\lambda - 2e_0\eta)/(4k_B T)] \quad (3\text{-}11)$$

$$k_{red} = A\exp[-(\lambda + 2e_0\eta)/(4k_B T)] \quad (3\text{-}12)$$

在高的过电位下,由于B-V理论并未考虑一个抛物线反应表面,公式3-11和3-12不再适用,而且,就像在Marcus理论中的,电位依赖电化学速率常数的二次方[33]。通过近似处理反应表面,如抛物线表面,可以观察到Marcus翻转区的电化学性质[34,35],与Marcus理论相关的速率关系将修订为公式3-13和3-14。

$$k_{ox}=A\exp[-(\lambda-2e_0\eta)^2/(4k_BT)] \tag{3-13}$$

$$k_{red}=A\exp[-(\lambda+2e_0\eta)^2/(4k_BT)] \tag{3-14}$$

如同$B-V$理论，将Marcus理论中的抛物线平面电势近似地处理为线性函数，当$\lambda/\eta\ll 1$时，公式3-13、3-14可简化为公式3-11和3-12。

氧化还原活性分子和金属电极之间电子传递的阴极和阳极速率常数可以通过D和A能级测量，对超出能量ε的区域进行数字拟合，D和A分子能级与金属的费米能级相关。在此，仅仅考虑了速率常数的阳极表达，因为阴极速率常数的表达仅仅在对称性上与阳极有区别。

$$k_{ox}=A\int_{-\infty}^{\infty}d\varepsilon D_{ox}(\varepsilon)\rho(\varepsilon)f(\varepsilon) \tag{3-15}$$

$$D_{ox}(\varepsilon)\frac{1}{\sqrt{4\pi\lambda k_BT}}\exp\left[-\frac{(\lambda-\varepsilon+e_0\eta)^2}{4\lambda k_BT}\right] \tag{3-16}$$

金属的费米函数$f(\varepsilon)$在一个给定的温度下，其给定的电子能级将被占有，金属的密度态$\rho(\varepsilon)$缓慢变化接近费米能级，并且假定超出了整体评价的能量范围，氧化还原中心电子受体能级的分布通过G函数$D_{ox}(\varepsilon)$来表达，其宽度定义为重组能。指前因子A包含的因素有电子耦合、通过电子能垒的隧穿以及氧化还原活性位点的表面覆盖[36,37]。

如果λ与应用电势在数量上相似，将出现一个明显的稳定区域，即随η的增加，速率不再增加。公式3-15和3-16假定了η接近λ（出现翻转区），正如在$B-V$理论中预言的，速率常数不再呈指数连续增长，但当$\eta=\pm\lambda$时，其将达到最大值[33]。

当$\eta\gg\lambda$时，氧化还原位点在电极上呈重叠态，增加驱动力(η)没有增加电子传递速率。当$\eta>\lambda$时，在翻转区，速率常数也并不随驱动力的增加而增加。事实上，正如Weber等人强调的，速率常数与驱动力无关，这些效应要求实验证实。

为了通过适当的方法对实验CV拟合检测K_{ET}，Eggers等人[38]对Fc修饰的SAM，使用公式3-17和3-18对实验CV进行了拟合。在这种方法中，电流i作为时间的函数，在扫描期间，由公式3-17给出，Γ_T是Fc的表面覆盖，θ_{Fc}是Fc的覆盖度(Γ_O/Γ_T)，对于Fc的氧化态，θ_{Fc}随时间变化，这可由公式3-18计算，k_O和k_R值分别由公式3-20和3-21来检测。

$$i(t)=F[d\theta_{Fc}(t)/dt]\Gamma_T \tag{3-17}$$

$$d\theta_{Fc}/dt=k_O[1-\theta_{Fc}(t)(1+e^{\frac{-\eta F}{RT}})] \tag{3-18}$$

$$i(t)=F\Gamma_T\{k_O[1-\theta_{Fc}(t)]-kR\theta_{Fc}(t)\} \tag{3-19}$$

氧化还原物质的氧化态k_O、还原态k_R的速率常数以Marcus理论来进行计算，产生的k_O和k_R值应用于公式3-17、3-18。应用这些公式，表面结合氧化还原物质的CV图可以给出E^θ、λ以及K_{ET}，拟合的峰电位与实验检测的峰电位相适应。如果实验数据是已知的，为了限制CV参数的数值，应用了实验数据的表观形式电位(E_{P_c}和E_{P_a}的平均值)，并且λ是固定的，改变k_O和k_R(通过改变E_P，$\eta=E_P-E^{\theta'}$)，直到拟合CV使E_P值与实验值适应，并产生了合理的值。然而，如果在系统中有明显的动力学异质性，这种途径将是不适用的。

$$k_O(\eta)=\frac{\rho|H_{AB}|^2}{h}\left(\frac{\pi}{kT\lambda}\right)^{1/2}\int_{-\infty}^{\infty}\exp\left[-\frac{(\varepsilon_F-\varepsilon+\eta-\lambda)^2}{4\lambda kT}\right]\times\left\{\frac{\exp[(\varepsilon-\varepsilon_F)/(kT)]}{1+\exp[(\varepsilon-\varepsilon_F)/(kT)]}\right\}d\varepsilon \tag{3-20}$$

$$k_R(\eta) = \frac{\rho|H_{AB}|^2}{h} \left(\frac{\pi}{kT\lambda}\right)^{1/2} \int_{-\infty}^{\infty} \exp\left[-\frac{(\varepsilon_F - \varepsilon + \eta - \lambda)^2}{4\lambda kT}\right] \times \left\{\frac{1}{1+\exp[(\varepsilon-\varepsilon_F)/(kT)]}\right\} d\varepsilon \quad (3-21)$$

4. 计时安培(CA)法

CA 法是一种电势阶跃方法,在单步阶跃实验中,同时检测工作电极的过电位以及随时间变化的电流衰减[16]。在双步阶跃实验中,电势围绕着氧化还原中心的形式电位施加小的增量,例如,如果氧化还原物质的形式电位是 0 V,第一和第二工作电位将是 +0.05 V 和 –0.05 V。这种方法必须考虑电解质和电极的电势限,另外,起始电位应接近于氧化还原中心的形式电位。

在电势阶跃步骤之间,一个重要的参数是时间施加的长度必须使电流完全衰减。足够长的施加时间可使法拉第电流从电荷电流中分离出来,这对于精确测量是非常重要的。只要施加的时间对于电荷电流比法拉第电流的速率常数小,就可以暂时从法拉第响应中分离。大的电势步骤将导致电荷电流比法拉第电流更大,使数据分析变得复杂。施加的适宜时间必须通过实验来确定,使电流返回基线水平。

测量的电流和工作电势随时间的变化可以分析相关实验数据,η 应当随 i_R 降进行修正,如公式 3-22,R_{SOL} 可以通过 EIS 法来检测。

$$\eta(t) = E(t) - E^{\theta} - i(t)R_{SOL} \quad (3-22)$$

电流 i_t 是法拉第电流和电荷电流(i_f 和 i_{ch})之和,电荷电流 i_{ch} 以公式 3-24 进行检测,C_{DL} 可以使用 EIS 法来检测。

$$i_t(t) = i_f(t) + i_{ch}(t) \quad (3-23)$$

$$i_{ch}(t) = C_{DL}(\Delta\eta/\Delta t) \quad (3-24)$$

电荷 Q_T 可以对整个时间的 i_f 进行积分获得

$$Q_T = \int i_f(t)dt \quad (3-25)$$

在任何给定的时间 t,电荷由公式 3-26 给出

$$Q(t) = Q_T - \int_0^t i_f(t)dt \quad (3-26)$$

最后,在给定时间的速率常数 K_{app} 可由公式 3-27 计算

$$K_{app}(t) = i_f(t)/Q(t) \quad (3-27)$$

作为时间函数的速率常数,$K_{app}(t)$ 并不是过电势的函数 $K(\eta)$,这归因于动力学异质性。如果动力学位点的分散非常明显,以 $\ln(i)$ 对时间作图,发现并不是预言的线性关系,则必须考虑系统的动力学异质性[39,40]。在依赖于时间通过量的电流衰减期间,可以测量不同的表观速率。因此,电流的衰减分为几个不同部分,在此,通过了不同的电流(和时间)量。

Tafel 曲线也可以通过每个过电势与测量的 $K_{app}(t)$ 作图来获得,实验 Tafel 曲线与应用 Marcus 密度态模型产生的理论曲线相适应,Marcus 密度态模型以适宜的参数 K_{ET}、λ 和 H_{AB} 来表示。这种方法已应用于各种系统,包括 Fc 和 Ru(NH$_3$)$_5$ 系统的电子传递动力学分析[39,41]。

5. 交互电流伏安(ACV)法

ACV 法与 CV 法相似,也是一种电势扫描法。循环模式如 Randles 环(图 3-3)所示,ACV 的起始和最终扫描电势特别指出了氧化还原物质的 E^{θ},此外,以正弦曲线振动的 AV 波在电势波形上有层理性,相比较于在电压上的全部变化,AC 的频率可以改变,其振动的幅度较

小,记录产生的交互电流并且其电化学响应明显呈一个单峰。

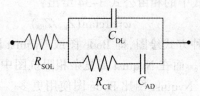

图 3-3　结合于 SAM 上的氧化还原物质的 Randles 环

Creager 和 Wooster 等人发展了氧化还原物质在 SAM 上的电子传递速率的检测方法。在 AC 频率范围内对一系列 AC 伏安变化作图,在此,峰电流 i_p 与背景电流 i_b 的比值由每一个变化的频率来检测。i_p/i_b 值对每个频率的对数值作图,这个图将与 Randles 的环模式相适宜[42]。在 ACV 法中,检测 K_{ET} 需要的参数是双电层电容(C_{DL})、电荷转移阻力(R_{CT})、电极表面积(A_{SUR})以及表面覆盖 Γ,另外,公式 3-28 至 3-31 给出 Randles 环的 4 个参数。

$$C_{DL} = C/A_{SUR} \tag{3-28}$$

$$C_{AD} = F^2 A \Gamma /(4RT) \tag{3-29}$$

$$R_{SOL} = 1/4\pi r_0 k \tag{3-30}$$

$$R_{CT} = 2RT/(F^2 A \Gamma K_{ET}) \tag{3-31}$$

应用上述方法可以在多肽[43-45]、短链覆盖[27,46]以及其他系统[47-50]中测量 K_{ET},这种方法检测 K_{ET} 有几个优点:首先,输出的变化(C_{DL},Γ,A_{SUR},R_{SOL})很容易从 CV 法或 EIS 法测量获得;第二,即使非常低的表面覆盖也可以测量 K_{ET},这归因于 ACV 法内在的灵敏性。此外,Creage 进一步发展了这种方法,在速率变化区域内,即在具有分散的速率系统中,或者在具有显著动力学异质性的系统中(归因于由 SAM 或表面缺陷引起的分子环境的变化),调整了 i_p/i_b 比率的各部分模拟的适应性。例如,在 5000~10000 s^{-1} 区域内,在高频区调整拟合 i_p/i_b 的适应性,而在 1~100 s^{-1} 范围内,调整了低频区的拟合数据。这种方法的不足是仅仅获得了 K_{ET},而没有获得 λ 和 H_{AB}。

6. 电化学交流阻抗(EIS)法

对于结合于 SAM 上的氧化还原物质的电子传递,Randles 环是最简单的模型,其将被应用于 EIS 法讨论[16,42,51]。EIS 法是通过测量阻抗 Z 来测量系统的频率响应,在一个特定的电势下,在频率范围内施加一个小的 AC 信号,频率的变化改变了 Randles 环中每一部分对整个阻抗的贡献。在一个宽范围的频率通过对阻抗的测量可以对 Randles 环的每个参数进行测量[16,52,53]。

R_{SOL} 和 R_{CT} 仅仅贡献于内相,或者是阻抗的真实部分 Z_{Re},电压和电流之间的电阻没有影响 φ 相,电压滞后电流 90°穿过电容器,所以电压和电流外相穿过 C_{DL} 和 C_{AD},即 C_{DL} 和 C_{AD} 贡献于外相,或者假定的部分(Z_{Im})[16]。

Z_{Re} 和 Z_{Im},Z 通常以复杂的公式 3-32 描述,ω 是 AC 信号的角频率。

$$Z(\omega) = Z_{Re} - jZ_{Im} \tag{3-32}$$

$$j = (-1)^{1/2}$$

通常,阻抗 Z 在复杂形式中的数值由公式 3-33 给出:

$$Z=(Z_{Re}^2+Z_{Im}^2)^{1/2} \tag{3-33}$$

而阻抗(φ)在复杂形式中的相由公式 3-34 给出：

$$\varphi=\arctan(Z_{Im}/Z_{Re}) \tag{3-34}$$

EIS 法数据通常以两种方式绘图，即 Bode 图或 Nyquist 图[16,52,53]。在 Bode 图中，$\lg Z$ 和 φ 相与频率的对数分离作图。而在 Nyquist 或复杂相阻抗图中，纵坐标是假设轴 Z_{Im}，横坐标是真实轴 Z_{Re}，从数据分析上，Nyquist 图比 Bode 图使用更多。在最简单的实验中，选择的电势没有发生电子传递(即电势偏离了结合于 SAM 上的氧化还原物质的 E^o)，在此，不考虑 Randles 环的法拉第部分，而仅仅考虑 R_{SOL} 和 C_{DL}。在高频区($\omega \rightarrow \infty$)没有对 C_{DL} 充电，曲线接近于 Z_{Re} 轴，当频率减小时，则有更多的时间对 C_{DL} 充电；而在低频区($\omega \rightarrow 0$)，对阻抗主要的贡献是 C_{DL}，其结果在图上是一条垂直线，R_{SOL} 贡献的阻抗没有受频率的影响[16,42]。

当选择一个特定的电势(例如，接近结合于 SAM 上的氧化还原物质的 E^o)时，发生了电子传递，必须考虑其对整个 Randles 环的贡献，对阻抗有贡献的 R_{SOL} 和 C_{DL} 使得 Nyquist 图的数据处理变得复杂[54-56]。在高频区，当 $\omega \rightarrow \infty$ 时，图形呈椭圆[42,56]，没有发生电子传递(R_{SOL} 和 C_{DL} 忽略不计)，即没有对 C_{DL} 充电的时间，当 $\omega \rightarrow 0$ 时，椭圆的高频部分 R_{SOL} 出现在 Z_{Re} 轴。

公式 3-35 表明了椭圆的中心定位在这点的真实轴上[56]

$$Ellipse_{center(Z_{re})}=R_{SOL}+R_{CT}/2[1+(C_{DL}/C_{AD})]^2 \tag{3-35}$$

公式 3-36 表明了在 Z_{Im} 轴上椭圆的最大值

$$Ellipse_{\max(Z_{Im})}=R_{CT}/2[1+(C_{DL}/C_{AD})] \tag{3-36}$$

公式 3-37 表明了椭圆上的点将在低频区($\omega \rightarrow 0$)穿过 Z_{Re} 轴

$$Ellipse_{low(\omega)((Z_{re})}=R_{SOL}+R_{CT}/[1+(C_{DL}/C_{AD})]^2 \tag{3-37}$$

然而，在低频区，Nyquist 图的椭圆部分并未接近 Z_{Re} 轴，实际上，随着更大 Z_{Im} 部分的增加，C_{DL} 和 C_{AD} 对阻抗的贡献将占优势，图形变成一条垂直线。垂直线对 Z_{Re} 轴贡献，相同点上的截距在 $\omega \rightarrow 0$ 时椭圆将穿过 Z_{Re} 轴。

在此，R_{SOL} 可以直接测量，应用恒电位器软件可以同时检测 C_{DL}、C_{AD} 和 R_{CT} 的值，一旦获得这些参数，应用公式 3-38 可以检测 K_{ET}[42]。

$$K_{ET}=1/R_{CT}C_{AD} \tag{3-38}$$

使用 EIS 法，测得十六烷基—SH/SAM 上的十五烷基—SH—Fc 乙酰胺基的 K_{ET} 是 9 s^{-1}[42]，在另一实验中，在 1-NH$_2$ 蒽醌衍生物通过 10-C 烷基链的单层膜在 0.1 mol/L H$_2$SO$_4$ 中测得的 K_{ET} 是 7.4 s^{-1}[57]，并且，在这个实验中，测得 1-NH$_2$ 蒽醌的 λ 具有一个较大的值 2.7 eV。

EIS 法因为可以同时获得几个不同的参数，所以应用非常广泛。但是，由于系统的非理想行为，也产生了许多问题。非理想行为可以改变 Randles 环模型分析获得的值，对于数据的精确性，必须强调这些非理想行为。处理非理想化的一种方式是使用外加参数，一般的方法是在环中合并为一个常数项元素(C_{PE})，来代替 Randles 环的一个或几个参数[42,58-60]。

7. 扫描电化学显微镜(SECM)

SECM 与其他微扫描技术，如 STM 很相似，这种方法一般是将一个修饰平面浸入电解质溶液，针尖呈微米~纳米尺寸的金属探针作为一个电极，或者应用一个超微电极(UME)。SECM 允许 UME 非常接近研究的表面，因此，通过扫描产生的表面反应可以给出形貌图，观察到的信号是溶液中电活性物质的法拉第电流[61,62]。

在 SECM 实验中,可将氧化还原分子结合于单层膜上或是将单层膜作为溶液中氧化还原物质和电极之间电子传递的壁垒。K_{ET} 是针尖与 SAM 覆盖电极之间距离与电流的函数,并且拟合理论曲线。SECM 实验的一个优点是双电层充电电流及抵抗电势降减少,因为所有测量是在稳态条件下进行的(例如,针尖保持在一个恒定的电偏压下),SECM 实验的另一个优点是有在通过电极表面的不同点获得多级速率常数的能力,并允许检测 SAM 上的空间异质性。

SECM 实验常常应用电子传递中介(图 3-4),电化学中介可以是非常小的氧化还原活性物质,如甲基紫精、$Ru(NH_3)_6^{3+}$ 或者 $Fe(CN)_6^{3-}$,中介也应用于许多蛋白电化学,但是直接测量通常是不可能的[63]。

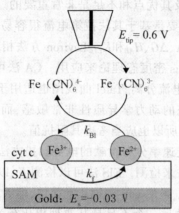

图 3-4 应用电子传递中介的 SECM 工作原理图[63]

对于 SAM 上没有结合的氧化还原物质,SECM 可用于测量溶液中的物质和穿过 SAM 电极之间的 K_{ET}[62,63]。然而,对于结合于 SAM 上的氧化还原物质,SECM 可以测量氧化还原物质穿越单层膜到电极之间的电子隧穿速率,而且,这种方法也可以检测氧化还原物质和溶液中介之间的生物分子电子传递速率。

SECM 法测量 K_{ET} 的首次报道,是以 $Ru(NH_3)_6^{3+}$ 作为溶液电子传递中介,测量了单层膜结合 $Fc[FcCO(CH_2)_7SH]^+$ 的 K_{ET},如以十六烷基—SH 作为混合 SAM 试剂,测定的标准速率常数是 7.0 s^{-1}[62],当使用壬烷—SH 作为混合 SAM 试剂时,测定的标准速率常数是 1.2×10^5 s^{-1}。而且,随烷基链亚甲基基团的增加,K_{ET} 呈指数减少,这是因为发生了隧穿,过程的 β 值是 1.0/亚甲基,这与其他类型单层膜的值基本一致[13,14,64]。电子传递中介 $Ru(NH_3)_6^{3+}$ 和结合于 SAM 上 Fc 之间的生物分子 K_{ET} 大于 4.5×10^{10} cm^3·mol^{-1}·s^{-1}。另外,研究者也测量了 $Ru(NH_3)_6^{3+}$ 穿越非电活性 SAM 的电子传递速率,在此,中介物质通过 SAM 的电子隧穿可以获得再生。

研究者以 $Fe(CN)_6^{3-}$ 作为中介,应用 SECM 测量了结合于 11—SH—COOH/SAM 上的 cyt-c 的 ET 动力学过程[63],研究表明,经由酰胺键形式在蛋白-lys 和羧基尾端 SAM 上共价结合的 cyt-c,与蛋白静电结合相比,具有较慢的隧穿速率,$K_{ET}=9.0$ s^{-1}。然而,如果蛋白是以静电形式结合,其 $K_{ET}=15$ s^{-1}。当使用 1:1 癸烷-SH 和 11-羧酸-SH 混合 SAM 时,可以观察到更快的隧穿速率——65 s^{-1},其原因是 cyt-c 在 SAM 表面的静电结合,增加了其迁移率,允许蛋白在电极上更有效地定位,使得 ET 更容易发生。测量的 cyt-c 和中介之间生物分子 ET 大

约是 2.2×10^8 $cm^3mol^{-1}\cdot s^{-1}$，这与相关文献值基本一致[63]。

8. 间接激光—诱导温度跳跃(ILIT)法

CV法和CA法通常受限于速率（$<10^4 s^{-1}$），对于更快的速率常数的测量容易产生较大误差。而ILIT技术可应用于更短烷基链接基团SAM速率常数的测量[26]。其方法是以短的激光脉冲撞击超薄金膜电极的后部，电极吸收的激光能（<1 ps）很快扩散，热穿过电极引起金膜的另一面温度上小的变化，使得SAM/电解质溶液界面的界面平衡发生了变化，并且电极的开环电势也随之改变。这种方法一般要求特定的装备，所以研究相对较少。

本节描述了几种常用的电化学技术，定量地测量了电子传递参数，如 λ、ΔG、H_{AB} 及 K_{ET}，了解这些技术的原理、数据处理及其优点和不足是非常重要的。

CV法是最常用的技术，主要是基于其法拉第电流很容易从其电荷电流中分离出来，CV法可以测量电子传递参数，如 λ、ΔG、H_{AB} 和 K_{ET}。Laviron方法相对简单，然而，它的应用存在一定的限制，所以经常结合Marcus密度态理论来应用。CA法可以控制 λ、H_{AB} 和 K_{ET}，在其测量中，法拉第电流可暂时与电荷电流分离，Tafel曲线图也适用于Marcus理论。另外，CA法测量的一个重要的方面是对系统的动力学异质性非常敏感，而且，在CV法和CA法测量中，溶液因可能产生反常低的 λ 值，所以也应当考虑其阻抗值。

ACV法是一种理想的确定速率分布贡献的电势扫描方法，但这种方法仅仅获得了 K_{ET}，λ、ΔG、H_{AB} 必须使用其他的方法来检测。EIS法可以检测 K_{ET} 以及 R_{SOL} 和 C_{DL}。与ACV法相似，EIS也不能测量 λ、ΔG 和 H_{AB} 值。虽然EIS实验非常简单，但因为必须考虑相的影响，而且，其数据分析也非常复杂；EIS法的另一个不足是其非理想化行为改变了ET参数的值。SECM可以在单个实验中同时测量吸附和溶解物质的 K_{ET}，小尺寸的SECM针尖可以获得SAM表面形貌的信息。因为SECM是在稳态条件下进行的测量，所以SECM的另一个优点是 C_{DL} 和溶液抵抗电势降非常小。

第二节　自组装膜上 K_{ET} 电化学测量的氧化还原体系

在应用电化学检测SAM中氧化还原物质的 K_{ET} 中，金属络合物显示了可逆的电化学响应。其高价和低价两种氧化态可以稳定存在，而且，这些络合物通常具有较高的配位数，并允许小型配体取代插入（图3-5），其氧化还原中心的内球 λ 值可以忽略，并且在能量上容易接近这些金属络合物的氧化还原电势（例如，SAM的稳定性以及电解质和溶剂决定了金属络合物的电势窗范围）。

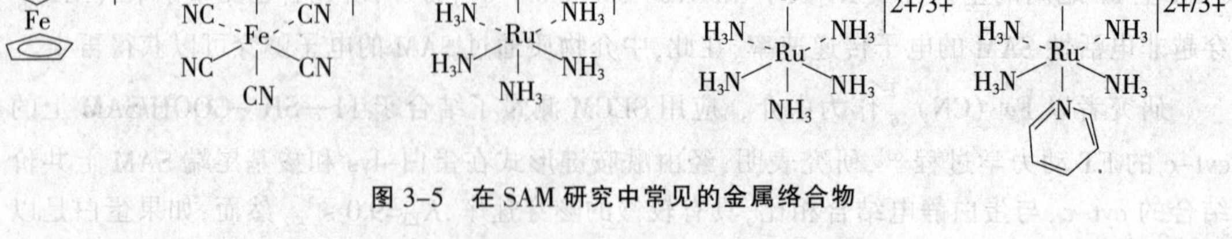

图3-5　在SAM研究中常见的金属络合物

1. 金属茂

1951年，Kealy和Pauson发现了有机金属络合物Fc，并且在1952年确定了它的结构，两个环戊二烯阴离子以η5与一个Fe(Ⅱ)配位,产生了完全中性的络合物[65]。络合物的几何尺寸很小,大约是 4.1 Å×3.3 Å,Fc⁻与Fc稍有不同,约为 4.1 Å×3.5 Å,测量结果表明其内部重组能非常低。Fc中环戊二烯的C—C键长为 1.4 Å,Fe离子和C原子的键长为 2.04 Å。这种夹心结构也可以应用其他过渡金属如Ru、Co、Ni来制备,但这些络合物并不应用于SAM的电化学研究。

Fc在电化学中最稳定的离子氧化态形式是$Fe^{3+/2+}$,Fe(Ⅱ)是橙色的,而Fe(Ⅲ)呈蓝色。Fc对于结合于环戊二烯环上功能基团的电子性质非常敏感[66-68]。Fc易溶于有机溶剂,不易溶于水,其具有芳环化合物的性质,能够通过Fe—C反应制备各种衍生物,如Fc—羧酸、Fc—二羧酸以及二甲基胺基甲基Fc等。另外,Fc很容易和四丁基锂反应生成锂化Fc,它是一种非常重要的对称化合物合成的前驱体[68-75]。

2. 钌—胺络合物

三氯六胺合钌络合物中的Ru中心以六个NH_3配体对称配位,呈八面体构型,$Ru(NH_3)_6^{3+}$中Ru—N键长是 2.104 Å,还原态Ru(Ⅱ)络合物中Ru—N键长微增至 2.144 Å[76]。和Fc中的Cp阴离子配体不同,钌络合物中性的NH_3配体并未中和电荷。络合物呈灰黄色,易溶于水,其良好的溶解性和稳定性使其作为电化学标准物得到了广泛应用[77]。

在无机化学中,很容易合成$[Ru(NH_3)_5Cl]^{2+}$,其他的配体也同时被研究,如 1965 年由 Allen 报道的络合物$[Ru(NH_3)_5N_2]^{2+}$[78]。1987 年,由于使用混合价态吡嗪(pz)吡啶桥联Ru的双核络合物$[Ru(NH_3)_5pzRu(NH_3)_5]^{5+}$在电子传递研究中的贡献,Taube 获得了诺贝尔奖,这种双核络合物称为C—T络合物[79]。

由$[Ru(NH_3)_5Cl]Cl_2$还原得到的五胺合钌络合物$[Ru(NH_3)_5H_2O]^{2+}$具有一个易变化的内部水分子配体,水分子配体可被各种N—给体配体(如—NH_2和N—杂环)替代,—NH_2配体络合物比N-杂环配体络合物的氧化还原电势更加灵敏[80,81],相关的λ参数可由$[Ru(NH_3)_5(py)]^{2+/3+}$转化获得,如氧化为Ru(Ⅲ)后,Ru—Npy键长在$[Ru(NH_3)_5(py)]^{2+/3+}$中从 2.058 Å变为 2.077 Å,表明其有小的λ_i[82]。各种稳定的四胺合钌络合物$[Ru(NH_3)_4L_2]^{2+}$也已经被分离并被表征[83]。

3. 锇—氮杂络合物

Os的配位化学性质接近于Ru[84],Os(Ⅱ)、Os(Ⅲ)和Os(Ⅳ)的N—杂环配体如双吡啶和邻二氮菲络合物很容易从K_2OsCl_2合成得到,并具有较高的稳定性和接近的氧化还原电势,广泛地应用于电化学研究[85](图 3-6)。

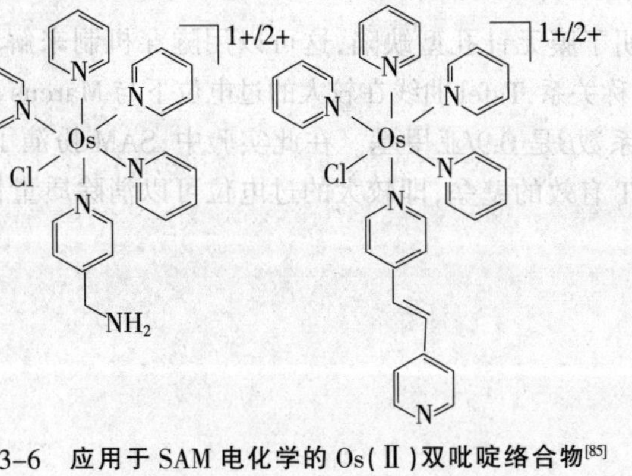

图 3-6 应用于SAM电化学的Os(Ⅱ)双吡啶络合物[85]

4. 富勒烯 (C_{60})

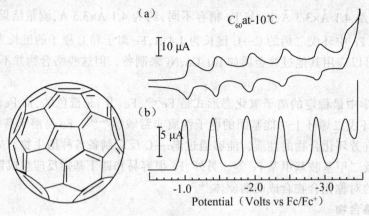

图 3-7 C_{60} 的结构及其电化学响应[86]

C_{60} 具有 8 个易接近的氧化态[86-88](图 3-7),并且具有非常小的内部 λ(约 0.06 eV)[89]。C_{60} 复合多卟啉结构作为光合成的模型[90-96],共价结合于 SAM 上[97-100],可以产生高效的光诱导长寿命电荷分离态。这种分子一般被固定于 ITO 表面,通过测量透明导电基底 C_{60}—SAM 光电流研究其光诱导过程。

C_{60} 也可以非共价形式结合于 SAM 上[86,101],例如,冠醚功能化的 C_{60} 通过冠醚基团与修饰电极上的—NH_2 基团相互作用进入 SAM 中[86]。大多数与 C_{60} 相关的光诱导电子传递研究使用了光谱技术[90,102-104],与 Fc 或其他金属络合物不同,电化学方法研究 C_{60} 电子传递动力学尚未见相关报道。

5. 溶液中的氧化还原物质

为了使用电化学方法测量溶液中氧化还原物质的 ET 参数,经常使用大的 η(≥1 V),较慢的 ET 速率是为了避免质量传递的影响[16]。在溶液中氧化还原物质满足较慢速率最简单的方法就是使电极 SAM 钝化,即 SAM 扮演了 ET 的壁垒,减缓了溶液中氧化还原物质和电极之间电子传递的速率,SAM 的厚度可以改变以调整 ET 速率,使之范围可应用于理想的 ET 研究。

Miller 等人[13]通过应用各种链长(从 2 个亚甲基到 16 个亚甲基)的羟基硫醇,在金电极上制备了不同厚度的单层膜,CV 响应由溶液中的氧化还原物质探针分子 $Fe(CN)_6^{3-}$ 和 $Fe(H_2O)_6^{3+}$ 来提供,电容测量证明了膜无针孔型缺陷,这可以用隧穿机制来解释实验现象。研究表明,K_{app} 和单层膜厚度呈对称关系,Tafel 曲线在较大的过电位下与 Marcus 理论预言的相一致[1,13,105],测量 ET 过程的隧穿系数 β 是 0.9/亚甲基。在此实验中,SAM 扮演了阻止了溶液中氧化还原物质与电极之间的 ET 有效的壁垒,即较大的过电位可以消除质量传递的影响。

6. 金属簇

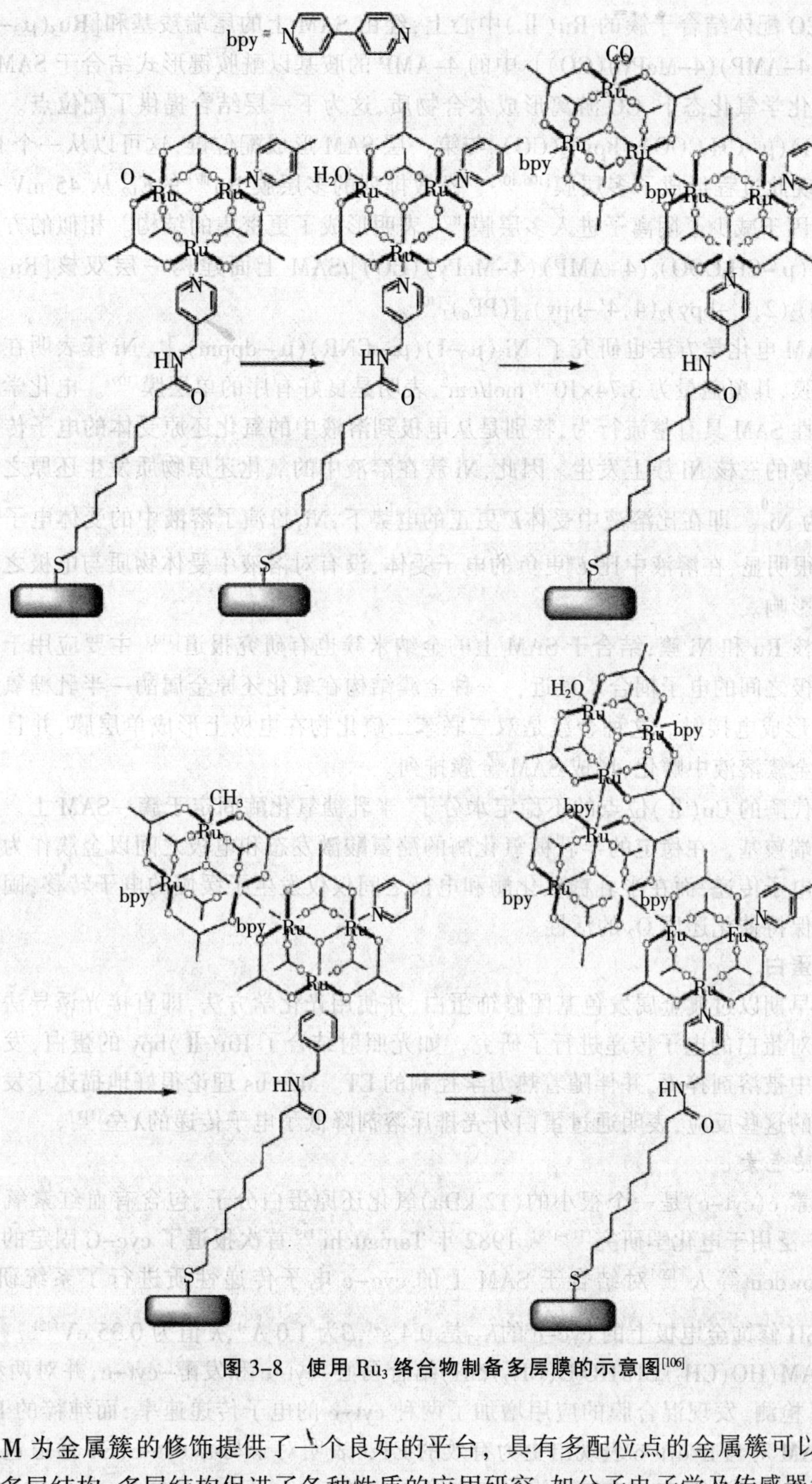

图 3-8 使用 Ru₃ 络合物制备多层膜的示意图[106]

SAM 为金属簇的修饰提供了一个良好的平台,具有多配位点的金属簇可以在 SAM 表面形成多层结构,多层结构促进了各种性质的应用研究,如分子电子学及传感器等[106-108]。三核 Ru 簇 [$Ru_3(\mu_3-O)(\mu-CH_3COO)_6(bpy)_2(CO)$] 是多核簇的一个典型形式,可应用于控制电活性多层形式的层—层沉积[106-108](图3-8)。这种络合物有两个Ru(Ⅲ)中心和一个Ru(Ⅱ)中心,一个CO配体结合于簇的Ru(Ⅱ)中心上,经由SAM上的尾端羧基和[$Ru_3(\mu_3-O)(\mu-CH_3COO)_6$(4-AMP)(4-MePy)(CO)]中的4-AMP的胺基以酰胺键形式结合于SAM上,在Ru(Ⅲ)的电化学氧化态下,CO游离形成水合物质,这为下一层结合提供了配位点。通过加入[$Ru_3(\mu_3-O)(\mu-CH_3COO)_6(bpy)_2(CO)$]与第一层SAM形成配位键,这可以从一个bpy—N簇的Ru,发现此过程形成了多层膜[106,107]。松散排列的多层膜$Ru^{Ⅲ/Ⅱ}$的$E_{1/2}$从45 mV迁移到90 mV,这归因于减少了阴离子进入多层膜[107],表明形成了更密集的结构。相似的方法是在[$Ru_3(\mu_3-O)(\mu-CH_3COO)_6$(4-AMP)(4-MePy)(CO)]/SAM上面建构一层双核[$Ru_2(\mu-O)(\mu-CH_3COO)_2(2,2'-bpy)_2(4,4'-bpy)_2](PF_6)_2$][108]。

使用SAM电化学方法也研究了[$Ni_3(\mu_3-I)(\mu_3-CNR)(\mu_2-dppm)_3$]$^+$。Ni簇表明在金表面形成了单层膜,其覆盖量为$3.74×10^{-10}$ mol/cm^2,表明是良好有序的单层膜[109]。电化学研究表明这种电活性SAM具有整流行为,特别是从电极到溶液中的氧化还原受体的电子传递仅仅在一个负电势的三核Ni簇上发生。因此,Ni簇在溶液中的氧化还原物质发生还原之前必须从Ni_3^+还原为Ni_3^0。即在比溶液中受体E^o更正的电势下,Ni_3^+扮演了溶液中的受体电子传递壁垒的角色。很明显,在溶液中比E^o更负的电子受体,没有对溶液中受体物质与电极之间的电子传递产生影响。

除了三核Ru和Ni簇,结合于SAM上的金纳米簇也有研究报道[110],主要应用于促进金属蛋白和电极之间的电子耦合。最近,一种金簇结构在氧化还原金属酶—半乳糖氧化酶的Cu(Ⅱ)之间形成电接触。这种金簇是双二联苯二硫化物在电极上形成单层膜,并且在巯基羧酸覆盖的金簇溶液中孵化,形成SAM金簇排列。

通过替代酶的Cu(Ⅱ)位点的不稳定水分子,半乳糖氧化酶固定于簇—SAM上。这种酶具有巯基尾端羧基,在稳定的半乳糖氧化酶的酪氨酸激发态和电极之间以金簇作为媒介发生了快速的电子传递,而在半乳糖氧化酶和电极之间仅仅发生了缓慢的电子转移,固定的半乳糖氧化酶保持催化还原O_2的活性。

7. 金属蛋白

研究者早期以过渡金属发色基团修饰蛋白,并使用光化学方法,即直接光诱导法以及闪光猝灭法等对蛋白的电子传递进行了研究。如光照射结合了Ru(Ⅱ)bpy的蛋白,发现其激发态在溶液中被溶剂猝灭,并伴随着热力学控制的ET。Marcus理论很好地描述了发生于蛋白活性位点的这些反应,表明通过蛋白外壳排斥溶剂降低了电子传递的λ垒[111]。

7.1 细胞色素c

细胞色素c(cyt-c)是一个很小的(12 kDa)氧化还原蛋白分子,包含有血红素氧化还原中心,已被广泛用于电化学研究[110-118]。1982年Taniguchi[119]首次报道了cyc-C固定的电极修饰,随后,Bowden等人[120]对结合于SAM上的cyc-c电子传递性质进行了系统研究,在$HO_2(CH_2)_{15}SH$修饰金电极上的cyc-c的K_{ET}=0.4 s^{-1},β为1.0 Å^{-1},λ值为0.35 eV[121]。研究者使用混合SAM($HO(CH_2)_nSH/HO_2C(CH_2)_nSH$)结合马心-cyt-c和发酵-cyt-c,并对两种SAM的K_{ET}进行了检测,发现混合膜的应用增加了两种cyt-c的电子传递速率,而纯粹的$HO_2C-(CH_2)_nSH$/SAM对马心cyt-c表现出更为有效的ET,表明对于马心cyt-c,ET通道包含离子相互作用,而对于发酵cyt-c,其ET途径是非离子化的。

氧化还原蛋白cyt-c通过羧基尾端[$HO_2C(CH_2)_nSH$]静电结合于Ag,对于短链(n<6),使

用电化学及SERRS法观察到蛋白的结构发生了变化[118],当$n=16$时,K_{ET}为0.073 s^{-1},当$n=11$时,K_{ET} =63 s^{-1},而对于$n\leq6$,K_{ET} =134 s^{-1}。研究者也研究了动力学同位素效应,表明系统的氢键重整及其质子传递(PT)亦受限于这种情况,PT随烷基链长的增加而减小[115,122-124]。

以上结果表明高电场效应(短的烷基链)增加了能量势垒,在一项蛋白结合于SAM的相关研究中,λ检测是0.26 eV,相比较于溶液中的0.6 eV较低,这归因于其溶解性存在差异,结合于SAM上的蛋白其氧化还原中心更难以接近。这个实验也表明短链SAM产生了结构控制,高电场增加了蛋白结构重构的活化能垒[115,122,123]。

使用吡啶尾端($pyCH_2NHCO(CH_2)_nSH$)/SAM,SERRS被应用于Ag和Au/SAM上cyt-c的研究[124],吡啶基团为cyt-c血红素提供了结合配体,配位平衡依赖于工作电位。其还原形式是五配位高自旋态,而其氧化态是六配位低自旋态,K_{ET} =760 s^{-1},这与之前测量的(780±40) s^{-1}具有很好的一致性[114,125]。

在使用相同的吡啶尾端SAM的另一个实验中,测得λ为(0.5±0.1) eV[126],对于短的烷基链,数据与模型保持了一致性,溶剂和蛋白结构减弱了质子化过程;而对于长的烷基链($n\geq12$),速率是非绝热的。这一研究结果与H_2OC—尾端SAM相比表明,短链的高电场效应限制了蛋白的构象化[115,122,123]。

结合于MUA上cyt-c的E^θ测量存在误差(范围从0.0 V到–0.06 V),其K_{ET}为(20~100 s^{-1}),Millo等人[115,116,122,127]认为存在误差的原因在于不同测量类型对不同的表面的要求是不同的,SERRS要求一个粗糙的Ag界面,而SEIRS使用粗糙的金界面,CV使用光滑的金界面,cyc静电结合的SAM吸附在四种界面(Au——粗糙、光滑;Ag——粗糙、光滑)上。CV研究表明其E^θ(–0.068 V,vsSCE)与金属或金属的形貌无关。然而,其K_{ET}值依赖于金属类型(16 s^{-1},Ag;33 s^{-1},Au)。

7.2 细菌氧化还原蛋白酶(Az)

研究者发现了Az在细菌中的电子传递链,并对其结构进行了彻底表征[111,117,128]。Az的活性位点包括存在于三个配体(两个his和一个cys)的平面上的单核Cu离子,一个轴向配体(M_{et}),并与一个酰胺键羰基O发生弱相互作用。受蛋白壳的限制,Az配合物的几何尺寸与其氧化态相比,仅有微小的变化,λ值约为0.7~1.0 eV。

Az的电化学研究是将Az吸附于不同链长的烷基硫醇/SAM金电极上[129,130],其理想行为是K_{ET} $(n=10)=(470±50)s^{-1}$,K_{ET} $(n=6)=(3200±300)s^{-1}$[129]。在另一项实验研究中,围绕Cu中心Az的疏水区域和烷基硫醇的甲基发生疏水相互作用被认为稳定了蛋白在SAM上的吸附[130]。Az在电极表面最可能的定位是其氧化还原中心与电极表面呈面对面排列,K_{ET}随烷基链的长度的增加呈指数衰减,$\beta=1.03±0.02$/亚甲基。当烷基链长$n>9$时,应用Laviron和EIS法测量了系统的表观速率常数K_{app},两种方法获得了相近的结果($n=11$,70 s^{-1};$n=17$,0.1 s^{-1})。然而,测得的λ值是0.3 eV,比SAM上的小分子氧化还原物质低得多,这可能归因于SAM环境疏水性的影响。

方波伏安法也应用于Az吸附在烷基硫醇SAM/Au电化学性质的检测[131],但没有观察到动力学同位素效应,也未发现pH值对K_{ET}产生影响。而且,溶液黏性对K_{ET}也无影响,其$K_{ET}=1.0\times10^3$ s^{-1},λ=0.7 eV。数据表明系统的电子传递与上述过程是不同的。

研究者在金基底上使用1:1的$H_3C(CH_2)_nSH/HO(CH_2)_nSH$($n=11,15$)混合SAM研究了Az的电化学性质,应用Az位点的定向变异去识别对电子传递非常重要的氨基酸残基。研究发现,仅仅对野生 Az 以及含有 Trp48 变异体的 Az 展示了一定的伏安响应。使用 Laviron 方法,对于长链(n=11)SAM,其 K_{ET} =63 s^{-1},lgK_{ET} 和链长间呈线性关系,其斜率约为 1.0/CH$_2$,表明在 SAM 头基(CH$_3$ 或 OH)和特定氨基酸残基之间的电子耦合对于电子传递非常重要。

在另一项相关研究中,发现离子强度并未对吸附于 CH$_3$/OH 尾端 SAM-Az 的 K_{ET} 产生影响[132],表明系统中的疏水相互作用比离子强度更为重要,K_{ET} 稳定在 1.0×10^3 s^{-1}(n<9),这归因于蛋白—SAM 界面动力学效应。

应用 1,2-二苯乙烯—SH(具有共轭链)和饱和烷基链的 SAM 也对 Az 的电子传递性质进行了比较[133]。1,2-二苯乙烯为基础的烷基链(15.8 Å)给出了 K_{ET} =1600 s^{-1},而脂肪族对比物分别是 481 s^{-1}(14.5 Å)和 600 s^{-1}(17 Å),显然,链的共轭性并未增加 K_{ET} [27,134]。

Az 蛋白结构中的尿素基团对其电化学性质也存在影响[135]。当吸附于 SAM 时,尿素基团可使蛋白变性,假定蛋白发生变性可观察到 λ 的改变,但奇怪的是,λ 对高浓度的尿素基团并不敏感,表明并未扰动其活性位点。研究表明,SAM 中长的烷基链产生了大的 λ 值,表明增加链长对其结构产生了较大的波动,而短链具有较高的电场强度,可以阻止这些波动[135]。

最近,研究者也使用电极上的纳米粒子膜对 Az 的电化学性质进行了研究[136]。使用烷基二硫醇结合金纳米粒子形成了具有功能性的纳米粒子膜,将 Az 吸附在其表面,电化学研究发现这种纳米膜的 K_{ET} 非常大(β≈0.01/CH$_2$),与之对照的 Az 直接吸附在 SAM 电极的 K_{ET} = 12~20 s^{-1},β=0.9/CH$_2$,表明这种系统的电子传递是 Hopping 机制而不是隧穿机制。

7.3 超氧化歧化酶(SOD)

研究者以三种小牛红血球,即(Cu/Zn—SOD)、(Fe—SOD)、(Mn—SOD),结合于 3-巯基丙酸 SAM/Au 表面,对依赖于 pH 的 K_{ET} 进行了研究,三种 SOD 的电化学性质(形式电位、电极反应的可逆性、动力学参数以及 pH 依赖性等)存在差异,表明其 ET 机制是不同的,Laviron 方法计算均表明在中性 pH 下 K_{ET} 最大[137]。

第三节 电子传递动力学的外球效应

离子对、范德华力、氢键以及溶剂极性对 λ_0 非共价相互作用的影响也对 ET 产生了重要影响,而且,其很难进行测量,其误差常常为±0.1 eV。很少有研究者关注对这个参数的测量,不考虑温度,获得的 λ_0 值与随驱动力实验获得的值并不一致,离子对效应、双电层效应、未补偿的溶液阴离子效应以及速率常数的异质性分布均对此差异产生了影响[39,40,112,113,138-141]。

从 Arrhenius 方程可以获得 λ 和指前因子 A [117],在一定温度范围内以 lnK_{ET} 对 1/T 作图可测量 K_{ET},活化能 E_A 由 Arrhenius 图的斜率获得,公式 3-39 可以计算表观 λ 值,Arrhenius 指前因子 A 及活化能 E_A 由公式 3-40 给出。

$$E_A=\lambda/4 \tag{3-39}$$

$$A=2\pi^{3/2}H_{AB}^2\rho/h \tag{3-40}$$

1. 溶剂及离子对效应

在通过 CA 法来研究[Os(bpy)$_2$Cl(pNp)]—SAM 的实验中,使用乙腈、丙酮、DMF、二氯甲烷、THF 以及氯仿溶剂,考察了溶剂对系统 ET 动力学和热力学的影响。使用相同的溶剂,在 p2p 膜上,K_{ET} 的范围从 7.4×10^3 s^{-1}(CHCl$_3$)至 1.1×10^5 s^{-1}(MeCN)变化;p3p 膜的 ET 值分别

是 $1.6×10^3 s^{-1}$ 和 $1.8×10^4 s^{-1}$。从这两种不同的膜上,可以观察到速率和溶剂相关,表明溶剂重组动力学强烈地影响了 ET。

研究者使用了 11 种溶剂:乙腈、丙酮、甲醇、丙醇、丁醇、DMF、DMSO、THF、乙醇、聚丙烯、碳酸盐,并以 $HO_2C(CH_2)_nSH$ 烷基硫醇稀释 $[Ru(NH_3)_5pyCH_2NHCO(CH_2)_nSH]^{2+}/SAM$ (n=10,15) 探索了 λ_0 的变化,λ 值是通过从 CA 数据产生的 Tafel 曲线来测量的[39]。水溶液中测得的 λ (0.8~0.9 eV)与 Marcus 理论预言的 λ 值相一致(0.9~1.0 eV),而在非水溶剂中,阴极 λ 比相同的阳极 λ 要大,研究者认为这是因为存在强烈的离子对效应。表 3-1 是应用公式 3-10 计算 $[Ru(NH_3)_5pyCH_2NHCO(CH_2)_{15}SH]^{2+}/HO_2C(CH_2)_{15}SH—SAM$ 在各种溶剂中的重组能 λ (eV)。

表 3-1 由公式 3-10 计算的 $[Ru(NH_3)_5pyCH_2NHCO(CH_2)_{15}SH]^{2+}/HO_2C(CH_2)_{15}SH—SAM$ 重组能

	阳极 λ/eV	阴极 λ/eV	计算 λ_0
H_2O	0.9	0.8	0.92
DMSO	0.9	0.7	0.75
乙腈	0.9	0.7	0.91
DMF	0.9	0.7	0.90
甲醇	0.9	0.7	0.92
乙醇	0.9	0.7	0.85
丙酮	0.9	0.7	0.84
丙醇	0.6~0.7	0.6	0.81
丁醇	0.6	0.5	0.78
THF	0.6	0.5	0.67

假设在 SAM 界面上局部存在高浓度水,且没有采取办法干燥有机溶剂,使用公式 3-10,λ_0 从 0.92 eV 变化到 0.75 eV,正如公式 3-10 预言的,非极性溶剂丙醇、丁醇、THF 有低的 λ_0 值,这意味着测量的值与水溶液条件下测量的值相似[39]。研究者认为精确测量 λ 的困难并不表明公式 3-10 存在错误,这可能是在 SAM 上的氧化还原中心估计 λ_0 的另一种方式。

研究者也测量了 Fc 酰胺键 SAM 的 K_{ET} 的离子对(BF_4^-, ClO_4^-, PF_6^-)效应[140,141],十二种非环状和环状的包含有酰胺键的 Fc 经由一个胱胺—硫基团固定于金微电极上,使用各种温度以 CV 法测量了 K_{ET} 和 λ,K_{ET} 的测量范围为 $(4.4~12)×10^3 s^{-1}$,在 BF_4^- 离子对中获得了最高的 λ 值,这是基于它与 Fc 阴离子有弱的结合,最刚性的环状多肽相比较于更灵活的非环状 Fc 多肽,发现其 λ 更小,其测量范围是从 0.3 eV 到 0.5 eV。

2. 质子参与的电子传递(PCET)

质子和电子传递的细节是理论和实验研究的热点问题,PCET 存在于许多生物过程(如酶促反应、光合成以及呼吸作用等)中,其应用领域包括能源(燃料和太阳能电池)、传感器以及分子电子学等[32,142,143]。研究者对 PCET 已进行了较为详尽的研究[139,144-149],这里仅仅对烷基硫醇 SAM 电子传递系统中的 PCET 进行简单讨论。

PCET 对质子和电子传递的可能机制分为分步及协同两种机制,在分步的机制中,ET 和 PT 分别发生,研究依赖于 ET 的 pH 可以阐明这个机制。在协同机制中,ET 和 PT 在相同的速率决定步骤中发生,这种机制一般通过动力学同位素效应来研究[150]。

对于 PCET,电化学方法可以对所有的 pH 依赖行为进行研究,如 Takeuchi 等人[151]研究

了 [Os(Ⅱ/Ⅲ)(terpy)(bpy)(H$_2$O)] 系统的 PCET 行为,Finklea 及其合作者[141]进一步对 [Os(Ⅱ)(bpy)$_2$(4-AMP)(H$_2$O)]系统的 PCET 行为进行了研究。结果表明,[Os(Ⅱ)(bpy)$_2$(4-AMP)(H$_2$O)]具有适当的电位,其 PCET 行为依赖于 pH 值。将 Os(Ⅱ)(bpy)$_2$(H$_2$O)(4-AMP)经由酰胺键结合于 HOOC—尾端烷基硫醇 SAM 上,从热力学角度看,结合的 Os(Ⅱ)水合系统展示了预期的电子行为,当 pH 小于 2 时(0.3 V vs SCE),其形式电位随 pH 变化而变化;而当 pH 大于 9 时(-0.11V vs SCE),形式电位保持稳定。相应的电势值和 pK_a 值与物质在溶液中具有一致性[151]。然而,从动力学的角度看,这种系统偏高于分步模式的预言,在低的和高的 pH 值下,阴极 λ 比阳极 λ 显著减小(0.6 eV vs 1.4 eV),以 lgK_{ET} 对 pH 作图,并未出现分步模式中预言的形状,而且,对于所有的 pH 值,假设 α 一直是 0.5,这种行为应归因于存在 PCET 的协同机制[148]。

结合 HOOC—尾端烷基硫醇 SAM 去比较其 PCET,形式电位并不随 pH 变化而变化,表明可以忽略任何双电层效应,阴极 λ(0.64 eV)和阳极 λ(0.5 eV)略有不同,其 K_{ET} 平均值为 (11±1) s^{-1}。

Finklea 及其合作者进一步使用 D$_2$O 对 Os(Ⅱ)水合系统进行了研究,调查了 PT/ET 机制中的动力学同位素效应。这种系统中不对称的 Tafel 曲线表明了对于 Os(Ⅱ)和 Os(Ⅲ)有不同的 λ。对于 Os(Ⅲ),λ 与 pH 无关(0.6~0.7 eV),而对于 Os(Ⅱ),λ 依赖于 pH 值(0.7~1.0 eV)(在一定 pH 范围内,具有更小的 λ,接近于络合物的 pK_a),依赖于 pH 的 λ 说明其电子传递依赖于 pH,研究者认为在这个系统中的 PCET 主要是协同机制。

最近,Costentin[152]对溶液中[Os(bpy)$_2$(py)(H$_2$O)]$^{2+/3+}$ 的 PCET 行为进行了研究,表明离子传递的速率受分步模式控制,结果与 Finklea 等人研究的 SAM 结合类似 Os 络合物相反,这表明了羧基在 SAM 溶液界面作为质子源,结合了 Finklea 络合物中心电子传递。

研究 PCET 机制最多的氧化还原活性有机分子是氢醌,它失去两个电子被氧化,并相应失去了两个质子。依赖于 pH 值的 K_{ET} 的氢醌可经由一个饱和烷基链 HQC-11 或者共轭链 OPV(HQ-OPV)(图 3-9)结合于电极表面(pH=8~12.6)[153]。在 SAM 上,氢醌基团以辛烷—SH 稀释,HQC-11 和 HQ-OPV 的电化学性质接近理想态,并且在一定 pH 值下是可逆的。对于 HQ-OPV 和 HQ-C11,在 pH=12.6 时,α 分别是 0.48 和 0.55。应用 Laviron 方程计算系统的 K_{ET},对于质子化形式的氢醌,链的结构显著影响其 K_{ET},HQ-C11 和 HQ-OPV 分别为 0.1 s^{-1} 和 77 s^{-1},而对于去质子化形式的氢醌,K_{ET} 链的结构缺乏灵敏性,HQ-C11 和 HQ-OPV 分别为 120 s^{-1} 和 268 s^{-1}。

将 2-甲基-1,4 萘醌衍生的包含有 5~12 个亚甲基的烷基链的硫醇作用于金基底上,在酸性条件下应用 Laviron 方法计算了 K_{app},lnK_{app} 对 12 C 链为 10.4,β 值为 0.89±0.16/亚甲基[154],这与之前描述的五胺基(吡啶)Ru—尾端 SAM 的值(1.06±0.04)基本一致[117]。

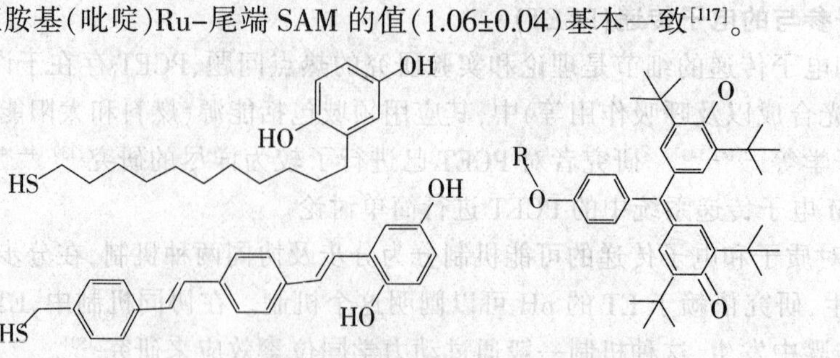

图 3-9 HQ-C11、HQ-OPV 以及 Ga 的分子结构

研究者应用PCET对烷基硫醇修饰Ga(图3-9)—SAM/Au的电化学性质进行了研究[147,148]，当pH<11时，可以预言CV图的不对称性。对于去质子化的Ga/Ga激发氧化还原电对，在pH=10~13时，K_{ET}是$4.5×10^3\ s^{-1}$，但没有Ga内部λ和外部λ的相关报道。这可与相似Fc-标记SAM系统对比，Fc系统以相同烷基链长的λ值约为0.8 eV。

以STM、EPR以及CV法对Au(111)表面的Ga层进行研究时发现两相具有不同的分子密度[149]，EPR证实了Ga的激发态被保护，失去一个电子的氧化以及相应的失去一个质子的还原由其氧化电位来决定，并且与其在溶液中的无激发行为相似，这表明吸附的激发态的电子性质并未剧烈地变化。

4-氨基-TEMPO结合于羧基尾端的HOOC-SAM/Au，应用CV法分析得到K_{ET}为$1.2\ s^{-1}$，TEMPO和TEMPO$^+$的λ分别为1.5 eV和1.3 eV[139]，这些值均是TEMPO在水溶液中的首次实验测量。

第四节　氧化还原自组装单层膜的结构

自组装膜的均一性对于精确测量K_{ET}、λ及H_{AB}非常重要，自组装膜的形成方法显著影响着自组装膜的结构。电极材料的选择、烷基硫醇稀释剂、自组装膜尾端基团以及与氧化还原中心联系的方式等均对其电化学性质产生了重要影响。电化学技术对于自组装膜完整性的表征具有高度灵敏性，这里仅仅对自组装膜的结构形式、自组装膜缺陷的类型、自组装膜的修饰以及自组装膜结构的电化学性质进行描述。

1. 自组装膜的结构形式

烷基硫醇和二硫化物在金上的自发吸附形成了良好有序的自组装膜，自组装膜的结构形式通常假定为分子在一个平坦界面上的均一排列[155-158]。然而，自组装膜表面也可能存在各种缺陷，如针孔、塌陷点、岛状物等[159]。扫描技术可表征其结构的变化，如人字形、阶梯形以及晶阶等[160-162]。基底金属的选择在检测自组装膜完整性方面是非常重要的因素，水银作为一种液态金属，具有无缺陷的表面[163]；而金、银以及其他金属无缺陷的聚晶表面倾向于形成原子尺寸的缺陷，使得自组装膜测量ET复杂化。研究者假设形成自组装膜的有机分子均一排列，烷基硫醇分子斜置角在自组装膜上有显著的变化[164]。在自组装膜形成的早期，烷基硫醇平躺在界面，随着时间的延长，大多数单层膜分子趋于平衡，以标准的分子轴排列在界面上。然而，一些分子链仍然平躺在界面上，在单层膜上的分子空白处称为针孔缺陷[165,166]。

2. 自组装膜缺陷的电化学分析

微探针扫描技术(如STM、AFM)很难认识自组装膜的缺陷[167,168]，而电化学技术对于纳米尺寸自组装膜缺陷具有高度的灵敏性，针孔缺陷的电化学识别可以作为超微电极分析的模型[169]。Lennox等人的研究表明CV法可以识别塌陷位点，如使用Fc烷基硫醇的电化学标记可以很快发现缺陷位点[170]。

在自组装膜结构和动力学研究中，CV法是最常使用的电化学方法。这是因为混乱无序的自组装膜可使电解质穿越自组装膜缺陷区到达电极表面，增加了电容电流。在早期自组装膜的ET研究中，羟基硫醇自组装膜的电容随链长的增加而减小，电容值从$12.6\ \mu F/cm^2 (2\ C)$减小到$1.36\ \mu F/cm^2 (16\ C)$，表明长链烷基硫醇形成的自组装膜无针孔缺陷[13]。

自组装膜多孔性控制的一个独特的实验是利用偶氮基团可逆的光致异构化(图3-10)，

Fc-偶氮基以各种方式结合于 ITO 电极上[171-173]。在 UV 照射下,Fc 尾端和偶氮基团的松散的自组装膜结构变得更加致密。应用 Laviron 方法对 UV 照射前后的 K_{ET} 值进行了测量,分别为 0.24 s^{-1} 和 0.11 s^{-1}。K_{ET} 的减小归因于围绕 Fc 的基团在自组装膜上微环境的变化,对应于偶氮基 N=C 键的 *trans-cis* 光致异构化[173],显然,单层膜上多孔性的减少增加了自组装膜对电子传递的阻塞能力。

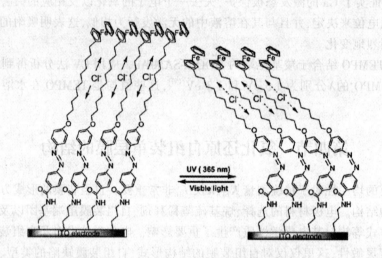

图 3-10 偶氮基团 Fc-自组装膜的光致异构化效应[171]

3. 自组装膜缺陷对电子传递测量的影响

自组装膜缺陷存在局部的微环境,使 ET 更易发生,并产生了明显的速率分布,如动力学分布和动力学异质性分布。对电化学响应的速率分布的影响已被计算模型化[174],如 Murray[40]报道了影响动力学分布的模型,给出了自组装膜表面 Fc 位点形式电位 $E^{\theta'}$ 的 ΔG 及 λ 分布值,相比较于均一相动力学位点的值,理想的 Tafel 曲线偏离 $E^{\theta'}$ 的 ΔG 分布,计算获得较低的 λ 值和更高的 K_{ET} 值。

对于不理想 CV 峰形的氧化还原位点和电极之间的距离也可通过动力学参数值的分布来描述[175],研究者也发展了空间多相性的数学模型,实际上,如果存在这种类型的动力学分布,K_{ET}、H_{AB} 和 λ 将不能被精确测量。

针孔缺陷可以应用电化学技术,通过估计单层膜有效的表面覆盖 θ 定量地测量,θ 值代表了区域被自组装膜覆盖,使得 ET 难以发生。$(1-\theta)$ 表示电极表面未被自组装膜覆盖的区域,可使溶液中的氧化还原物质接近电极表面,使用公式 3-41,可以计算 θ 值[176],R_{CT}^{0} 是裸电极的界面电阻,R_{CT} 是实验测量的电荷传递阻力。

$$\theta = (R_{CT} - R_{CT}^{0})/R_{CT} \tag{3-41}$$

4. 自组装膜的形式

在许多情况下,烷基硫醇在自组装膜形成之前已结合了氧化还原基团,这种结合经常使用 C—C 键、酰胺键或者酯键的形式[177,178]。使用低浓度(1 mmol/L)的烷基硫醇,长的烷基链,延长吸附时间,增加热力学退火过程,通过多次沉浸或是将生成的自组装膜进一步沉浸在第二种硫醇溶液中可以有效减少缺陷的发生[178,179]。

例如,Fc-自组装膜的形成,是将电极沉浸在 Fc 烷基硫醇稀溶液(0.1 mmol/L)和稀的烷基硫醇(1.0 mmol/L)溶液中,研究者测量了烷基链长含 5~18 个 CH_2 连接基团、各种硫醇稀释溶液、Fc 和烷基链之间三种不同的连接键结构(C—C,酰胺键,酯键)的 Fc-自组装膜系统的 K_{ET} 值。

$[Ru(NH_3)_5(4-AMP)]^{2+}$以酰胺键与尾端羧基硫醇结合形成自组装膜[180],这种自组装膜在 CV 图中具有多重信号。使用$[Os(bpy)_2(dipy)Cl]^+$也可有效地将 Os 结合于自组装膜,这些 Os—自组装膜在 Pt 电极的 CV 表明了在高氯酸阴离子和其阳离子头基之间以离子对形式强烈影响系统的能量,应用 Laviron 方法计算系统的K_{ET}的低限是 10^5 s^{-1}[181]。

以三种不同方式将 Os(Ⅱ)(bpy)络合物结合于金电极上,可应用 CV 法和 Laviron 法进行探索。不同链长的烷基硫醇($n=3,11,16$)通过酰胺键结合的 Os 络合物,使用 Laviron 处理 C_{11},C_{16}($\alpha=0.5$),K_{ET}值分别为 1870 s^{-1} 和 8 s^{-1}[182]。

研究者也使用$[Os(bpy)_2(bpe)Cl]^+$比较了共轭链和饱和链的影响[183],如不考虑链的共轭性以及短的电子传递距离,伏安响应是研究穿越空间的隧穿过程的最佳模式。对于共轭链,测定的K_{ET}是 9.4×10^4 s^{-1},约是饱和链(1,2-二吡啶基乙烷)的 1/30。这种差异可能的原因归于饱和链的灵活性,它使系统中的氧化还原中心更容易移动接近电极表面;相反,结合了氧化还原中心的共轭链是刚性的,阻止了氧化还原中心接近电极。所以,与共轭系统相比,就对饱和链产生了较大的耦合和更短的电子传递距离,产生了更快的穿越空间的 ET 反应。

5. 自组装膜形成之后氧化还原中心的结合

氧化还原中心与烷基硫醇的共价结合在合成上存在一定困难,或是仅仅限于有限的几种氧化还原物质(如金属蛋白),自组装膜形成之后氧化还原中心的进一步结合可有效地解决上述问题。氧化还原中心在自组装膜上的连接有两种途径:尾端结合配合物配体到自组装膜,然后到络合金属中心;自组装膜上尾端功能基团与金属中心配体的功能基团之间形成化学键,两种途径均应用于金属蛋白以及小分子氧化还原中心在自组装膜上的固定[184-186]。

使用自组装膜的尾端基团作为金属键合配体,研究者报道了几种可置换配体。Van Ryswyk 等人[186]以 H_2O 分子作为$[Ru(NH_3)_5(H_2O)]^{2+}$标记配体,水分子可被自组装膜中 11-SH 异烟碱胺的吡啶功能基团置换。该反应是在 Ar 气氛下,将 11-SH 异烟碱胺自组装膜电极浸入$[Ru(NH_3)_5(H_2O)](PF_6)_2$的 THF 溶液中。如果将修饰电极浸入 H_2O 中,则仅仅存在少量的或没有替代的水合配体,即$[Ru(NH_3)_5(H_2O)]^{2+}$发生了去水化过程并与吡啶基团反应,使用非配位试剂如 THF 减缓了去水化过程。置换反应的另外一种解释是自组装膜在 THF 中比在水中更具流动性,增加了自由基团吡啶与去水化过程的竞争[185,186]。

以相似的方式,将一个尾端吡啶 L_1 基团结合于预先形成的自组装膜上,尾端吡啶或咪唑基团与 $trans$-$[Ru(Ⅱ)(NH_3)_4(SO_3)(H_2O)]$的配体发生置换反应可将 Ru 络合物结合于自组装膜上,反应可在水溶液中进行,SO_3 配体转变为 SO_4,并且氧化 Ru(Ⅱ)为 Ru(Ⅲ),电化学还原这种配合物产生了 $trans$-$[Ru(Ⅱ)(NH_3)_4(H_2O)(L_1)]$—自组装膜形式。在水溶液中形成高产率 $trans$-$[Ru(Ⅱ)(NH_3)_4(H_2O)(L_1)]$—自组装膜的原因是因为 $trans$-$[Ru(Ⅱ)(NH_3)_4(SO_3)(H_2O)]$是一个中性络合物,比带电荷的配合物$[Ru(NH_3)_5(H_2O)]^{2+}$更容易结合亲水自组装膜界面[185]。

在 EDC 存在下,自组装膜尾端羧基与组氨酸氨基在自组装膜上修饰了 $trans$-$[Ru(Ⅲ)$—$(NH_3)_4(SO_4)(L)]$,实验表明 ET 速率与氧化还原中心修饰烷基—SH 的形式相似,但并未见有 ET 速率的报道。

多项研究是通过共价作用将氧化还原分子结合于自组装膜上，其中研究最多的途径是使用酰胺键形式。如应用不同烷基链长羧基修饰[$Ru(NH_3)_5(4-AMP)$]$^{2+}$，酰胺化反应是将羧基尾端自组装膜修饰电极浸入 EDC 和[$Ru(NH_3)_5(4-AMP)$]$^{2+}$溶液[185]。

Fc 通过酰胺键或 C—C 键的形式结合于自组装膜上，K_{ET}测量比较了两种键合对 ET 的影响，发现对于同一链长的键，通过酰胺键结合与通过 C—C 键结合 Fc 的 K_{ET}几乎完全一样。另外，将[$Os(II)(bpy)_2(4-AMP)(H_2O)$]$^{2+}$共价结合于 16-COOH/12-SH 或 16-SH 混合自组装膜上，可以研究 K_{ET} 对 pH 的依赖性，在 pH=7.2 时，K_{ET} 是 9.9 s^{-1}[117]。

第五节 氧化还原自组装膜烷基链结构

烷基链是自组装膜的关键部分，可以有效控制电极和氧化还原中心的距离，进而控制 H_{AB}和 K_{ET}，增加烷基链长减小电子传递速率值。另外，从控制电子耦合以及从 β 衰减的角度评价烷基链的结构也是非常关键的，公式 3-42 反映了 K_{ET}、β 以及 d 之间的关系，A 是指前因子，d 是烷基链长。

$$K_{ET} = A e^{-\beta d} \tag{3-42}$$

1. 烷基链的结构

研究表明，与短链形式相比，长链烷基硫醇形成了更加密集、高度有序的自组装膜。膜中烷基链之间范德华相互作用对于疏水性的长链烷基—SH 非常强，另外，长链硫醇在溶剂中缺乏溶解性，这通常使自组装膜更易沉积，并形成密集堆积、良好有序的 SAM。

K_{ET} 随链长增加而减小，例如，Fc 在一系列烷基链上与电极之间的 K_{ET} (5~13CH$_2$) 变化从最长链的 10^2 s^{-1} 增加到最短链的 3.0×10^7 s^{-1}[179]。

通常，在金基底上穿越硫醇 SAM 的 β（隧穿系数）实验测量是 1.1[178,179]。然而，隧穿系数 β 随氧化还原物质的位置以及自组装膜上稀释剂的不同而发生变化，这可由一系列[$Ru(NH_3)_5pyCH_2NHCO-(CH_2)_nSH$]$^{2+/3+}$络合物和稀释剂 HS(CH$_2$)$_m$COOH/SAM 来证明。当 $n<m$ 时，Ru 深埋于自组装膜中，β=0.16；当 $n=m$ 时，β=0.97；当 $n>m$ 时，Ru 在 SAM 上凸出，β=0.83。β 值的变化表示穿越 SAM 有不同的电子隧穿途径，这表明依赖于自组装膜上氧化还原中心对稀释物质的位置存在差异[41]。

如果烷基硫醇 SAM 的内部链结构是烯基或炔基[187]，则穿越 SAM 的 ET 速率将减小，原因是 H_{AB} 的减小。氧化还原物质和电子之间的 H_{AB} 依赖于烷基链碳—氢键最接近的 2~3 个原子之间 σ 和 σ^* 轨道的重叠，而烯基与炔基中断了烷基链中心 σ 和 σ^* 轨道的重叠，与未修饰的饱和烷基链相比，使其氧化还原中心和电子之间缺乏有效的电子耦合[187,188]。

烷基链中亚甲基数目的奇偶性是烷基硫醇 SAM 稳定性方面的一项重要研究内容。不考虑链长，烷基硫醇在金表面呈 30° 的斜置角。对于 CH$_3$(CH$_2$)$_n$S—Au 系统，当 n 是奇数时，SAM 尾端 C—C 键垂直于电极表面，SAM 表面仅有 CH$_3$ 一种基团；当 n 是偶数时，尾端 C—C 与电极表面呈一角度，因此，这种 SAM 表面具有 CH$_2$ 和 CH$_3$ 两种基团[164]（图 3-11）。

当链长增加时，烷基硫醇 SAM 在电极上的斜置角将趋于减小[189]，超交换机制的 β 受 SAM 上烷基链的斜置影响，并不是恒定的 1.2/CH$_2$。作为奇—偶效应的结果，奇数 CH$_2$/SAM 比偶数 CH$_2$/SAM 有较慢的电子传递速率。关于 SAM 上的奇—偶效应可参考相关文献[164]。

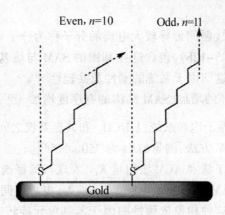

图 3-11　SAM 结构上的奇—偶结构示意图[164]

2. 功能基团

Au/SAM 上的尾端功能基团最常见的有—CH_3、—OH 和—COOH（图 3-12a），—OH 尾端基团在 SAM 上的斜置角为 28°，与—CH_3 尾端基团烷基硫醇相似，尾端—OH/SAM 提供了良好的亲水界面，β 检测是 0.9/CH_2，这也与—CH_3 尾端基团烷基链的 SAM 相似[190]。但与烷基硫醇相比，发现—OH 尾端 SAM 存在一定缺陷和渗透性，另外，极性—OH 基团引入 SAM 界面影响了系统的表观电位 $E^{0\prime}$ 和表观 K_{ET} 值，表现在 CV 图上的这种 SAM 的双电层电容增加引起 K_{ET} 值比其实际值明显偏低[38]。

—COOH 尾端 SAM 表明比—OH 尾端 SAM 有更多的缺陷，更利于水和离子的渗透，这种行为归因于在尾端—COOH 基团之间存在氢键形式[191]。—COOH 尾端硫醇也可以在基底形成良好有序密集的 SAM（归因于烷基链的疏水相互作用），其斜置角为 32°，与—CH_3、—OH 尾端 SAM 相似。—COOH 尾端 SAM 很容易进行化学修饰，这是这类 SAM 最吸引研究者的性质[141]。除了形成酰胺键，—COOH 也可以形成酯键，研究表明 SAM 上的酯功能基团存在一定的无序性，尾端酯键膜的斜置角测量是 35°。另外，研究发现酯键在 SAM 上的水解比酯在溶液中的水解稍慢[186]。

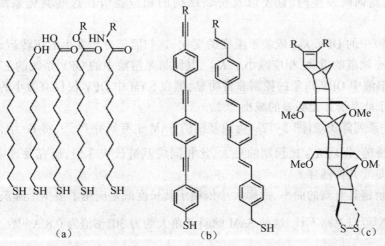

R 是结合氧化还原中心的基团

图 3-12　形成 SAM 的分子[191]

3. 共轭烷基链

具有高度共轭并可微接间传导较大电流的分子称为分子导线[192],研究最多的分子导线有 OPEs 和 OPVs[191](图 3-12b),包含这些基团的 SAM 与烷基硫醇 SAM 相比,具有相似的有序性,其斜置角较小,这归因于系统的刚性及共轭性[159]。

在 OPEs 中随共轭芳环的增加,SAM 结构的有序度增加[193],OPEs/SAM 系统的 β 测量在 0.36 到 0.57 $Å^{-1}$ 之间,β 值界于饱和系统(1.1/CH_2)和共轭系统之间(0.2/CH_2)。对于包含 6 个苯环单元的 SAM,使用 ACV 方法,测量的 K_{ET} 为 350 s^{-1}[194,195]。

在 OPE 系统中,以 ILIT 法和 ACV 法测量 K_{ET},发现与稀释剂无关[27],当 OPE 被稀释时,它对电子穿越烷基硫醇 SAM 没有产生影响。Fc 和 Au 电极之间穿越 OPE/SAM 的 ET 依赖于 OPE 分子的构象形式,这种构象依赖性归因于苯基单元旋转态较低的能垒,即减少了系统的共轭性,这也就解释了 β 为什么是一个中间值[27,117]。

使用 ILIT 法与 ACV 法测量相似的分子,发现存在明显的差异,表明 AC 信号可能不准确,这归因于测量池的时间常数 R_uC_{film},R_u 是未补偿的溶液阻抗,C_{film} 是 SAM 的电容。另外,ACV 法测量的表面覆盖率也比 ILIT 法更低。

OPV 比 OPE 共轭性更强,通常 OPV 具有更快的 ET,即使分子长度大于 35 Å,测量的 K_{ET} 仍大于 10^4 s^{-1},这归因于其高度共轭的结构特点[196]。快速的电子传递表明氧化还原中心和电极之间的耦合对超过 28 Å 长的 OPV,长度并非一个重要的限制因素,较大的耦合性使得 K_{ET} 随 SAM 厚度的增加,没有发生大的变化,穿越这种系统的 ET 机制目前并不明确(跳跃 vs 隧穿?)[197]。

为了推测可能的机制,研究者测量了 OPV 的 LUMO 和给体 HOMO 的能级水平[197],将一个电子注射到 OPV(跳跃机制),给体的 HOMO 能级必须与 OPV 的 LUMO 能级重叠。研究发现 Fc 共价连接于 OPV 的 HOMO 和 LUMO 之间有较大的能隙,研究者认为发生了跳跃机制并且测量了此系统中发生的 ET。溶液中的并四苯作为给体,通过 OPV 桥联结合于一个受体,发现给体的 HOMO 能级与 OPV 的 LUMO 相似,决定 ET 的可能是跳跃机制。当跳跃和隧穿两种机制同时发生时,D/A 以及桥联结构的相应能级在这些共轭系统中是非常重要的。

对于溶液中的 OPV,K_{ET} 依赖于温度的变化,这归因于分子发生的扭转运动。扭转运动减弱了 OPVπ 轨道的重叠,相应减小了 K_{ET},其结果是溶液中的 OPV 系统的 ET 受其构象控制。尽管在溶液中 OPV 构象的控制非常明显,但在 SAM 中 OPV 仅仅有微小的构象变化,并且其 K_{ET} 对此系统并无明显的减小[198]。

最近,一系列降莰烷(图 3-12c)为烷基链的 SAM 也有研究[199-201],将 Fc 结合于这种链长为 21.3 Å 的桥链,发现 K_{ET} 比预期的更大,比相同烷基链长快 3 倍,研究发现电子穿越这种系统的机制是非共振隧穿。

降莰烷桥链具有高的曲率,曲率大小依赖于其长度或是应用了多少个降莰烷单元,短的降莰烷链曲率可以忽略不计,这种 SAM 的斜置角大约为 30°,β 值为 0.8 $Å^{-1}$[199]。

4. 肽键

蛋白的长程电子转移对于生命过程研究非常重要[202],大多数进行的工作是利用闪光光解作用法去检测 ET[203-206]。最近,电化学方法也应用于研究 ET 穿过 SAM 上(相对)短的多肽链。研究表明氧化还原蛋白之间的耦合效率由介入多肽基质的 3D 结构来决定,蛋白结构是调控 ΔG、λ、H_{AB} 以及 ET 的基础[207-212]。

研究者测量了通过亮氨酸结合的 Fc 尾端螺旋状多肽在金表面上的量[209,211,212]。通常,分子在表面的垂直定位以及更紧密的 SAM 降低了电子传递速率,但在螺旋状多肽 SAM 上的 ET 可能与分子的运动相联系,存在三种可能的机制:通过螺旋收缩的电子隧穿;结合了螺旋交换从 R-螺旋到 3_{10} 螺旋的电子隧传;电子沿着骨架经由酰胺键的跳跃[44,45]。

Mandal 和 Kraatz[208]研究了多肽螺旋双极子瞬间对 ET 的影响,他们使用 Fc 亮氨酸多肽,双极子瞬间的多肽被 SAM 排列,也在双极子相反形成,RAIRS 光谱法揭示了双极子—相反多肽比双极子—排列多肽位置更加垂直。而且,它们的 ET 性质显著不同。两种膜中的分子动力学的差异性是合理的,使用 B-V 方法由 CV 数据获得了 K_{ET},双极子—相反 SAM 比双极子—排列 SAM 展示了更小的 K_{ET}(K_{ET}=1.2×10^{-3} s^{-1} vs 1.5×10^{-2} s^{-1})。

研究者使用 CV 法和 ACV 法对 FcCO(Gly)$_n$NH(CH$_2$)$_2$SH/CH$_3$(CH$_2$)$_m$SH(n=2~6)混合 SAM 进行了研究[210],对于短链衍生物,穿越这些桥链的 ET 速率快速减小,K_{ET} 范围从 9000 s^{-1} 到 1 s^{-1},在更长的桥链上(n≥5)可以观察到对距离依赖性不强的 ET。这种肽桥链的第二结构或是从隧穿到跳跃机制的差异被认为是速率常数不同的可能原因。

改变谷氨酸和亮氨酸的多肽序列并且在组氨酸末端修饰 Ru(NH$_3$)$_5$,并且经由胱氨酸巯基基团在金表面上形成 SAM,肽被设计为依赖 pH 的第二结构,可以研究其结构和 ET 的相关性[213,214]。在低的 pH 值下,羧酸基团质子化,多肽具有紧密的螺旋结构;在高的 pH 值下,羧酸基团去质子化产生了延伸的构造,归因于羧基间的静电相互作用,分子动力学模型预言了对 pH 敏感的多肽构造,并经由电化学测量得到证实。在低的 pH 下,ET 更快(7.8 s^{-1},110 s^{-1},230 s^{-1}分别对应于 pH=3.0,2.0,1.0),随 pH 增加,K_{ET} 逐渐减小,在 pH=6.9 时 ET 不再发生(禁止)。

应用短的 Fc-肽胱氨酸形成密集性不好的 SAM,其厚度为 7 Å,而且通过 Cu 的去电位沉积研究,表明有 10%~15%金表面裸露,使用 B-V 法估计其 K_{ET} 值为 5~8×10^3 s^{-1},如以己烷—SH 稀释这种 SAM 将产生更快的氧化还原反应,这归因于 SAM 的"变硬"[207]。

5. 核酸桥联

应用 SAM 对 DNA 中的电荷传递也得到了广泛研究[215-221],Meade 及合作者[222]将金属络合物共价结合于 DNA 中,发展了灵敏的电子检测系统,组合 Fc 衍生物于 DNA 寡聚核苷酸中可以检测特定位点的突变[215,216]。Barton 及合作者[217,218]以共价或非共价形式在 SAM 上结合 DNA,并将氧化还原有机物质正定霉素和亚甲基蓝插入其中,这可以灵敏地扰乱 DNA 结构,表明在这些系统中碱基堆砌是电荷传递的基础。使用 Fc-尾端 ds-DNA(20 碱基对),如果 Fc 是在目标片段 3′处,Kraatz 及合作者报道了系统的电荷传递速率为 25 s^{-1},如果是在 5′处,其电子传递速率为 115 s^{-1}。研究者认为 5′-Fc 比 3′-Fc 更容易接近碱基对[219]。Anne 及其合作者[220,221]将 Fc 结合于 ssDNA,发现 DNA 发生弯曲使 Fc 可以更容易接近电极表面,存在着构造效应。

肽核酸(PNA)低聚物可在 SAM 中检测核酸,其方法是将 PNA 片段固定在一个氨基乙基胺基乙酸骨架上,并且经由 Watson-Crick 氢键形成复制,骨架是中性的,片段在表面上彼此并不排斥,这为核酸的检测提供了一个良好的平台。

单链 PNA(ss—PNA)包含 3~7 个胸腺嘧啶核苷,1 个尾端 cys,1 个 N-尾端的 Fc 基团,在金电极上形成了SAM，这种SAM的β检测为0.9/Å，K_{ET}范围从3个碱基的2000 s^{-1}到7个碱基的0.02 s^{-1}。

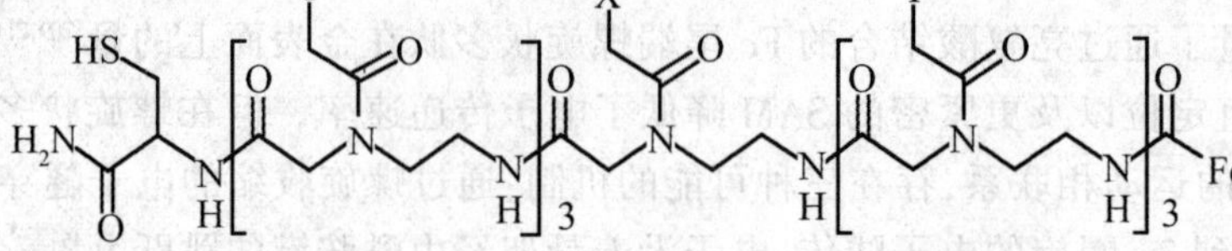

图3-13 具有T_3—X—T_3(X=C,T,A,G,CH$_3$)序列的多肽核酸低聚物[223]

研究者也对具有T_3—X—T_3序列的ss—PNA/SAM进行了研究[223](图3-13)，X=C、T、A、G、CH$_3$，这种SAM的电荷传递速率从0.014 s^{-1}到0.067 s^{-1}，CH$_3$基团给出了最慢的电荷传递速率(<0.005 s^{-1})，这依赖于核碱基X的同一性。计算表明PNA/SAM的电荷传递是通过中空媒介超交换机制发生的，电荷传递速率常数和X氧化电势之间具有相关性。

最近，研究者应用CV法对既含有ss—cys—Tn—Fc、ss—cys—An—Fc，也含有ds—cys—(AT)$_n$—Fc的PNA/SAM的电荷传递性质进行了研究[224]。对于短链ss-PNA/SAM，电荷传递速率常数随PNA链长的增加快速减小。根据超交换中介隧穿机制，对于长的ss-PNA和ds-PNA，电荷传递速率具有弱的距离依赖性，并且与核碱基的氧化电势相关，表明发生了跳跃机制。在超交换和跳跃机制之间的PNA链长依赖性，由Rarner提出的紧密结合模型传递性可以进行合理的解释[225]。

第六节 氧化还原自组装膜电子传递研究中的电极材料

电极材料的选择对于K_{ET}的测量非常重要，这归因于Marcus理论电子态密度在电极材料上的应用。以硫醇为基础的SAM电子传递研究应用最广泛的电极材料是Au和Ag，然而，研究者也对在其他金属如Ni、Cu、Pd、Hg、Pt表面的SAM形式进行了研究，—SH也可与这些金属以高度亲合性形成密集的SAM。但很少有研究对这些金属SAM表面的ET动力学进行直接比较。此外，半导体材料也作为SAM的基底应用于K_{ET}的测量[173,226,227]。

1. 电极材料的影响——理论和实验

Gosavi和Marcus[228]对金属(Pt,Au)的费米能级上的K_{ET}电子态密度的理论效应(DOS)进行了研究。对于金属，s电子支配DOS。然而，接近费米能级的DOS，Pt比Au要高7.5倍，这归因于d能带的重叠。与sp带相比，Pt费米能级每种电子态的电子耦合对于d带的影响非常弱，这种差异可能是因为d带更易离域化。因此，仅仅对结合在SAM上的氧化还原中心有弱的结合。

Finklea等人[229]使用Ru(NH$_3$)$_5$/SAM，并应用Gosavi's理论对发生在Au、Pt、Ag表面SAM的K_{ET}进行了比较研究，由CA法产生的Tafel曲线，测得Au、Ag、Pt的K_{ET}分别为1.0 s^{-1}、0.6 s^{-1}和1.7 s^{-1}，K_{Pt}/K_{Au}为1.7，与其接近费米能级的DOS比率的7.5相比明显偏低。Au和Ag之间的差异与电子热常数一致，并与接近金属费米能级的密度态(0.65 vs Ag, 0.73 vs Au)成比例，其结果与理论预言高度一致，即d能态仅仅与Ru(NH$_3$)$_5$中心有弱的结合。

[Os(OMe-bpy)$_2$(p3p)Cl]$^+$/SAM结合于碳纤维、Hg、Pt、Au、Cu以及Ag微电极上，用电化

学方法研究了 DOS 对 K_{ET} 的影响[230]。对于每一种电极材料,均使用 AC 法产生了 Tafel 曲线。曲线与 Marcus–DOS 理论预言的高度一致,$\lambda=0.27$ eV,K_{ET} 对于 Pt、Au 和 C 分别是 4.0×10^4 s^{-1}、1.8×10^4 s^{-1} 和 3.0×10^3 s^{-1}。

实验测量的指前因子与离域在电极上的金属态和定位在 Os 络合物氧化还原态之间的弱的电子耦合相一致。相比较 DOS 比率的 7.5,Pt 和 Au 的指前因子的比值是 2.9±0.7,这个比率比 $Ru(NH_3)_5$ 的 K_{ET} 比率高,证实了电子传递和电极 DOS 的关系并非简单的线性关系。

2. 电极材料

2.1 汞(Hg)

Hg 在室温下呈液态,它可以提供无缺陷的界面,与其他金属相比,Hg 和—SH 具有更高的亲合性,并且形成高度密集的 SAM,阻止了亲水和疏水氧化还原探针到达金属表面[163]。大量的实验研究主要是调查了溶液中的氧化还原物质穿越 SAM 的异质 K_{ET},很少有研究涉及测量汞表面氧化还原活性 SAM 的 K_{ET}[163,174,231,232]。研究者曾尝试在 Hg 表面形成 $HO_2C(CH_2)_{15}SH/$ SAM,随后结合 $[Ru(NH_3)_5(4-AMP)]^{2+}$,测量 K_{ET},但未获得成功[229]。研究发现 $HO_2C(CH_2)_{15}SH$ 在 Hg 表面形成的是多层膜而不是单层膜,出现的还原峰归因于—SH 在 -0.6 V 的脱附以及 $Ru(II/III)$ 电对的非理想行为。

研究者应用各种方法对 Hg 上烷基硫醇 SAM 可能存在的缺陷进行了研究。基于 SAM 的缺陷可产生电势诱导离子门效应,Demox 等人使用十六烷基—SH,报道了高度密集的 SAM/Hg 阻止了 $Ru(NH_3)_6^{3+}$ 的渗透,C_{60}—SH/SAM 在 Hg 上的电化学研究表明溶液中的氧化还原物质被电化学阻塞,其表面具有高度疏水性[163]。研究者也应用 SECM 去评价了烷基硫醇 $CH_3(CH_2)_nSH(n=8,10,11,15)$ 在 Au 和 Hg 表面 SAM 的缺陷。另外,利用 $FcCH_2NHCO(CH_2)_{12}SH$ 和尾端聚硼 FcNB–SH 在 SAM 上的针孔效应,通过 Pd 纳米粒子的生长可以识别针孔,这发展了 K_{ET} 在针孔存在下的模型,FcNB 物质给出的 K_{ET} 为 189 s^{-1},这个值与 $FcCH_2NHCO(CH_2)_{12}SH$ 在相同长度测量时相似[61]。

2.2 银(Ag)和铜(Cu)

在真空条件下,良好有序的 SAM 可以在新鲜蒸镀的 Ag 和 Cu 膜上形成,在 Ag 和 Cu 表面烷基硫醇的斜置角比其在 Au 表面要小,并且没有观察到奇 偶效应[233]。

相对于 Au,Ag 和 Cu 暴露在空气中易形成一层氧化膜,烷基硫醇 SAM 在这些氧化表面上形成,然而,其结构和性质与未氧化表面形成的 SAM 有差别。研究已经表明氧化还原反应在金属氧化态和烷基硫醇分子之间发生,即烷基硫醇被氧化为硫化物,伴随着 Ag 和 Cu 氧化态的还原,这些还原表面与等量的烷基硫醇继续反应形成 SAM。Cu 一般不被应用于 ET/SAM 研究,而 Ag 则广泛地应用于 SERRS 研究,并且金属蛋白 K_{ET} 测量也常常使用 Ag-SAM 表面[234,235]。

2.3 镍(Ni)

烷基硫醇在 Ni 上形成 SAM 的均相性可通过 XPS、俄歇电子能谱以及电化学表面覆盖来确定[236,237]。在常压和室温下,Ni 表面立即形成一层氧化层,在最近的报道中,Ni 表面形成的 SAM,在最佳工作电势下,通过还原 NiO 产生的还原电流可以测量 $Fc(CH_2)_{11}SH$ 的表面覆盖[238-240]。

2.4 钯(Pd)

对于在晶体 Pd 上的 SAM,长的烷基链与 Ag 和 Au 上的 SAM 有序性相似。烷基硫醇

SAM 提高了 Pd 的抗腐蚀性,并且与烷基链长无关,XPS 数据揭示了复杂硫化 Pd 相 SAM 的稳定性和抗腐蚀性,SAM 在 Pd 上可以更好地应用于微触印刷术。烷基硫醇在 Pd 上的还原脱附电势与碳氢的链长无关,这与 Au、Ag、Pt 和 Ni 的金属—硫化物界面不同[241]。

在 Pd 上的 OEG 尾端 SAM 可以抵抗蛋白以及孵化细胞的非特定吸附[242],OEG-SAM 抵抗细胞入侵的时间可达到四周,而以相同的方式,SAM 在 Au 上仅保持两周[243]。

在各种电解质中(碱性、中性、酸性),使用两种 Pd 氯络合物($PdCl_2$,Na_2PdCl_4)将高粗糙的 Pd 表面沉积于 GC 电极上,癸烷—SH 和丁烷—SH 在这些 Pd 表面的 SAM 阻塞了溶液中氧化还原物质的电化学反应[244]。

2.5 半导体材料

在硅表面高稳定的 SAM 可以通过硅烷化反应以强的 Si—C 键形成,这比以 Au—S 键形式的同类 SAM 有更大的稳定性[245],理想的硅表面化学对微电子装置的快速发展是非常关键的[246]。

电极材料的电子结构影响其界面的化学性质,并且影响其电化学反应。半导体的导带和禁带被带隙分开,半导体带隙约为 0.5~2.0 eV,表现的电子传导性非常低[36]。通过掺杂可以加强半导体的传导性,半导体表面和电解质之间的电势差异归因于半导体(低的传导)和电解质溶液(相对高的传导)之间传导性的差异,而且,半导体掺杂之前的电势差异发生在电极的边界层,而不是发生在界面—溶液层[33]。

半导体界面电势与金属电极相反[33],相比较于半导体,金属电极电势随反应的摩尔 ΔG 的变化而改变。掺杂的半导体静电电势发生变化使带隙弯曲,当半导体电极电势变化时,在半导体表面带隙的位置并未改变,这归因于半导体固有的低传导性。

杂化分子/硅组装相比于同类化合物在金表面的排列,具有两个突出的优点:首先,硅的电性质很容易通过选择适当的掺杂物或掺杂浓度来改变,或者经由光照射下电子中空对的再生;第二,Si—C 键和 Si—O 键提供了超级稳定性界面,Si(111)表面可以以原子级平坦单氢—尾端表面反应形成,并可以进一步修饰[247]。

在诸如传感器及生物活性界面应用中,界面性质的控制非常重要[159]。导体和掺杂半导体界面与生物分子的相互作用,可以通过阻抗谱进行详细研究。生物活性物质,如酶、抗体/抗原或 DNA,在电极或半导体界面的固定,也可以通过电容或界面电子传递阻抗来控制。对在电极界面上发生的生物识别,阻抗谱应用于其界面电荷(变化)的研究[51]。

Fc 和卟啉衍生物经由一定连接体作用于 Si(100)表面,对于 Zn 卟啉 SAM,氧化还原的 ET 速率与同类巯基卟啉衍生物在金表面很相似[226]。然而,与相应的烷基硫醇在金表面相比,发现 Fc/SAM 在 Si(100)表面上的 K_{ET} 更小,这种现象归因于氧化还原中心与表面距离的变化,即连接链的定位发生了调整。

研究者改变连接原子(O、S 和 Se)及连接体结构,应用 XPS、FITR 以及电化学方法对两类 Zn 卟啉结合在 Si(100)表面的电子传递性质进行了研究[227]。不考虑结合原子,氧化还原中心与 Si(100)结合可以通过短的(苯基)和非常短的(亚甲基)连接体,显然,这些卟啉基团在 Si(100)上的定位可以通过选择连接体基团来控制。相比较于亚甲基连接体,苯基连接体为卟啉基团提供了更直立的空间定位,而且,其 K_{ET} 依赖于表面覆盖(对于 $\Gamma=(0.25~3.0)\times10^{-11}$ mol/cm^2,$K_{ET}=10^4~10^5$ s^{-1}),也比亚甲基连接体小得多,并且改变结合原子并未显著地影响 K_{ET}。

将乙烯基 Fc 经由 Si—C 键结合于单晶 Si(100)表面,研究者应用 XPS 和电容-V 对其电化学性质进行了研究[248],但并未检测其K_{ET}。Fc以尾端烷基修饰于Si(111)表面,其表面覆盖可以通过稀释剂n-葵烷来控制,K_{ET}在50 s^{-1}时,发现与氧化还原中心Fc的表面覆盖无关[49]。

2.6 ITO

ITO作为一种透明传导材料,常作为基底去研究光—电化学过程,如人工光合成、光诱导电子传递等[249],然而,很少有研究应用ITO键联的氧化还原中心用电化学方法检测K_{ET}。

研究者以不同的方式化学预处理ITO去检测其电化学效应。如将Fc-三氯硅烷结合于含有ITO、ZITO、CdO以及ITO衍生物的离子束沉积的半导体电极上(IAD-ITO),最大的Fc表面覆盖($7.9×10^{-10}$ mol/cm^2)和最大的电子传递速率(9.23 s^{-1})与用O$_2$等离子处理的ITO保持了一致。CV法表明对IAD-In$_2$O$_3$产生了最大的ΔE_p和$FWHM$,表明在IAD-In$_2$O$_3$表面的SAM更加无序[250]。

Laviron方法被应用于K_{ET} (s^{-1})的计算,7.12(ZITO)>6.6(作为标准的ITO)>5.07(IAD-ITO)>0.42(CdO)>0.03(IAD-In$_2$O$_3$),IAD-In$_2$O$_3$的ET速率比作为标准的ITO小两个数量级,这归因于其相对低的传导性(1000 S/cm),在源电压和电极表面之间发生了明显的电势滞后现象,所有ITO表面均产生了这种滞后。对CV施加一个更极端的电势,相比较于电活性SAM在贵金属电极上,ITO的CV图将产生一个伸展的外观形状。

第七节 自组装结构的模型体系研究

1946年Zisman等人[251]报道了SAM通过自组装在金属表面的实验制备,1966年,研究者应用MC方法对单层和双层膜进行了理论研究,极大地推动了生物膜性质和结构研究的发展[252-254]。对SAM的理论研究,目前主要包括自组装基本过程的理解、界面现象、结构与功能的相互关系以及与分子电子学相关的SAM上的电荷传递等[255,256]。

SAM的结构对于电化学测量非常重要,这里主要讨论SAM的结构模型,这些模型系统包括退火、斜置角、奇—偶效应以及头基效应等。

1. 早期的工作——分子动力学

20世纪70年代后期和80年代初期随着超级计算机的发展,应用计算方法如MC法及分子动力学模型(MD)对多粒子系统进行了广泛研究。SAM动力学MD研究首次报道于20世纪70年代后期。MD方法产生的以简单相互作用为基础的完整结构和动力学细节通常被认为比MC法更为有力,然而,MD模拟仍然比较原始[257,258]。

Toxvaerd's[258]使用成对电势描述了密集C链之间的相互作用,研究了SAM的密度态方程。大量的动力学计算是使用粗糙的SAM模型,单独分子被处理为纯粹的2维物质(例如,分子仅仅允许在SAM的平板上移动)。这种处理忽略了分子链与大量可能的立体构象之间的相互作用,并且没有考虑SAM的晶态分布以及分子的溶解行为。

1980年,Weigel及其合作者[259]进一步报道了SAM的MD模型,有序相中的液膜是从有序液态到无序气态的过渡。不久,van der Ploeg和Berdensen[260]提出了有代表性的双层液膜作为生物膜的模型,这种双层液膜模型包括Lennard-Jones、双平面以及当键长一定时键角与电势的相互作用。

1988年,Harris和Rice[261]应用MD法研究了水中五—羧酸—SH单分子膜的热力学性质,在这种模型中,水被处理为极性连续体,五—羧酸—SH被处理为内部键约束的、角弯曲和扭曲分子内相互作用以及 Lennard-Jones 原子—原子分子内相互作用的 15 个假原子链。当温度在 300~400 K 时，这些模拟的结果显示了低压相，即低密度的气相和良好有序的浓缩相，这与实验研究相矛盾，表明有稳定的液态延伸相存在。与此同时，研究者也对分子构象、分子内相互作用与单层膜或多层系统结构之间的相互关系进行了研究。

2. 在表面上的自组装膜模型

1989 年，Hautman 和 Klien[262]首次应用 MD 模型对金属表面形成的硫醇 SAM 结构和动力学的分子分散性进行了研究。对于 $HS(CH_2)_{15}CH_3$ 分子在金基底形成 SAM 层的吸附分子—表面相互作用，研究者探索了两种模型：第一种模型要求在室温下，S—C 键与基底表面接近平行；第二种模型要求产生的 SAM 链一个接一个排列，并且与表面呈标准斜置。两种模型产生了不同的斜置角，不考虑斜置角的差异，两种 SAM 模型的厚度应该是相同的，并且膜结构修饰头基的影响被限制于接近金属表面的区域。在链旋转动力学方面，两种模型表现出显著的差异。

1993 年，新的力场参数，即 ab 从头算被引入烷基硫醇与 Au 和 Ag 表面结合的模型系统[263]。使用这种新的力场参数，第一分子机制(MM)能量最小化，以 ab 从头算，研究者对 HS 和 CH_3S 在 Au(111)、Au(100)、Ag(111)以及 Ag(100)表面的最佳模型化进行了研究，并检测了 MD 力场参数。这些计算的结果表明，硫醇分子在金属表面存在两种化学吸附模式，非常接近 SH 在 Au(111)表面的能量，在第一种化学吸附模式中，表面—S—C 夹角呈 180°，这归因于 sp 杂化；第二种模式表明表面—S—C 夹角是 104°，这归因于 sp^3 杂化，两种模式揭示了烷基硫醇 SAM 的退火的可能机制。

3. 尾端基团效应

20 世纪 80 年代后期，n-烷基硫醇开始作为模型系统的分子被广泛应用于 SAM 分子结构、表面和表面性质相关性的实验探索[264,265]，研究者尽管使用了各种技术如 AFM、STM、FTIR、SERS、X-RAY、电子和氦原子衍射对n-烷基硫醇在 Au(111)进行了研究，但烷基硫醇的化学吸附模式仍然不清晰[266-268]。

为了解决化学吸附模式及实验观察到的奇—偶效应问题，几何优化和 MV 模拟的计算研究被应用于 n-烷基硫醇(R—SH)以及 $ROC_{12}H_8SH(R=C_{16}H_{33}, C_{17}H_{35})$的 Au(111)表面系统。1987 年，Li[269]提出了完整的原子力场，结合 MD 模拟可以解释奇—偶效应。通过 IR 对长链 n-烷基硫醇和 4-烷氧基-二-苯基-SH 的观察，研究者建立了化学头基结构与—SH 在 Au(111)上聚集结构之间的相关性。

2005 年，Harding 等人[270,271]对羧基尾端 SAM 性质的模型进行了研究，其目的是验证实验工作，表明链分子(C 原子是奇数或偶数)在长链羧酸 SAM 中是否定位生长。研究者也应用 MD 模型考察了奇—偶效应，在 300 K，水分子存在下头基聚集结构的差异证实了头基效应，即奇—偶效应的存在。

2005 年，Goddard 等人[272]以计算方法研究了 BPDT-SAM 在分子电子装置中的电荷传递性质，他们从 MD 力场和退火模拟的 BPDT-SAM 证实了能量适宜的人字形 SAM 堆积构造，并从聚集定位和斜置角的角度对三种分子电子装置模型的差异性进行了比较。应用 Green's 函数方法计算，与由 Landauer-Büttiker 公式产生的 i—V 曲线具有一致的电荷传递性质。

在低压下，从 i—V 曲线可以发现人字形 SAM 有 30°的斜置角，这与平行定位 SAM30°的斜置角相似。当人字形模型呈 15°斜置角时，其电流值要比 30°斜置角或者平行结构模型 30°

斜置角小。而对于高偏压区域，三种模型的性质有显著的差异，这归因于其苯环的键结构。在低偏压区域 BPDT/SAM 的 $i—V$ 性质主要通过分子结或是单个分子与电极接触的 Si—Au 相互作用来决定，而在高偏压区域，分子内构象和相互作用对 BPDT/SAM 的 $i—V$ 性质产生了影响。

实验表明 OEG 尾端烷基硫醇 SAM 阻止了蛋白吸附[273]，然而，OEG 结构的理论研究表明仅仅是其螺旋构象阻止了蛋白吸附，而最密集的平面对蛋白吸附阻力相对很小[274]。

为了解释 SAM 结构对蛋白吸附的影响，Grunze 等人应用 ab 从头算 HF，解释了 OEG 基团在 Ag 上，一个 7/2 螺旋构象（螺旋 SAM）和平面上的 trans-构象（trans-SAM）对蛋白吸附性质的差异。在其研究中，对接近于 SAM 表面的水以包含有 20 个水分子的分子簇模型化，同时对超过 12 个刚性 OEG 片段的六边形结构模型化。计算表明单螺旋 OEG 片段直立，并通过 OEG 片段中两个连续 O 原子与水分子形成强的氢键，而平面上的 trans-构象仅仅允许 OEG 的一个 O 原子与水分子以单氢键的形式结合。

与 ab 从头算 HF 不同，对接近于灵活 OEG-尾端 SAM 表面的水，可以使用刚性模式的 MC 模型来解释。应用一个 TIP4P 模型，OEG 尾端烷基硫醇 SAM 表面的水被模型化，对接近于螺旋和 trans-OEG 尾端烷基硫醇 SAM 的水的模拟揭示了 SAM 结构在接近水层的短程效应。

相比于 trans-SAM，水分子能够更深地渗透进入螺旋 SAM 中，并且形成更多的氢键，这与实验结果相一致。水的存在影响 OEG 基团的无序性，蛋白分子在 OEG-SAM 表面上的吸附可推测为水分子是从其表面上发生取代，因此，具有更高的表面水分子密度以及氢键的 SAM 更加阻止了蛋白吸附。

研究者对 pH 响应的多肽 SAM 也应用 MD 和电化学方法进行了研究[214]，在 pH 为 2 和 7 时，使用 MD 法对多肽序列 HELELELELELC 的第二结构进行了研究。当 pH=2 时，MD 结果表明螺旋构象稳定超过 200 ns，在 pH=7 时，70 ns 后产生了一个随机不可逆的盘绕，这归因于谷氨酸残基的—COO—基团的负电荷存在相斥性。研究者使用这种构象模型来引导实验，在不同 pH 下，对多肽 SAM 的 ET 的性质进行了研究。

4. 自组装膜表面 Fc 的模型研究

研究者也对包含 $Fc(CH_2)_{12}S—Au/C_{10}S—Au$ 混合 SAM 的分子动力学模拟进行了研究[170]，结构和能量性质的模拟计算是为了探索 SAM 中可能存在的中性 Fc 的多相性，结构多相性被认为是引起非理想电化学响应的主要原因。

使用理论计算，研究者对 Fc 烷基硫醇的五种系统进行了研究，从相对贡献的角度描述了二元 SAM 中 Fc 隔离岛和 Fc 簇规则的分布，Fc 基团更倾向于和烷基链发生疏水相互作用，而不是界面区域的亲水相互作用，每一种相互作用（Fc—Fc，Fc—烷基，Fc—H_2O）的能量贡献可以推测 Fc 是呈孤立态还是簇态。

氧化还原活性 SAM 为电子传递动力学研究提供了良好的平台，通过应用一定的空间链长，氧化还原中心与电极之间的距离可以被有效控制。将氧化还原物质，包括过渡金属络合物（如 Fe、Ru、Os、金属簇等）以及有机分子（如 C_{60} 等）结合于 SAM 上，各种电化学方法，如 CV、ACV、EIS 以及 CA 可用来研究电子传递动力学过程，检测氧化还原活性 SAM 的电子传递速率。同时，SAM 也为研究氧化还原物质的外球相互作用、单层膜的组成、结构以及电极材料对其电子传递速率的影响等提供了理想的环境。与实验研究几乎同步，关于 SAM 结构的理论模型也获得了快速发展[275]。

目前,应用电化学方法研究氧化还原活性 SAM,在长程生物电子转移、人工光合成以及分子电子领域引起了研究者的广泛关注,分子开关、分子整流器以及分子晶体管等结合了 SAM 的电子器件的纳米制造均通过其 ET 性质来表征。D 和 A 之间连接基团的性质是一个非常活跃的研究领域,SAM 电化学研究提供了大量关于 ET 中心连接基团电子效应以及分子环境的信息[38,48]。

本章描述的各种电化学方法和表面化学方法可使研究者去检测特定的 ET 参数,SAM 的结构对于精确测量 K_{ET} 非常关键。该领域未来的工作将是通过共价和非共价形式电子传递参数的测量,继续探索氧化还原活性 SAM 的结构与性质的相关性。

参考文献

[1] Marcus R A. Theoretical study of electron transfer reactions of solvated electrons. Adv. Chem. Ser., 1965, 50: 138–140.

[2] Yu C J, Wan Y, Yowanto H, et al. Electronic detection of single-base mismatches in DNA with ferrocene-modified probes. J. Am. Chem. Soc., 2001, 123: 11155–11161.

[3] Bowler B E, Meade T J, Mayo S L, et al. Long-range electron transfer in structurally engineered pentaammineruthenium (histidine-62) cytochrome c. J. Am. Chem. Soc., 1989, 111: 8757–8759.

[4] Meade T J, Gray H B, Winkler J R. Driving-force effects on the rate of long-range electron transfer in ruthenium-modified cytochrome c. J. Am. Chem. Soc., 1989, 111: 4353–4356.

[5] D'Alessandro D M, Keene F R. Current trends and future challenges in the experimental, theoretical and computational analysis of intervalence charge transfer (IVCT) transitions. Chem. Soc. Rev., 2006, 35: 424–440.

[6] D'Alessandro D M, Topley A C, Davies M S, et al. Probing the transition between the localised ("class ii") and localised-to-delocalised ("class ii–iii") regimes using intervalence charge transfer (IVCT) solvatochromism in a series of mixed-valence dinuclear ruthenium complexes. Chem. Eur. J., 2006, 12: 4873–4884.

[7] Gray H B, Winkler J R. Long-range electron transfer. Proc. Natl. Acad. Sci. U.S.A., 2005, 102: 3534–3539.

[8] Winkler J R, Gray H B, Prytkova T R, et al. Electron transfer through proteins. Bioelectronics, 2005, 15–33.

[9] Imahori H, Kashiwagi Y, Hasobe T, et al. Porphyrin and fullerene-based artificial photosynthetic materials for photovoltaics. Thin Solid Films, 2004, (451-452): 580–588.

[10] Bertin P A, Georganopoulou D, Liang T, et al. Electroactive self-assembled monolayers on gold via bipodal dithiazepane anchoring groups. Langmuir, 2008, 24: 9096–9101.

[11] Trammell S A, Moore M, Lowy D, et al. surface reactivity of the quinone/hydroquinone redox center tethered to gold: comparison of delocalized and saturated bridges. J. Am. Chem. Soc., 2008, 130: 5579–5585.

[12] Chidsey C E D, Loiacono D N. Chemical functionality in self-assembled monolayers: structural and electrochemical properties. Langmuir, 1990, 6: 682–691.

[13] Miller C, Cuendet P, Graetzel M. Adsorbed omega-hydroxy thiol monolayers on gold electrodes: evidence for electron tunneling to redox species in solution. J. Phys. Chem., 1991, 95: 877-886.

[14] Becka A M, Miller C J. Electrochemistry at omega-hydroxy thiol coated electrodes. 3. Voltage independence of the electron tunneling barrier and measurements of redox kinetics at large overpotentials. J. Phys. Chem., 1992, 96: 2657-2668.

[15] Sikes H D, Smalley J F, Dudek S P, et al. Rapid electron tunneling through oligophenylenevinylene bridges. Science, 2001, 291: 1519-1523.

[16] Bard A J, Larry F R. Electrochemical methods: Fundamentals and applications. 2nd ed.. New York: John Wiley & Sons Inc., 2001.

[17] Wohnrath K, dos Santos P M, Sandrino B. A Novel Binuclear Ruthenium Complex: Spectroscopic and Electrochemical Characterization, and Formation of Langmuir and Langmuir-Blodgett Films. J. Braz. Chem. Soc., 2006, 17(8): 1634-1641.

[18] Evans D H, O'Connell K M, Petersen R A, et al. Cyclic voltammetry. J. Chem. Ed., 1983, 60: 290.

[19] Brown A P, Anson F C. Cyclic and differential pulse voltammetric behaviour of reactants confined to electrode surface. Anal. Chem., 1977, 49: 1589-1595.

[20] Laviron E. The use of linear potential sweep voltammetry and of a.c. voltammetry for the study of the surface electrochemical reaction of strongly adsorbed systems and of redox modified electrodes. J. Electroanal. Chem. Interfacial Electrochem., 1979, 100: 263-270.

[21] Laviron E. General expression of the linear potential sweep voltammorgam in the case of diffusionless electrochemical systems. J. Electroanal. Chem. Interfacial Electrochem., 1979, 101: 19-28.

[22] Hartnig C, Koper M T M. Molecular dynamics simulations of solvent reorganization in electron-transfer reactions. J. Chem. Phys., 2001, 115: 8540-8546.

[23] Marcus R A. electrostatic free energy and other properties of states having nonequilibrium polarization. J. Chem. Phys., 1956, 24: 979-980.

[24] Marcus R A. Chemical and electrochemical electron-transfer theory. Annu. Rev. Phys. Chem., 1964, 15: 155-157.

[25] Marcus R A, Sutin N. Electron transfers in chemistry and biology. Biochim. Biophys. Acta., 1985, 811: 265-266.

[26] Smalley J F, Feldberg S W, Chidsey C E D, et al. The kinetics of electron transfer through ferrocene-terminated alkanethiol monolayers on gold. J. Phys. Chem., 1995, 99: 13141.

[27] Smalley J F, Sachs S B, Chidsey C E D, et al. Interfacial electron-transfer kinetics of ferrocene through oligophenyleneethynylene bridges attached to gold electrodes as constituents of self-assembled monolayers: Observation of a nonmonotonic distance dependence. J. Am. Chem. Soc., 2004, 126: 14620.

[28] Closs G L, Miller J R. Intramolecular long-distance electron transfer in organic molecules. Science, 1988, 240: 440-447.

[29] Marcus R A. relation between charge transfer absorption and fluorescence spectra and the inverted region. J. Phys. Chem. A, 1989 93: 3078-3080.

[30] Chen P, Duesing R, Graff D K, et al. Intramolecular electron transfer in the inverted

region. J. Phys. Chem. 1991, 95: 5850-5858.

[31] Chen P, Meyer T J. Medium effects on charge transfer in metal complexes. Chem. Rev., 1998, 98: 1439-1477.

[32] Huynh M H V, Meyer T J. Proton-coupled electron transfer. Chem. Rev., 2007, 107: 5004-5064.

[33] Schmickler W. Interfacial Electrochemistry. Oxford: Oxford University Press, 1996.

[34] Tender L, Carter M T, Murray R W. Cyclic voltammetric analysis of ferrocene alkanethiol. Monolayer electrode kinetics based on marcus theory. Anal. Chem., 1994, 66: 3173-3181.

[35] Weber K, Creager S E. Voltammetry of redox-active groups irreversibly adsorbed onto electrodes-treatment using the marcus relation between rate and overpotential. Anal. Chem., 1994, 66: 3164-3172.

[36] C Kittel. Introduction to Solid State Physics. 4th ed.. New York: John Wiley & Sons, 1971.

[37] Smalley J F. Prediction of standard interfacial electron-transfer rate constants for the $Ru(NH_3)_6^{3+/2+}$ couple through omega-hydroxyalkanethiol self-assembled monolayers on gold electrodes. J. Phys. Chem. B., 2007, 111: 6798-6806.

[38] Eggers P K, Hibbert D B, Paddon-Row M N, et al. Molecular dynamics simulations of solvent reorganization in electron-transfer reactions. J. Phys. Chem. C., 2009, 113: 8964-8965.

[39] Ravenscroft M S, Finklea H O. Kinetics of electron transfer to attached redox centers on gold electrodes in nonaqueous electrolytes. J. Phys. Chem., 1994, 98: 3843-3850.

[40] Rowe G K, Carter M T, Richardson J N, et al. Consequences of kinetic dispersion on the electrochemistry of an adsorbed redox-active monolayer. Langmuir, 1995, 11: 1797-1806.

[41] Finklea H O, Liu L, Ravenscroft M S, et al. Multiple Electron Tunneling Paths across Self-Assembled Monolayers of Alkanethiols with Attached Ruthenium (II/III) Redox Centers. J. Phys. Chem., 1996, 100: 18852-11858.

[42] Creager S E, Wooster T T. A new way of using ac voltammetry to study redox kinetics in electroactive monolayers. Anal. Chem., 1998, 70: 4257-4263.

[43] Arikuma Y, Takeda K, Morita T, et al. linker effects on monolayer formation and long-range electron transfer in helical peptide monolayers. J. Phys. Chem. B., 2009, 113: 6256-6266.

[44] Okamoto S, Morita T, Kimura S. Electron transfer through a self-assebled monolayer of a double-helix peptide with linking the terminals by ferrocene. Langmuir, 2009, 25: 3297-3304.

[45] Takeda K, Morita T, Kimura S. Effects of monolayer structures on long-range electron transfer in helical peptide monolayer. J. Phys. Chem. B., 2008, 112: 12840-12850.

[46] Brevnov D A, Finklea H O, Van Ryswyk H. Ac voltammetry studies of electron transfer kinetics for a redox couple attached via short alkanethiols to a gold electrode. J. Electroanal. Chem., 2001, 500: 100-107.

[47] Eckermann A L, Shaw J A, Meade T J. Kinetic dispersion in redox active dithiocarbamate monolayers. Langmuir, 2010, 26(4): 2904-2913.

[48] Eggers P K, Zareie H M, Paddon-Row M N, et al. Structure and properties of redox active self-assembled monolayers formed from norbornylogous bridges. Langmuir, 2009, 25: 11090-11096.

[49] Fabre B, Hauquier F. Single-component and mixed ferrocene-terminated alkyl monolayers covalently bound to Si(111) surfaces. J. Phys. Chem. B., 2006, 110: 6848-6855.

[50] de Araujo M P, de Figueiredo A T, Bogado A L, et al. Ruthenium Phosphine/Diimine Complexes: Syntheses, Characterization, Reactivity with Carbon Monoxide, and Catalytic Hydrogenation of Ketones. Organometallics, 2005, 24 (25): 6159 - 6168.

[51] Katz E, Willner I. Probing biomolecular interactions at conductive and semiconductive surfaces by impedance spectroscopy: Routes to impedimetric immunosensors, DNA-sensors, and enzyme biosensors. Electroanalysis (New York), 2003, 15: 913-947.

[52] Park S-M, Yoo J S. Electrochemical impedance spectroscopy for better electrochemical Measurements. Anal. Chem., 2003, 75: 455A-461A.

[53] Macdonald D D. Reflections on the history of electrochemical impedance spectroscopy. Electrochim. Acta, 2006, 51: 1376-1388.

[54] Bowden E F. The distribution of standard rate constants for electron transfer between thiol-modified gold electrodes and adsorbed cytochrome c. Journal of Electroanalytical Chemistry, 1996, 410(7): 9-13.

[55] Song S H, Clark R A, Bowden E F, et al. Characterization of cytochrome c/alkanethiolate structures prepared by self-assembly on gold. J. Phys. Chem., 1993, 97 (24): 6564-6572.

[56] Creager S, Yu C J, Bamdad C, et al. Electron transfer at electrodes through conjugated "molecular wire" Bridges. J. Am. Chem. Soc., 1999, 121 (5): 1059-1064.

[57] Abhayawardhana A D, Sutherland T C. donator acceptor map of psittacofulvins and anthocyanins: Are they good antioxidant substances? J. Phys. Chem. C., 2009 113: 4915-4921.

[58] Brug G J, Van den Eeden A L G, Sluyters R M, et al. The analysis of electrode impedances complicated by the presence of a constant phase element. J. Electroanal. Chem. Interfacial Electrochem., 1984, 176: 275-295.

[59] Nahir T M, Bowden E F. The distribution of standard rate constants for electron transfer between thiol-modified gold electrodes and adsorbed cytochrome c. J. Electroanal. Chem., 1996, 410: 9-13.

[60] Jorcin J B, M Orazem E, Pebere N, et al. CPE analysis by local electrochemical impedance spectroscopy. Electrochim. Acta, 2006, 51: 1473.

[61] Kiani A, Alpuche-Aviles M A, Eggers P K, et al. Scanning electrochemical microscopy. 59. Effect of defects and structure on electron transfer through self-assembled monolayers. Langmuir, 2008, 24: 2841-2849.

[62] Liu B, Bard A J, Mirkin M V, et al. Electron transfer at self-assembled monolayers measured by scanning electrochemical microscopy. J. Am. Chem. Soc., 2004, 126: 1485-1492.

[63] Holt K B. Using scanning electrochemical microscopy (SECM) to measure the electron-transfer kinetics of cytochrome c immobilized on a COOH-terminated alkanethiol monolayer on a gold electrode. Langmuir, 2006, 22: 4298-4304.

[64] Edwards P P, Gray H B, Lodge M T J, et al. Electron transfer and electronic conduction through an intervening medium. Angew. Chem. Int. Ed. Engl., 2008, 47: 6758-

6765.

[65] Wilkinson G, Rosenblum M, Whiting M C, et al. The structure of iron bis-cyclopentadienyl. J. Am. Chem. Soc., 1952, 74: 2125-2126.

[66] Nguyen P, Gomez-Elipe P, Manners I. Organometallic polymers with transition metals in the main chain. Chem. Rev., 1999, 99: 1515-1548.

[67] Colacot T J. A concise update on the applications of chiral ferrocenyl phosphines in homogeneous catalysis leading to organic synthesis. Chem. Rev., 2003, 103: 3101-3118.

[68] Osakada K, Sakano T, Horie M, et al. Unique properties based on electronic communication between amino group of the ligand and Fe center. Coord. Chem. Rev., 2006, 250: 1012-1022.

[69] Nijhuis C A, Ravoo B J, Huskens J, et al. Redox-active supramolecular systems. Coord. Chem. Rev., 2007, 251: 1761-1780.

[70] Bonini B F, Fochi M, Ricci A. Synthesis and chemistry of new central and planar chiral sulfur-containing ferrocenyl compounds. Synth. Lett., 2007, 360-373.

[71] Wagner M. A new dimension in multinuclear metallocene complexes. Angew. Chem. Int. Ed. Engl., 2006, 45: 5916-5918.

[72] Berlinguette C P, Dragulescu-Andrasi A, Sieber A, et al. A charge-transfer-induced spin transition in a discrete complex: the role of extrinsic factors in stabilizing three electronic isomeric forms of a cyanide-bridged Co/Fe cluster. J. Am. Chem. Soc., 2005, 127 (18): 6766-6779.

[73] Nakamura E. Bucky ferrocene and bucky ruthenocene. serendipity and discoveries. J. Organomet. Chem., 2004, 689: 4630.

[74] Herrmann R, Huebener G, Ugi I. Chiral sulfoxides from (R)-α-dimethylaminoethylferrocene. Tetrahedron, 1985, 41: 941-947.

[75] Marquarding D, Klusacek H, Gokel G, et al. Correlation of central and planar elements of chirality in ferrocene derivatives. J. Am. Chem. Soc., 1970, 92: 5389-5393.

[76] Stynes H C, Ibers J A. Effect of metal-ligand bond distances on rates of electron-transfer reactions. Crystal structures of hexaammineruthenium(II) iodide, [Ru(NH3)6]I2, and hexaammineruthenium(III) tetrafluoroborate, [Ru(NH3)6]3. Inorg. Chem., 1971, 10: 2304-2308.

[77] Endicott J F, Taube H. Studies on oxidation-reduction reactions of ruthenium ammines. Inorg. Chem., 1965, 4: 437-445.

[78] Creutz C, Ford P C T, Meyer J. Henry Taube: Inorganic chemist extraordinaire. Inorg. Chem., 2006, 45 (18): 7059-7068.

[79] Creutz C, Taube H. Direct approach to measuring the Franck-Condon barrier complexes of ruthenium ammines. J. Am. Chem. Soc., 1973, 95: 1086-1094.

[80] Ford P C, Rudd D P, Gaunder R, et al. Synthesis and properties of pentaamminepyridineruthenium (ii) and related pentaammineruthenium complexes of aromatic nitrogen heterocycles. J. Am. Chem. Soc., 1968, 90: 1187-1194.

[81] Matsubara T, Ford P C. Some applications of cyclic voltammetry reduction potentials and rate studies. Inorg. Chem., 1976, 15: 1107-1110.

[82] Shin Y G K, Szalda D J, Brunschwig B S, et al. Electronic and molecular structures

of pentaammineruthenium pyridine and benzonitrile complexes as a function of oxidation state. Inorg. Chem., 1997, 36: 3190–3197.

[83] Isied S S, Taube H. Rates of substitution in cis and trans ruthenium (ii) aquotetraammines. Inorg. Chem., 1976, 15: 3070–3075.

[84] Che C M, Yam V W W. High valent compounds of Ruthenium and Osmium. Adv. Inorg. Chem., 1992, 39: 233–325.

[85] Haddox R M, Finklea H O. Proton-coupled electron transfer of an osmium aquo complex on a self-assembled monolayer on gold. J. Phys. Chem. B., 2004, 108: 1694–1700.

[86] Echegoyen L, Echegoyen L E. Electrochemistry of fullerenes and their derivatives. Acc. Chem. Res., 1998, 31: 593–601.

[87] Yang Y, Arias F, Echegoyen L, et al. Reversible fullerene electrochemistry: Correlation with the HOMO-LUMO energy difference for C_{60}, C_{70}, C_{76}, C_{78}, and C_{84}. J. Am. Chem. Soc., 1995, 117: 7801–7804.

[88] Suzuki T, Maruyama Y, Akasaka T, et al. Redox properties of organofullerenes. J. Am. Chem. Soc., 1994, 116: 1359–1363.

[89] Fukuzumi S, Nakanishi I, Suenobu T, et al. Electron-transfer properties of C_{60} and tert-butyl-C_{60} radical. J. Am. Chem. Soc., 1999, 121: 3468–3474.

[90] Imahori H, Sakata Y. Donor-linked fullerenes: photoinduced electron transfer and its potential application. Adv. Mater. (Weinheim, Ger.) 1997, 9: 537–546.

[91] Imahori H, Yamada H, Nishimura Y, et al. Vectorial multistep electron transfer at the gold electrodes modified with self-assembled monolayers of ferrocene-porphyrin-fullerene triads. J. Phys. Chem. B., 2000, 104: 2099–2018.

[92] Yamada H, Imahori H, Fukuzumi S. Photocurrent generation using gold electrodes modified with self-assembled monolayers of fullerene-porphyrin dyad. J. Mater. Chem., 2002, 12: 2034–2040.

[93] Yamada H, Imahori H, Nishimura Y, et al. Photovoltaic properties of self-assembled monolayers of porphyrins and porphyrin-fullerene dyads on ito and gold surfaces. J. Am. Chem. Soc., 2003, 125: 9129–9139.

[94] Cho Y J, Ahn T K, Song H, et al. Unusually high performance photovoltaic cell based on a [60]fullerene metal cluster-porphyrin dyad SAM on an ITO electrode. J. Am. Chem. Soc., 2005, 127: 2380–2381.

[95] Chukharev V, Vuorinen T, Efimov A, et al. Photoinduced electron transfer in self-assembled monolayers of porphyrin-fullerene dyads on ITO. Langmuir, 2005, 21: 6385–6391.

[96] Isosomppi M, Tkachenko N V, Efimov A, et al. Photoinduced electron transfer in multilayer self-assembled structures of porphyrins and porphyrin-fullerene dyads on ITO. J. Mater. Chem., 2005, 15: 4546–4554.

[97] Caldwell W B, Chen K, Mirkin C A, et al. Self-assembled monolayer films of C_{60} on cysteamine-modified gold. Langmuir, 1993, 9: 1945–1947.

[98] Song F, Zhang S, Bonifazi D, et al. Self-assembly of [60]fullerene-thiol derivatives on mercury surfaces. Langmuir, 2005, 21: 9246–9250.

[99] Arias F, Godinez L A, Wilson S R, et al. Asymmetric catalytic synthesis of-aryloxy alcohols: kinetic resolution of terminal epoxides via highly enantioselective ring opening with

phenols. J. Am. Chem. Soc., 1996, 118: 6086-6087.

[100] Shi X, Caldwell W B, Chen K, et al. A well-defined surface-confinable fullerene-monolayer self-assembly on Au(111). J. Am. Chem. Soc. 1994, 116: 11598-11599.

[101] Vuorimaa E, Vuorinen T, Tkachenko N, et al. Photoinduced electron transfer between self-assembled resorcinarene-[60]fullerene complex and poly (3-hexylthiophene) in Langmuir-Blodgett films. Langmuir, 2001, 17: 7327-7331.

[102] Fukuzumi S, Honda T, Ohkubo K, et al. Charge separation in metallomacrocycle complexes linked with electron acceptors by axial coordination. Dalton Trans., 2009, 3880-3889.

[103] Guldi D M, Illescas B M, Atienza C M, et al. Fullerene for organic electronics. Chem. Soc. Rev., 2009, 38: 1587-1597.

[104] Sakata Y, Imahori H, Tsue H, et al. Control of electron transfer and its utilization. Pure Appl. Chem., 1997, 69: 1951-1956.

[105] Marcus R A. electron transfer reactions in chemistry: Theory and experiment (Nobel Lecture). Angew. Chem. Int. Ed. Engl., 1993, 32: 1111-1121.

[106] Abe M, Michi T, Sato A, et al. Electrochemically controlled layer-by-layer deposition of metal-cluster molecular multilayers on gold. Angew. Chem. Int. Ed. Engl., 2003, 42: 2912-2915.

[107] Michi T, Abe M, Takakusagi S, et al. Spontaneous rapid growth of triruthenium cluster multilayers on gold surface. Cyclic voltammetric in situ monitoring and AFM characterization. Chem. Lett., 2008, 37: 576-577.

[108] Uehara H, Inomata T, Abe M, et al. Multicomponent molecular layers that exhibit electrochemical potential flip on Au (111) by use of proton-coupled electron-transfer reactions. Chem. Lett., 2008, 37: 684-685.

[109] Henderson J I, Feng S, Ferrence G M, et al. Self assembled monolayers (SAMs) of dithiols, di-isocyanides, and isocyanothiols on gold (111). "Chemically sticky" surfaces for covalent attachment of metal clusters and studies of interfacial electron transfer. Inorg. Chim. Acta, 1996, 242: 115-124.

[110] Jensen P S, Chi Q, Grumsen F B, et al. Gold nanoparticle assisted assembly of a heme protein for enhancement of long-range interfacial electron transfer. J. Phys. Chem. C., 2007, 111: 6124-6132.

[111] Gray H B, Winkler J R. Electron tunneling through proteins. Q. Rev. Biophys., 2003, 36: 341-372.

[112] Blankman J I, Shahzad N, Miller C J, et al. Direct voltammetric investigation of the electrochemical properties of human hemoglobin: relevance to physiological redox chemistry. Biochemistry, 2000, 39: 14806-14812.

[113] Blankman J I, Shahzad N, Dangi B, et al. Voltammetric probes of cytochrome electroreactivity: the effect of the protein matrix on outer-sphere reorganization energy and electronic coupling probed through comparisons with the behavior of porphyrin complexes. Biochemistry, 2000, 39: 14799-14805.

[114] Wei J, Liu H, Dick A R, et al. Direct wiring of cytochrome c's heme unit to an electrode: electrochemical studies. J. Am. Chem. Soc., 2002, 124: 9591-9599.

[115] Murgida D H, Hildebrandt P. Proton-coupled electron transfer of cytochrome c. J. Am. Chem. Soc., 2001, 123: 4062-4068.

[116] Clark R A, Bowden E F. Voltammetric peak broadening for cytochrome c/alkanethiolate monolayer structures: dispersion of formal potentials. Langmuir, 1997, 13: 559-565.

[117] Bjerrum M J, Casimiro D R, Chang I J, et al. Electron transfer in ruthenium-modified proteins. J. Bioenerg. Biomembr., 1995, 27: 295-302.

[118] Cotton T M, Schultz S G, Van Duyne R P. Surface-enhanced resonance Raman scattering from cytochrome c and myoglobin adsorbed on a silver electrode. J. Am. Chem. Soc., 1980, 102: 7960-7962.

[119] Taniguchi I, Toyosawa K, Yamaguchi H, et al. Reversible electrochemical reduction and oxidation of cytochrome c at a bis (4-pyridyl) disulphide-modified gold electrode. J. Chem. Soc. Chem. Commun., 1982, 1032-1033.

[120] Song S, Clark R A, Bowden E F, et al. Characterization of cytochrome c/alkanethiolate structures prepared by self-assembly on gold. J. Phys. Chem., 1993, 97: 6564-6572.

[121] Kasmi A E, Wallace J M, Bowden E F, et al. Controlling interfacial electron transfer kinetics of cytochrome c with mixed self-assembled monolayers. J. Am. Chem. Soc., 1998, 120: 225-226.

[122] Murgida D H, Hildebrandt P. Heterogeneous electron transfer of cytochrome c on coated silver electrodes. Electric field effects on structure and redox potential. J. Phys. Chem. B., 2001, 105: 1578-1586.

[123] Murgida D H, Hildebrandt P. Electrostatic-field dependent activation energies modulate electron transfer of cytochrome c. J. Phys. Chem. B., 2002, 106: 12814-12819.

[124] Murgida D H, Hildebrandt P, Wei J, et al. Surface-enhanced resonance raman spectroscopic and electrochemical study of cytochrome c bound on electrodes through coordination with pyridinyl-terminated self-assembled monolayers. J. Phys. Chem. B., 2004, 108: 2261-2269.

[125] Wei J, Liu H, Khoshtariya D E, et al. Electron-transfer dynamics of cytochrome c: A change in the reaction mechanism with distance. Angew. Chem. Int. Ed. Engl., 2002, 41: 4700-4703.

[126] Yue H, Khoshtariya D, Waldeck D H, et al. On the electron transfer mechanism between cytochrome c and metal electrodes. evidence for dynamic control at short distances. J. Phys. Chem. B., 2006, 110: 19906-19913

[127] Millo D, Bonifacio A, Ranieri A, et al. pH induced changes in absorbed cytochrome c voltammetric and surface-enhanced raman (SERRS) characterization performed simultaneously at chemically modified silver electrodes. Langmuir, 2007, 23: 9898-9904.

[128] Miller J E, Di Bilio A J, Wehbi W A, et al. Electron tunneling in rhenium-modified pseudomonas aeruginosa azurins. Biochim. Biophys. Acta: Bioenerg., 2004, 1655: 59-63.

[129] Jeuken L J C, Armstrong F A. Investigation of non-ideal electrochemical behavior of adsorbed proteins: Contrasting behavior of the "blue"copper protein, azurin, adsorbed on pyrolytic graphite and modified gold electrodes. J. Phys. Chem. B., 2001, 105: 5271-5282.

[130] Chi Q, Zhang J, Andersen J E T, et al. Ordered assembly and controlled electron transfer of the blue copper protein azurin at gold (111) single-crystal substrates. J. Phys. Chem.

B., 2001, 105: 4669-4679.

[131] Jeuken L J C, McEvoy J P, Armstrong F A. Insights into gated electron-transfer kinetics at the electrode-protein interface: A square wave voltammetry study of the blue copper protein azurin. J. Phys. Chem. B., 2002, 106: 2304-2313.

[132] Yokoyama K, Leigh B S, Sheng Y, et al. electron tunneling through pseudomonas aeruginosa azurins on sam gold electrodes. Inorg. Chim. Acta, 2008, 361: 1095-1099.

[133] Armstrong F A, Barlow N L, Burn P L, et al. Fast, long-range electron-transfer reactions of a "blue" copper protein coupled non-covalently to an electrode through a stilbenyl thiolate monolayer. Chem. Commun., 2004, 316-317.

[134] Sachs S B, Dudek S P, Hsung R P, et al. Rates of interfacial electron transfer through pi-conjugated spacers. J. Am. Chem. Soc., 1997, 119: 10563-10564.

[135] Guo Y, Zhao J, Yin X, et al. Electrochemistry investigation on protein protection by alkanethiol self-assembled monolayers against urea impact. J. Phys. Chem. C., 2008, 112: 6013-6021.

[136] Vargo M L, Gulka C P, Gerig J K, et al. Distance dependence of electron transfer kinetics for azurin protein adsorbed to monolayer protected nanoparticle film assemblies. Langmuir, 2110, 26(1): 560-569.

[137] Tian Y, Mao L, Okajima T, et al. Electrochemistry and electrocatalyticactivities of superoxide dismutases at gold electrodes modifiedwith a self-assembled monolayer. Anal. Chem., 2004, 76: 4162-4168.

[138] Shafiey H, Ghourchian H, Mogharrab N. How does reorganization energy change upon protein unfolding? Monitoring the structural perturbations in the heme cavity of cytochrome c. Biophys. Chem., 2008, 134: 225-231.

[139] Swiech O, Hrynkiewicz-Sudnik N, Palys B, et al. Gold nanoparticles tethered to gold surfaces using nitroxyl radicals. J. Phys. Chem. C, 2011, 115 (15): 7347-7354.

[140] Orlowski G A, Chowdhury S, Kraatz H B. Reorganization energies of ferrocene-peptide monolayers. Langmuir, 2007, 23: 12765-12770.

[141] Madhiri N, Finklea H O. Potential-, pH-, and isotope-dependence of proton-coupled electron transfer of an osmium aquo complex attached to an electrode. Langmuir, 2006, 22: 10643-10651.

[142] Hammes-Schiffer S, Soudackov A V. Proton-coupled electron transfer in solution, proteins, and electrochemistry. J. Phys. Chem. B., 2008, 112: 14108-14123.

[143] Costentin C. Electrochemical approach to the mechanistic study of proton-coupled electron transfer. Chem. Rev., 2008, 108: 2145-2179.

[144] Hammes-Schiffer S, Hatcher E, Ishikita H, et al. Theoretical studies of proton-coupled electron transfer: models and concepts relevant to bioenergetics. Coord. Chem. Rev., 2008, 252: 384-394.

[145] Jones M R. The petite purple photosynthetic powerpack. Biochem. Soc. Trans., 2009, 37: 400-407.

[146] Reece S Y, Nocera D G. Proton-coupled electron transfer in biology: results from synergistic studies in natural and model systems. Annu. Rev. Biochem., 2009, 78: 673-699.

[147] Finklea H O, Haddox R M. Coupled electron/proton transfer of galvinol attached to

SAMs on gold electrodes. Phys. Chem. Chem. Phys., 2001, 3: 3431-3436.

[148] Haddox R M, Finklea H O. Proton coupled electron transfer of galvinol in self-assembled monolayers. J. Electroanal. Chem., 2003, (550 - 551): 351-358.

[149] Niermann N, Degefa T H, Walder L, et al. Galvinoxyl monolayers on Au(111) studied by STM, EPR, and cyclic voltammetry. Phys. Rev. B: Condens. Matter Mater. Phys., 2006, 74: 235424/1-13.

[150] Hammes-Schiffer S. Isot Eff. Kinetic isotope effects for proton-coupled electron transfer reactions. Chem. Biol., 2006, 499-519.

[151] Takeuchi K J, Samuels G J, Gersten S W, et al. Multiple oxidation states of ruthenium and osmium based on dioxo/diaqua couples. Inorg. Chem., 1983, 22: 1407-1409.

[152] Costentin C, Robert M, Saveant J M, et al. Concerted and stepwise proton-coupled electron transfers in aquo/hydroxo complex couples in water: oxidative electrochemistry of $[Os^{II}(bpy)_2(py)(OH)_2]^{2+}$. Chem. Phys. Chem., 2009, 10: 191-198.

[153] Trammell S A, Lowy D A, Seferos D S, et al. Heterogeneous electron transfer of quinone-hydroquinone in alkaline solutions at gold electrode surfaces: Comparison of saturated and unsaturated bridges. J. Electroanal. Chem., 2007, 606: 33-38.

[154] Kazemekaite M, Bulovas A, Talaikyte Z, et al. Synthesis and self-assembling properties on gold of 2-methyl-1,4-naphthoquinone derivatives containing w-mercaptoalkylalkanoate group. Tetrahedron Lett., 2004, 45: 3551.

[155] Nuzzo R G, Allara D L. Adsorption of bifunctional organic disulfides on gold surfaces. J. Am. Chem. Soc., 1983, 105: 4481-4483.

[156] Nuzzo R G, Fusco F A, Allara D L. Spontaneously organized molecular assemblies. 3. Preparation and properties of solution adsorbed monolayers of organic disulfides on gold surfaces. J. Am. Chem. Soc., 1987, 109: 2358-2368.

[157] Bain C D, Whitesides G M. Molecular-level control over surface order in self-assembled monolayer films of thiols on gold. Science, 1988, 240: 62-63.

[158] Bain C D, Whitesides G M. Formation of monolayers by the coadsorption of thiols on gold: variation in the length of the alkyl chain. J. Am. Chem. Soc., 1989, 111: 7164-7175.

[159] Love J C, Estroff L A, Kriebel J K, et al. Self-assembled monolayers of thiolates on metals as a form of nanotechnology. Chem. Rev., 2005, 105: 1103-1170.

[160] Poirier G E. Characterization of organosulfur molecular monolayers on Au(111) using scanning tunneling microscopy. Chem. Rev., 1997, 97: 1117-1128.

[161] Benitez G, Vericat C, Tanco S, et al. Role of surface heterogeneity and molecular interactions in the charge-transfer process through self-assembled thiolate monolayers on Au (111). Langmuir, 2004, 20: 5030-5037.

[162] Sondag-Huethorst J A M, Schonenberger C, Fokkink L G J. Formation of holes in alkanethiol monolayers on gold. J. Phys. Chem., 1994, 98: 6826-6834.

[163] Demoz A, Harrison D J. Characterization and extremely low defect density hexadecanethiol monolayers on mercury surfaces. Langmuir, 1993, 9: 1046-1050.

[164] Tao F, Bernasek S L. Understanding odd?even effects in organic self-assembled monolayers. Chem. Rev., 2007, 107: 1408-1453.

[165] Poirier G E, Pylant E D. The Self-assembly mechanism of alkanethiols on Au(111). Science, 1996, 272: 1145-1148.

[166] Xu S, Cruchon-Dupeyrat S J N, Garno J C, et al. SFM observations of the two-stage growth process in solution. Direct proof that the striped phase consists of molecules lying down. J. Chem. Phys., 1998, 108: 5002-5012.

[167] Chambers R C, Inman C E, Hutchison J E. Electrochemical detection of nanoscale phase separation in binary self assembled monolayers. Langmuir, 2005, 21: 4615-4621.

[168] Collard D M, Fox M A. Use of electroactive thiols to study the formation and exchange of alkanethiol monolayers on gold. Langmuir, 1991, 7: 1192-1197.

[169] Vericat C, Vela M E, Salvarezza R C. Self-assembled monolayers of alkanethiols on Au(111): surface structures, defects and dynamics. Phys. Chem. Chem. Phys., 2005, 7: 3258-3268.

[170] Lee L Y S, Lennox R B. Ferrocenylalkylthiolate labeling of defects in alkylthiol self-assembled monolayers on gold. Phys. Chem. Chem. Phys., 2007, 9: 1013-1020.

[171] Campbell D J, Herr B R, Hulteen J C, et al. Ion-gated electron transfer in self-assembled monolayer films. J. Am. Chem. Soc., 1996, 118: 10211-10219.

[172] Herr B R, Mirkin C A. Self-assembled monolayers of ferrocenylazobenzenes: monolayer structure vs. response. J. Am. Chem. Soc., 1994, 116: 1157-1158.

[173] Li C, Ren B, Zhang Y, et al. A novel ferroceneylazobenzene self-assembled monolayer on an ITO electrode: Photochemical and electrochemical behaviors. Langmuir, 2008, 24: 12911-12918.

[174] Calvente J J, Lopez-Perez G, Ramirez P, et al. Experimental study of the interplay between long-range electron transfer and redox probe permeation at self-assembled monolayers: Evidence for potential-induced ion gating. J. Am. Chem. Soc., 2005, 127: 6476-6486.

[175] Calvente J J, Andreu R, Molero M, et al. Influence of spacial redox distribution on the electrochemical behaviour of electroactive self-assembled monolayers. J. Phys. Chem. B., 2001, 105: 9557-9568.

[176] Sabatani E, Rubinstein I. Monolayer-based ultra-microelectrodes for the study of very rapid electrode kinetics. J. Phys. Chem., 1987, 91: 6663-6669.

[177] Hockett L A, Creager S E. Redox kinetics for ferrocene groups immobilized in impermeable and permeable self-assembled monolayers. Langmuir, 1995, 11: 2318-2321.

[178] Weber K, Hockett L, Creager S. Long-range electronic coupling between ferrocene and gold in alkanethiolate-based monolayers on electrodes. J. Phys. Chem. B., 1997, 101: 8286-8291.

[179] Robinson D B, Chidsey C E D. Submicrosecond electron transfer to monolayer-bound redox species on gold electrodes at large overpotentials. J. Phys. Chem. B., 2002, 106: 10706-10713.

[180] Rowe G K, Creager S E. Interfacial solvation and double-layer effects on redox reactions in organized assemblies. J. Phys. Chem., 1994, 98 (21): 5500-5507.

[181] Acevedo D, Abruna H D. Electron-transfer study and solvent effects on the formal potential of a redox-active self-assembling monolayer. J. Phys. Chem., 1991, 95: 9590-9594.

[182] Ricci A, Rolli C, Rothacher S, et al. Electron transfer at Au surfaces modified by

Tethered Osmium bipyridine-pyridine complexes. J. Solid State Electrochem., 2007, 11: 1511–1520.

[183] Forster R J, Figgemeier E, Loughman P, et al. Conjugated vs nonconjugated bridges: hetereogeneous electron transfer dynamics of osmium polypyridyl monolayers. Langmuir, 2000, 16: 7871–7875.

[184] Devaraj N K, Decreau R A, Ebina W, et al. Rate of interfacial electron transfer through the 1,2,3-triazole linkage. J. Phys. Chem. B, 2006, 110: 15955–15962.

[185] Luo J, Isied S S. Ruthenium tetraammine chemistry of self-assembled monolayers on gold surfaces: Substitution and reactivity at the monolayer interface. Langmuir, 1998, 14: 3602–3606.

[186] Van Ryswyk H, Turtle E D, Watson-Clark R, et al. Reactivity of ester linkages and pentaammineruthenium (iii) at the monolayer assembly/solution interface. Langmuir, 1996, 12: 6143–6150.

[187] Cheng J, Saghi-Szabo G, Tossell J A, et al. Modulation of electronic coupling through self-assembled monolayers via internal chemical modification. J. Am. Chem. Soc., 1996, 118: 680–684.

[188] Liang C, Newton M D. Ab initio studies of electron transfer: Pathway analysis of effective transfer integrals. J. Phys. Chem., 1992, 96: 2855–2866.

[189] Yamamoto H, Waldeck D H. Effect of tilt-angle on electron tunneling through organic monolayer films. J. Phys. Chem. B, 2002, 106: 7469–7473.

[190] Nuzzo R G, Dubois L H, Allara D L. Fundamental studies of microscopic wetting on organic surfaces. 1. Formation and structural characterization of a self-consistent series of polyfunctional organic monolayers. J. Am. Chem. Soc., 1990, 112: 558–569.

[191] Schreiber F. Structure and growth of self-assembling monolayers. Prog. Surf. Sci., 2000, 65: 151–256.

[192] Kim B, Beebe J M, Olivier C, et al. Temperature and length dependence of charge transport in redox-active molecular wires incorporating ruthenium (II) Bis(s-arylacetylide) complexes. J. Phys. Chem. C., 2007, 111: 7521–7526.

[193] Adams D M, Brus L, Chidsey C E D, et al. Charge transfer on the nanoscale: Current status. J. Phys. Chem. B., 2003, 107: 6668–6697.

[194] Woitellier S, Launay J P, Spangler C W. Intervalence transfer in pentaammineruthenium complexes of alpha-omega-dipyridylpolyenes. Inorg. Chem., 1989, 28: 758–762.

[195] Adams D M, Brus L, Chidsey C E D, et al. Charge transfer on the nanoscale: Current status. J. Phys. Chem. B., 2003, 107: 6668–6697.

[196] Dudek S P, Sikes H D, Chidsey C E D. Synthesis of ferrocenethiols containing oligo (phenylenevinylene) bridges and their characterization on gold electrodes. J. Am. Chem. Soc., 2001, 123: 8033–8038.

[197] Sikes H D, Sun Y, Dudek S P, et al. Photoelectron spectroscopy to probe the mechanism of electron transfer through oligo (phenylene vinylene) bridges. J. Phys. Chem. B., 2003, 107: 1170–1173.

[198] Davis W B, Ratner M A, Wasielewski M R. Conformational gating of long distance electron transfer through wire-like bridges in donor-bridge-acceptor molecules. J. Am. Chem.

Soc., 2001, 123: 7877-7886.

[199] Yang W R, Jones M W, Li X, et al. Single molecule conductance through rigid norbornylogous bridges with zero average curvature. J. Phys. Chem. C., 2008, 112: 9072-9080.

[200] Liu J, Gooding J J, Paddon-Row M N. Unusually rapid heterogeneous electron transfer through a saturated bridge 18 bonds in length. Chem. Commun., 2005, 631-633.

[201] Beebe J M, Engelkes V B, Liu J, et al. Testing the molecular wire concept: Direct evidence for through-bond tunneling in molecular junctions incorporating rigid, curved molecules. J. Phys. Chem. B., 2005, 109: 5207-5215.

[202] Gray H B, Winkler J R. Electron transfer in metalloproteins. Electron Transfer Chem., 2001, 3: 3-23.

[203] McGourty J L, Blough N V, Hoffman B M. Electron transfer at crystallographically known long distances (25 Angstrom) in hybrid hemoglobin. J. Am. Chem. Soc., 1983, 105: 4470-4472.

[204] M J Therien, M Selman, H B Gray, et al. Long-range electron transfer in ruthenium-modified cytochrome c: evaluation of porphyrin-ruthenium electronic couplings in the Candida krusei and horse heart proteins. J. Am. Chem. Soc., 1990, 112 (6): 2420 - 2422

[205] Conrad D W, Zhang H, Stewart D E, et al. Distance dependence of long-range electron transfer in cytochrome c derivatives containing covalently attached cobalt cage complexes. J. Am. Chem. Soc., 1992, 114: 9909-9915.

[206] Wang K, Zhen Y, Sadoski R, et al. Definition of the interaction domain for cytochrome c on cytochrome c oxidase. Ii. Rapid kinetic analysis of electron transfer from cytochrome c to Rhodobacter sphaeroides cytochrome oxidase surface mutants. J. Biol. Chem., 1999, 274: 38042-38050.

[207] Bediako-Amoa I, Sutherland T C, Li C Z, et al. Electrochemical and surface study of ferrocenoyl oligopeptides. J. Phys. Chem. B., 2004, 108: 704-714.

[208] Mandal H S, Kraatz H B. Electron transfer across α-helical peptides: potential influence of molecular dynamics. Chem. Phys., 2006, 326: 246-251.

[209] Morita T, Kimura S. Long-range electron transfer over 4 nm governed by an inelastic hopping mechanism in self-assembled monolayers of helical peptides. J. Am. Chem. Soc., 2003, 125: 8732-8733.

[210] Sek S, Sepiol A, Tolak A, et al. distance dependence of the electron transfer rate through oligoglycine spacers introduced into self-assembled monolayers. J. Phys. Chem. B., 2004, 108: 8102-8105.

[211] Takeda K, Morita T, Kimura S. Effects of different linkers on electron transfer through helical peptide monolayers. Peptide Sci., 2006, 43: 77.

[212] Watanabe J, Morita T, Kimura S. Effects of dipole moment, linkers, and chromophores at side chains on long-range electron transfer through helical peptides. J. Phys. Chem. B., 2005, 109: 14416-11425.

[213] Devillers C H, Boturyn D, Bucher C, et al. Redox-active biomolecular architectures and self-assembled monolayers based on a cyclodecapeptide regioselectively addressable functional template. Langmuir, 2006, 22: 8134-8143.

[214] Doneux T, Bouffier L, Mello L V, et al. Molecular dynamics and electrochemical

investigations of a pH-responsive peptide monolayer. J. Phys. Chem. C, 2009, 113: 6792–6799.

[215] Meade T J, Kayyem J F, Fraser S E. Nucleic acid mediated electron transfer. Patent., 1994, 12: 9, 515, 971.

[216] Yu C J, Wang H, Wan Y, et al. 2'-Ribose-ferrocene oligonucleotides for electronic detection of nucleic acids. Journal of Organic Chemistry., 2001, 66: 2937–2942.

[217] Kelley S O, Jackson N M, Hill M G, et al. Long-range electron transfer through DNA films. Angew. Chem. Int. Ed. Engl., 1999, 38: 941.

[218] Liu T, Barton J K. DNA electrochemistry through the base pairs not the sugar? phosphate backbone. J. Am. Chem. Soc., 2005, 127: 10160–10161.

[219] Long Y T, Li C Z, Sutherland T C, et al. A comparison of electron transfer rates of ferrocenoyl linked DNA. J. Am. Chem. Soc., 2003, 125: 8724–8725.

[220] Anne A, Demaille C. Dynamics of electron transport by elastic bending of short DNA duplexes. Experimental study and quantitative modeling of the cyclic voltammetric behavior of 3'-ferrocenyl DNA end-grafted on gold. J. Am. Chem. Soc., 2006, 128: 542–557.

[221] Anne A, Demaille C. Electron transport by molecular motion of redox-DNA strands: Unexpectedly slow rotational dynamics of 20-mer ds-DNA chains end-grafted onto surfaces via C_6 linkers. J. Am. Chem. Soc., 2008, 130: 9812–9823.

[222] Frank N L, Meade T J. Modification of duplex DNA with aruthenium electron donor-acceptor pair using solid-phase DNA synthesis. Inorg. Chem., 2003, 42: 1039–1044.

[223] Paul A, Bezer S, Venkatramani R, et al. Role of nucleobase energetics in single stranded peptide nucleic acid charge transfer. J. Am. Chem. Soc., 2009, 131: 6498–6507.

[224] Paul A, Watson R M, Wierzbinski E, et al. Distance dependence of the charge transfer rate for peptide nucleic acid monolayers. J. Phys. Chem. B., 2009, (ACS ASAP).

[225] Berlin Y A, Burin A L, Ratner M A. Elementary steps for charge transport in DNA: Thermal activation vs. tunneling. Chem. Phys., 2002, 275: 61–74.

[226] Roth K M, Yasseri A A, Liu Z, et al. Measurement of electron-transfer rates of charge-storage molecular monolayers on Si (100). towards hybrid molecular/semiconductor information storage devices. J. Am. Chem. Soc., 2003, 125: 505–517.

[227] Yasseri A A, Syomin D, Loewe R S, et al. structural and electron-transfer characteristics of o-, s-, and se-tethered porphyrin monolayers on Si (100). J. Am. Chem. Soc., 2004, 126: 15603–15612.

[228] Gosavi S, Marcus R A. Nonadiabatic electron transfer at metal surfaces. J. Phys. Chem. B., 2000, 104: 2067–2072.

[229] Finklea H O, Yoon K, Chamberlain E, et al. Effect of the metal on electron transfer across self-assembled monolayers. J. Phys. Chem. B., 2001, 105: 3088–3092.

[230] Forster R J, Loughman P, Keyes T E. Effect of electrode density of states on the heterogeneous elactron-transfer dynamics of osmium containing monolayers. J. Am. Chem. Soc., 2000, 122: 11948–11955.

[231] Cohen-Atiya M, Mandler D. Studying electron transfer through alkanethiol self-assembled monolayers on a hanging mercury drop electrode using potentiometric measurements. Phys. Chem. Chem. Phys., 2006, 8: 4405–4409.

[232] Slowinski K, Slowinska K U, Majda M. Electron tunneling across hexadecanethiolate monolayers on mercury electrodes. Reorganization energy, structure and permeability of the alkane/water interface. J. Phys. Chem. B., 1999, 103: 8544-8551.

[233] Laibinis P E, Whitesides G M, Allara D L, et al. A comparison of the structures and wetting properties of self-assembled monolayers of n-alkanethiols on the coinage metal surfaces, Cu, Ag, Au. J. Am. Chem. Soc., 1991, 113: 7152-7167.

[234] Himmelhaus M, Gauss I, Buck M, et al. Core-level spectroscopy in early-transition-metal compounds. J. Electron Spectrosc. Relat. Phenom. 1998, 92: 139-149.

[235] Ziegler K J, Doty R C, Johnston K P, et al. Synthesis of organic monolayer-stabilized copper nanocrystals in supercritical water. J. Am. Chem. Soc., 2001, 123: 7797-7803.

[236] Vogt A D, Han T, Beebe Jr T P. Adsorption of 11-mercaptoundecanoic acid on Ni (111) and its interaction with probe molecules. Langmuir, 1997, 13: 3397-3403.

[237] Mekhalif Z, Laffineur F, Couturier N, et al. Elaboration of self-assembled monolayers of n-alkanethiols on nickel polycrystalline substrates: Time, concentration, and solvent effects. Langmuir, 2003, 19: 637-645.

[238] Mekhalif Z, Riga J, Pireaux J J, et al. Self-assembled monolayers of n-dodecanethiol on electrochemically modified polycrystalline nickel surfaces. Langmuir, 1997, 13: 2285-2290.

[239] Bengio S, Fonticelli M, Benitez G, et al. Electrochemical self-assembly of alkanethiolate molecules on Ni(111) and polycrystalline Ni surfaces. J. Phys. Chem. B., 2005, 109: 23450-23460.

[240] Hoertz P G, Niskala J R, Dai P, et al. Comprehensive investigation of self-assembled monolayer formation on ferromagnetic thin film surfaces. J. Am. Chem. Soc., 2008, 130: 9763-9772.

[241] Love J C, Wolfe D B, Haasch R, et al. Formation and structure of self-assembled monolayers of alkanethiolates on palladium. J. Am. Chem. Soc., 2003, 125: 2597-2609.

[242] Jiang X, Bruzewicz D A, Thant M M, et al. Palladium as a substrate for self-assembled monolayers used in biotechnology. Anal. Chem., 2004, 76: 6116-6121.

[243] Love J C, Wolfe D B, Chabinyc M L, et al. Self-assembled monolayers of alkanethiolates on palladium are good etch resists. J. Am. Chem. Soc., 2002, 124: 1576-1577.

[244] Soreta T R, Strutwolf J, O'Sullivan C K. Electrochemically deposited palladium as a substrate for self-assembled monolayers. Langmuir, 2007, 23: 10823-10830.

[245] Linford M R, Chidsey C E D. Alkyl monolayers covalently bonded to silicon surfaces. J. Am. Chem. Soc., 1993, 115: 12631-12632.

[246] Sailor M J, Lee E J. ChemInform abstract: Surface chemistry of luminescent silicon nanocrystallites. Adv. Mater. Weinheim, Ger., 1997, 9: 783-793.

[247] Choi C H, Liu D J, Evans J W, et al. Passive and active oxidation of Si (100) by atomic oxygen: A theoretical study of possible reaction mechanisms. J. Am. Chem. Soc., 2002, 124: 8730-8740.

[248] Dalchiele E A, Aurora A, Bernardini G, et al. XPS and electrochemical studies of ferrocene derivatives anchored on n-and p Si (100) by Si—O or Si—C bonds. J. Electroanal. Chem., 2005, 579: 133-142.

[249] Hasobe T, Imahori H, Ohkubo K, et al. Structure and photoelectrochemical properties

of ITO electrodes modified with self-assembled monolayers of meso, meso-linked porphyrin oligomers. J. Porphyrins Phthalocyanines, 2003, 7: 296-312.

[250]Li J, Wang L, Liu J, et al. Characterization of transparent conducting oxide surfaces using self-assembled electroactive monolayers. Langmuir, 2008, 24: 5755-5765.

[251]Bigelow W C, Pickett D L, Zisman W A. Adsorbed from solution in non-polar liquids. J. Colloid Sci., 1946, 1: 513-517.

[252]Whittington S G, Chapman D. Effect of density on configurational properties of long-chain molecules using a Monte Carlo method. Trans. Faraday Soc., 1966, 62: 3319-3324.

[253]Scott Jr H L. Monte Carlo studies of the hydrocarbon region of lipid bilayersBiochim. Biophys. Acta, 1977, 469: 264-271.

[254]Belle J, Bothorel P. Theoretical study of spin-labeled aliphatic chains in bilayers. Biochem. Biophys. Res. Commun., 1974, 58: 433-436.

[255]Duffy D M, Harding J H. Modeling the properties of self-assembled monolayers terminated by carboxylic acids. Langmuir, 2005, 21: 3850-3857.

[256]Haran M, Góose J E, Clote N P, et al. Multiscale modeling of self-assembled monolayers of thiophenes on electronic material surfaces. Langmuir, 2007, 23: 4897-4909.

[257]Cotterill R M J. Computer simulation of model lipid membrane dynamics. Biochim. Biophys. Acta, 1976, 433: 264-270.

[258]Toxvaerd S. Conditional lifetimes in geminate recombination. J. Chem. Phys., 1977, 67: 2056-2060.

[259]Kox A J, Michels J P J, Wiegel F W. Simulation of a lipid monolayer using molecular dynamics. Nature, 1980, 287: 317-319.

[260]Van der Ploeg P, Berendsen H J C. Molecular dynamics simulation of a bilayer membrane. J. Chem. Phys., 1982, 76: 3271-3276.

[261]Tupper K J, Colton R J, Brenner D W. Simulations of self-assembled monolayers under compression: Effect of surface asperities. Langmuir, 1994, 10: 2041-2043.

[262]Hautman J, Klein M L. Simulation of a monolayer of alkyl thiol chains. J. Chem. Phys., 1989, 91: 4994.

[263]Sellers H, Ulman A, Shnidman Y, et al. Structure and binding of alkanethiolates on gold and silver surfaces: implications for self-assembled monolayers. J. Am. Chem. Soc., 1993, 115: 9389-9401.

[264]Bain C D, Whitesides G M. Modeling organic surfaces with self-assembled monolayers. Angew. Chem., 1989, 101: 522-528.

[265]Prime K L, Whitesides G M. Self-assembled organic monolayers: Model system for studying adsorption of proteins at surfaces. Science, 2009, 252: 1164-1167.

[266]Alves C A, Smith E L, Porter M D. Atomic scale imaging of alkanethiolate monolayers at gold surfaces with atomic force microscopy. J. Am. Chem. Soc., 1992, 114: 1222-1227.

[267]Bryant M A, Pemberton J E. Surface Raman scattering of self-assembled monolayers formed from 1-alkanethiols: behavior of films at gold and comparison to films at silver. J. Am. Chem. Soc., 1991, 113: 8284-8293.

[268]Camillone III N, Chidsey C E D, Eisenberger P, et al. Structural defects in self-assembled organic monolayers via combined atomic beam and X-ray diffraction. J. Chem. Phys.,

1993, 99: 744–747.

[269] Li T W, Chao I, Tao Y T. The relationship between packing structures and head groups of self-assembled monolayers on Au (111): Bridging experimental observations through computer simulations. J. Phys. Chem. B., 1998, 102: 2935–2942.

[270] Bond A D. On the crystal structures and melting point alternation of the n-alkyl carboxylic acids. New J. Chem., 2004, 28: 104–114.

[271] Wenzl I, Yam C M, Barriet D, et al. Structure and wettability of methoxy-terminated self-assembled monolayers on gold. Langmuir, 2003, 19: 10217–10224.

[272] Kim Y H, Seung S J, William A G. Conformations and charge transport characteristics of biphenyldithiol self-assembled-monolayer molecular electronic devices: A multiscale computational study. J. Chem. Phys., 2005, 122: 244703–244704.

[273] Prime K L, Whitesides G M. Self-assembled organic monolayers: model systems for studying adsorption of proteins at surfaces. Science, 1991, 252: 1164–1167.

[274] Harder P, Grunze M, Dahint R, et al. Molecular conformation in oligo(ethylene glycol)-terminated self-assembled monolayers on gold and silver surfaces determines their ability to resist protein adsorption. J. Phys. Chem. B., 1998, 102: 426–436.

[275] Eckermann A L, Feld D J, Shaw J A, et al. Electrochemistry of redox-active self-assembled monolayers. Coordination Chemistry Reviews, 2010, 254(15–16): 1769–1802.

第四章 基于硅基自组装膜的分子电子器件研究

微电子器件诞生于 20 世纪 50 年代，半导体材料如 Si、Ge、Ga—As 等是微电子工业的基础。当前，微电子技术已经改变了人类的活动方式，其在人类活动的几乎所有领域(如汽车、家电、医疗及科学仪器等)都发挥了显而易见的作用，并且，伴随其性质的继续开拓也发展了新的应用领域，驱动了当今世界现代技术的发展。

微电子器件的特点之一是它的微型化，微型化意味着装置的尺寸越来越小，但其工作效率更高，价格也更低。微型化的概念由 Moore 定律所控制，Moore 定律预言传导器的尺寸将不小于 20 nm，这是因为低于这个尺寸，传导器将面临诸如电流泄漏、能量消耗以及在电荷—构成方面随机出现的波动等。这就促使研究者去发展新的微电子系统，并从材料、结构、制备方法以及原理等方面进行系统研究[1]。

第一节 分子电子器件的出现及其目前的地位

当整个传导器的尺寸降至几个纳米时，表明在一个传导器中仅仅有少量原子在工作，由此，就出现了微电子器件的制备和组成(装配)问题，并且电子出现的信号要求独立的或群体分子的设计具有特定的电子功能，有机分子由于其尺寸机械加工的灵活性以及化学耐受性，被期望可以在分子电子器件中发挥关键作用[2]。有机分子的电学性质在 1971 年首次由 Mann 和 Kahn 进行了测量[3]，这项工作也促进了有机分子经由自组装的有序单层膜沉积技术，即自组装膜的发展。单个有机分子作为整流器的理论由 Aviram 和 Ratner 于 1974 年首次提出[4]，这种分子由一个易氧化的基团和一个易还原的基团相联系。LUMO 和 HOMO 的定向排列将使得这样的电子传导易于在一个方向上进行，这促进分子二极管的出现。随后，研究者提出了几种基于分子逻辑结构装置的概念。但在当时，由于实验条件及理论发展的限制，以及以硅为基础的半导体材料在微电子器件所占据的主导地位，发展分子电子技术被认为是不可能的。然而，伴随着 SPM 技术的出现，分子电子研究在上个世纪 80 年代得到了较快发展[5]。应用 SPM 不仅使我们能够看到和操纵有机分子，并可以测量它们的电学性质。目前，我们仍处于分子电子研究的初始阶段，主要的研究集中于具有不同电子功能的新有机分子的合成，并且在测量"单个分子"或"群体分子"的电学性质中发展新的方法[6]。

自组装代表了一种先进的表面功能化策略，其可以创造良好定位、具有可剪裁性质以及预设功能的纳米结构，在固体表面分子的自组装是通过分子—分子以及分子—基底相互作用来控制的。目前，应用的 SAM 通常有两种类型。

1. **以 Au 为代表的金属基底的烷基硫醇单层膜**

这类自组装膜研究最为广泛和深入，通过金属与功能有机分子结合的应用，获得了分子良好有序、规则定位的排列。描述以 Au 等金属作为基底与不同烷基硫醇形成的 SAM 目前已有多篇综述性文献报道[7-9]。

2. **有机分子经由自组装过程结合于 SiO_2 或 Si 基底上的单层膜**

Bigelow 等人[10]发现三氯硅烷可以自溶液中在氧化硅表面形成自组装有机单层膜。后来,Maoz 和 Sagiv 等[11]进一步发展了这项技术。由 Si 提供的能量间隙可以为设计新的一类共振隧穿装置提供可能,并可以有效地发展组合分子装置及杂化装置[12]。

近年来,将功能化的有机单分子层和以硅为基体的器件相结合引起了人们极大的关注,并促进了有机硅—杂化材料及器件的设计和制备。有机分子一般是通过 Si—C、Si—O 或 Si—N 键嫁接在活性硅表面上,将有机分子接枝到硅表面将赋予传统的硅材料更多新的功能,它具有许多其他表面难以比拟的优点:

(1)硅半导体是微电子工业的基础,研究比较深入,成熟的微电子集成技术和工艺可以充分利用;

(2)可以通过掺杂程度调节导电状态;

(3)硅表面平整,比金属更易于得到原子级平整度的单晶平面;

(4)Si—C 键比 S—Au 键的键能大,可以获得更稳定的表面;

(5)硅化学反应中副反应少,容易得到均一的有机层。

因此,将有机分子嫁接在硅表面上,通过单层有机分子精细调节硅表面的性质,可以获得功能化、智能化的表面。研究单层有机分子修饰硅表面的宏观物理、化学性质,并开展纳米/分子尺度的光、电等特性及器件化的研究,发展相关理论,将有力地推动分子电子器件和分子电子学的发展。

第二节 基于硅基自组装膜的电子传导

脂肪链 SAM 由于其 HOMO 和 LUMO 具有较大的间隙而被期望用于介电材料,理论计算表明,脂肪链的 HOMO—LUMO 间隙最大可达到 9 eV,实际上,三氯硅烷 SAM($n>8$C)的研究表明通过它们的裂分电流非常低($<10^8$ A/cm^2),因此,有机薄膜晶体管(OTFT)可被用于介电材料[13]。关于有机自组装膜(SAM)中的电子传导性目前已发展了各种理论模型,这些理论模型可以在一个介电 SAM 中去更好地理解电子传导性,并且可以应用这些模型去分析一些实验结果。

表 4-1　M—I—M 结构中各种可能的电子传导机制(与 J 的关系)示意图

传导机制	电子传导跨越能垒示意图	与施加电压的关系	温度依赖性
隧穿 (1)直接隧穿		(1)$eV\ll\Phi$ 或 $V\approx 0$, $J\propto V$	无关
(2)FN 隧穿	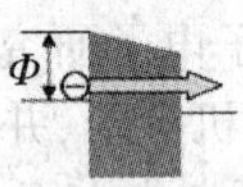	(2)$eV<\Phi/2$, $J\propto V+V^3$	无关
		(3)$eV>\Phi/2$, $\ln(J/V^2)\propto 1/V$	无关
热电子激发	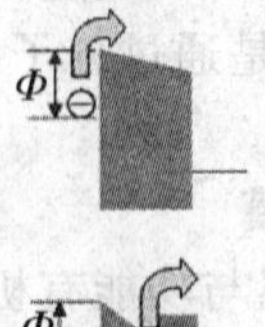	$\ln(J)\propto V^{1/2}$	$\ln(J/T^2)\propto 1/T$
Poole-Frankel 激发	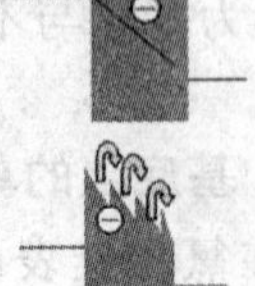	$\ln(J/V)\propto V^{1/2}$	$\ln(J)\propto 1/T$
跨越传导		$J\propto V$	$\ln(J/V)\propto 1/T$

1. 理论模型

两个金属电极间的脂肪链 SAM 三明治结构可以通过一个金属(1)—绝缘体—金属(2)(M_1—I—M_2)的能量带描述,在一个 M—I—M 结构中,电子传导可由 4 种不同的机制,即隧穿、热电子激发、Poole-Frankel 激发以及跨越传导所控制,表 4-1 列出了各种传导机制。

1.1 隧穿

M—I—M 结构中简单的隧穿模型由电子产生的电流来描述,在金属—绝缘体界面,有一个起始电势能垒Φ,即使电子能量远低于Φ,在其界面上电流波形并未完全消失,而是随能垒增加成比例降低。这表明了存在一种可能性,即电子可从两个电极间穿越一个短的距离进入绝缘体,而不考虑其能量水平,电流波形的重叠产生非共振的隧穿电流密度,Simmon 指出了直接隧穿电流密度(J_{DT})与V对能垒的理论关系[14],

$$J_{DT} = (\alpha/d^2)\{\Phi\exp[-Ad(\Phi)^{1/2}] - (\Phi+eV)\exp[-Ad(\Phi+eV)^{1/2}]\} \quad (4-1)$$

$\alpha=e/(4\pi^2\beta'^2\hbar)$,$A=2\beta'(2m^*/\hbar^2)^{1/2}$ (e为电子电荷,m^*为电子在绝缘体中的有效质量,β'为一常数,其值≈1,h为普朗克常数),Φ为平均能垒高度,d为能垒宽度,V为两电极间的电压。

研究者在非常低的电压下,如$eV<<\Phi$,或是$V\approx 0$,导出电流密度[14],

$$J_{DT} = (\gamma\Phi^{1/2}/d)\exp(-AD\Phi^{1/2})V \quad (4-2)$$

这里$\gamma=e(2m^*)^{1/2}/(4\pi^2\beta'^2\hbar^2)$,因为$eV<<\Phi$,可以认为$\Phi$并不依赖于$V$,因此,这里$J$与$V$成比例。

对于中间V值,如$eV<\Phi/2$,电流密度以下式表达[14],

$$J_{DT} = (\gamma\Phi^{1/2}/d)\exp(-AD\Phi^{1/2})(V+\sigma V^3) \quad (4-3)$$

其中$\sigma=(Ae)^2/(96\Phi d^2)-Ae^2/(32d\Phi^{3/2})$。

对于高V,如$eV>\Phi$,电势能垒具有不规则的形状,电流密度关系式简化为通常的 Fowler-Nordheim(FN)形式[14],

$$J_{FN} = BE^2\exp(-C/E) \quad (4-4)$$

$E=V/d$是穿越单层膜的电场,$B=e^3/(16\pi^2m^*h\Phi)$,$C=[4(2m^*)^{1/2}/3eh]\Phi^{3/2}$,穿越介电层的电流传递将由 FN 隧穿控制。

除了以上讨论的直接隧穿和 Fowler-Nordheim 非共振隧穿机制,如果 HOMO—LUMO 的间隙相对较小,如对于 π 共轭分子,一个共振隧穿也可以通过 SAM 的分子轨道发生。

1.2 热电子激发

在金属中,电子的能量分布由 Fermi-Dirac 分散式决定,在一定的温度下,金属中部分电子将获得足够的能量越过绝缘体的能量势垒,此过程称为热电子激发,其电流密度由下式给出[15]:

$$J_{TE} = T^2\exp\{[\Phi-e(eV/4\pi\varepsilon d)^{1/2}]/(KT)\} \quad (4-5)$$

K在这里是波尔兹曼常数。

1.3 Poole-Frankel 激发

如果绝缘层包含非理想性结构,如不纯原子导致的缺陷,那么这些缺陷将扮演电子陷阱的作用,诱陷电子的场加强热激发将产生电流,其电流密度以下式表达[15]:

$$J_{PF} = V\exp\{-[\Phi_B-e(eV/4\pi\varepsilon d)^{1/2}]/(KT)\} \quad (4-6)$$

Φ_B是陷阱势垒的高度,此过程称为 Poole-Frankel(P-F)激发,是一个热力学活化过程。

1.4 跨越传导[15]

如果绝缘体的缺陷很大,电子的传导将直接由跨越机制控制,其电流密度为$J=V\exp-(\Delta E/KT)$,ΔE为电子活化能。

2. 通过脂肪短链的电子传导机制

隧穿机制与温度无关,而与施加的偏压有关,直接隧穿在低偏压下发生,而Fowler-Nordheim隧穿发生在较高的偏压下。热电子激发、Poole-Frankel激发以及跨越传导与温度有关,因此,为了在一个特定SAM中了解传导的机制,在一个V值较大范围内变化的以及对温度依赖的J—V数据的测量是非常重要的,温度依赖于J—V数据的结果可以通过上面讨论的理论模型来分析。

最近,研究者对脂肪短链SAM电子传导机制进行了研究,选择p$^+$—Si(SiO$_x$)/SH—(CH$_2$)$_3$—Si—(OCH$_3$)$_3$(MPTMS)/Au结构是基于以下原因:SAM脂肪链中仅包含3个C原子,链长很短(0.8 nm),因此,可望获得较高的直接隧穿电流;SAM的外围基团是—SH,由于强烈的Au—S化学键的相互作用,可以有效阻止对电极金的扩散。研究表明,这种SAM结构的J—V曲线是对称的,且J—V数据与温度无关。另外,Simmons理论的定量比较也表明MPTMS单分子膜具有良好的绝缘行为,直接隧穿是其主要传导机制[16]。

对于MPTMS单分子膜,在低偏压下其逆向电流比其正向要高。然而,在偏压超过2.45 eV时,整流的方向发生反转,即正向电流高于其逆向电流。在中间偏压($V=0\sim1.3$ V)时,其逆向和正向具有不同的Φ值,逆向偏压产生$\Phi_1=(2.14\pm0.01)$ eV,这与MPTMS/Si界面电子能量势垒的高度相一致,而电子能量势垒$\Phi_2=(2.56\pm0.01)$ eV是通过正向偏压数据获得的,势垒高度的差值($\Delta\Phi=0.42$ eV)几乎等于两个电极工作函数的差值,这个结果与Simmons理论相一致。

3. 电子传导性质与距离间的关系

当偏压居中(0.2~1.0 eV)时,研究者分析了以Al作为对电极的以—CH$_3$为尾端的脂肪链SAM的J—V数据($n=8,11,14,18$),研究表明,随链长增加,电流降低,检测的Φ值也随链长的增加而减小,到$n=18$时,降至1.5 eV。另外,当对电极与单分子膜的表面基团以化学键的形式作用时,将减小烷基链HOMO—LUMO间隙,即降低了能垒高度。获得的J值比隧穿传导机制要高,显然,其电子传导并不完全由隧穿机制控制[17]。增大的J值存在两种可能性:当尾端—CH$_3$基团与Al不能形成化学键时,对电极金属可以扩散进入单分子膜,可能导致额外的细丝传导通道;电子传导的机制随链长的增加可能发生从隧穿到其他机制的转变,解决这一问题最好的方式是分子尾端为—SH,以Au为对电极,并独立于温度进行J—V关系研究。

同样,从J—d数据可以获得隧穿衰减参数β,其变化范围0.4~>1 Å^{-1}(包括烷基硫醇在Au基底和烷基链在Si基底上的单层膜),理论J—d获得的β值为0.8 Å^{-1},较低的β值表示较高的隧穿效率。然而,如上所述,额外的细丝传导通道以及其他传导机制(不同于隧穿)的存在,可能获得一个更低的β值。但如果传导实验是在高质量—SH尾端烷基链SAM($n>8$)上进行的,获得的β值就具有一致性。

4. 基于自组装膜的介电衰减

通过烷基SAM的裂分电流非常低,然而,应用它们作为电子装置中的介电材料,低的裂分电流并非唯一选择,在高电场压力下,绝缘膜将渐渐衰减,这将引起自组装膜不可逆的毁坏,导致裂分电流增加,并最终使装置失效[15]。衰减的物理意义是裂分结内部原子键的毁坏,这种情况通常与电荷构筑有关，主要是因为它增加了电场，因此，一个绝缘介电材料也可以通过电场的衰减来进行表征。

在绝缘体中研究电子衰减最简单的方式是在高电压下记录J—V数据，直到观察到衰减，一旦单层膜发生强烈的衰减，随电压的增加将产生高的J值，并与V呈线性关系，这说明单层膜在衰减过程中产生的压力场将产生持续的缺陷。

一旦强烈的衰减发生，由于产生了持久的缺陷，单层膜将变成导体。考虑到这种因素，在高电压下，母体的氧化将不能维持其介电性。计算衰减区域$E_{BD}=V_{BD}/d_{SAM}$，对于 MPTMS 单层膜可以达到约 50 mV/cm，这个值比使用不同系统，如八烷基三氯硅烷在 SiO_2/Si[18]、烷基硫醇在 Au 上[19]、烷基在 H—尾端的 Si 上的长链烷基链报道的值要大得多(12~20 mV/cm)[20]，说明短链具有更高的衰减区域。此外，支持的母体氧化物是一个优良的介电材料，考虑到它的厚度，其衰减区域$E_{BD}=V_{BD}/(d_{ox}+d_{SAM})$是约 16 mV/cm，这个值仍然比相同厚度的$SiO_2$(约 2 nm)的衰减区域要高，说明这种单层膜适用于杂化电子器件[21]。

第三节 基于硅基自组装膜的分子电子装置

烷基链(σ分子)具有良好的介电性质是基于较大的 HOMO—LUMO 间隙，并具有高度的电子衰减区域，而共轭π分子由于电子的离域而显示良好的电子传导性。将σ和π分子体系相结合，并通过自组装结合于 Si 基底表面，可以制备一些分子电子器件。

1. 分子二极管

Aviram 和 Ratner 首次提出了分子整流二极管的概念，如图 4-1 所示，它是基于电子受体和电子给体部分以短的桥键相连，A—b—D 分子的桥键具有σ和π两种形式，迄今分子整流二极管中研究最多的分子是$C_{16}H_{33}$—Q—3CNQ[22]。研究表明这种分子中通过π桥相连的 D 和 A 部分并不完全孤立，分子轨道在整个 Q—3CNQ 单元中是高度离域的，整流的效率由接近于某一电极的 Q—3CNQ 单元长链的主体对称性决定。

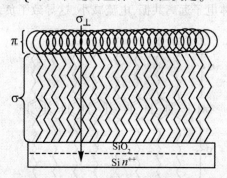

图 4-1 以 Si 为基底的σ—π分子整流器结构示意图[4]

在应用σ—π单层膜制备分子二极管中，Al-对电极通过电子蒸发沉积(10^{-4} cm^2，10 nm 厚)，选择 Al 作为对电极是为了避免两个电极间工作函数整流效率的差异(4.2 eV 对 Al，4.1 eV 对n-Si)。

在对π-尾端 SAM 整流性能的研究中，HETS $(SiCl_3—(CH_2)_{15}—CH=CH_2)$、OETS$(SiCl_3—(CH_2)_6—$

CH=CH$_2$)和TETS(SiCl$_3$—(CH$_2$)$_{12}$=CH=CH$_2$)是作为脂肪链整流行为研究的典型分子,在-1 V测量的电流密度均较1 V要高,并均观察到了相似的整流行为。但当脂肪链缺乏π基团时,无整流行为发生。OETS、TETS和HETSSAM的J—V曲线是对称的,对于每一个π-功能基团单层膜,整流率(RR)被定义为-1 V时的电流密度与1 V时电流密度的比值($RR=|J_{-1V}|/|J_{1V}|$),对于功能性单层膜,RR的平均值是3~13,最佳的RR可以达到37,RR不因末端基团化学性质的变化显著变化,也不需随空间链长的改变而进行校正。

研究表明,σ—π SAM的整流行为是基于π基团HOMO的共振隧穿,整流行为随施加于Al电极的负偏压而产生,这是因为Si的费米能级和π轨道的HOMO之间的能量差值比其π轨道LUMO低。假设通过HOMO存在共振效应,$E_0<0$,电流密度为[23]:

$$J=\frac{2J_0}{\pi}\{\arctan[(|E_0|)+\eta eV]-\arctan[(|E_0|)+(1-\eta)eV]\} \quad (4-7)$$

V为施加于金属电极上的工作电势,e为电子电荷,η为π单元的电势部分,J_0为基态电流,θ为电极/分子联结参数,η值应用简单的介电模型来检测,SAM的σ和π部分厚度分别为d_σ和d_π并且介电常数分别为ε_σ和ε_π[23]。

$$\eta=1-\frac{1}{2}\frac{1}{1+(\varepsilon_\pi d_\sigma)/(\varepsilon_\sigma d_\pi)} \quad (4-8)$$

OETS、TETS和HETS/SAM的测量值[24]分别为0.83、0.87和0.9,E_0与烷基链的长度无关,门电压V_T也不依赖于π基团,这些证明了二极管中电子传递的理论计算和实验测试结果的一致性。

2. 分子共振隧穿二极管(MRTD)

如图4-2所示,MRTD是一个双能垒异相结构,它主要包括σ—π—σ分子在硅基底上形成的自组装单层膜。σ—π—σSAM的J—V行为,是由共振隧穿机制而产生的,当分子的HOMO—LUMO间隙比π分子更大时,增加应用偏压,如$V>0$,其电流由于非共振隧穿模式而增加。当偏压低于临界工作偏压V_R时,样品的电势通过LUMO校准,并出现了共振隧穿;当偏压略高于临界值$>V_R$时,导体电子远离共振,电流减小,这导致了负差分阻抗(NDR)的产生。

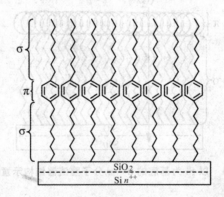

图4-2 以σ—π—σ分子自组装膜为基础的MRTD结构示意图[25]

根据NDR的作用机制,化学吸附于Si上的单π分子的MRTD,其结果可以通过分离的有机分子来观察,对于n型Si(100)上的苯乙烯分子,NDR仅仅观察到了负样品偏压,而正样品偏压将导致电子受激脱附。对于TEMPO分子,在每一个偏压极化下均未观察到电子的受激脱附,在这种情况下,NDR仅仅在n型Si(100)的负样品偏压以及p型Si(100)的正样品偏压区观察到[25]。独特的行为与本体Si能带结构和吸附分子离散轨道间的共振隧穿机制相一致,这为以Si作为基底的分子电子器件研究和应用提供了可能。

3. 分子记忆

具有氧化还原的活性分子,如金属茂、卟啉、三层三明治结构的功能配合物均可以作为电荷储存的元件,这些分子可以通过自组装过程作用于以Si为基底的装置上,如图4-3所示。分子储存与去储存电荷为基础的分子储存是基于分子具有不同的氧化还原态,氧化态表示信息的写入,还原态表示擦除或是读出信息[26,27]。

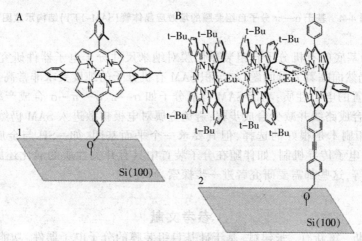

A 为卟啉单体;B 为三层三明治形卟啉/酞菁结构

图4-3 以Si为基底形成自组装膜的卟啉分子结构图[28]

研究表明,卟啉分子在相对低的电压下(<1.6 V)可以提供多级信息存储;可以经历数十亿次写/读循环;相比较于其他一半导体材料,展示了更长的电荷保留时间(数分钟);在极端条件下运行的稳定性较强(如共价作用于Si基底的卟啉SAM于400 ℃运行半小时,也保持稳定,因此,适合于半导体器件的加工和运行)[28]。第一代分子信息存储器件可与半导体材料进行杂化设计,目前,在此领域是进一步研究其他类型的氧化还原活性分子,并去理解和控制其作用参数,如电子的传递速率,因为它限制了写/读及电荷保留时间,并决定着信息更新的速率[29]。

4. 分子晶体管

场效应晶体管(FET)是三终端装置,基于SAM-FET的装置如图4-4,源(S)和沟渠(D)电极间隔距离(L),通过π基团沉积于SAM上,烷基链的长度(t)作为门电极(G)介电层。FET的作用原理是基于一个门场效应来操纵半导体电子隧穿传导来改变装置的开和关[30,31],最近,Kagan等人[32]成功设计和制备了具有化学和物理性能的SAM-FET,并应用电子隧穿理论分析了装置的$L>2.5\sim3.0$ nm,门介电厚度$t=L/1.5$。

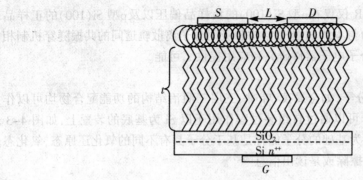

图 4-4 基于 σ—π 分子自组装膜的场效应晶体管(SAM-FET)结构示意图[28]

以 Si 作为基底的有机分子自组装单层膜对纳米尺寸分子电子器件研究发挥了极大的推动作用,但仍然面临着大量问题。如沉积 SAM 有机分子的纯度要求非常高,不纯性将产生缺陷并影响装置的电子性质;功能 SAM 有机分子如 σ—π、σ—π—σ 合成产率必须提高,这需要探索新的合成路线并减少合成步骤;避免金属对电极扩散进入 SAM 仍然是需要解决的问题,软平板印刷术是很好的选择,但其要求一个平面基团,如—SH 与金属的结合;许多 SAM 存在新的电子传导机制,如伴随在分子装置中具有开关性质的氧化还原,分子系统构造将发生变化等,这些都需要研究者进一步探索、研究。

参考文献

[1] 左国防, 雷新有, 张建斌. 基于硅基自组装膜的分子电子器件. 功能材料与器件学报, 2010, 16(5): 429–436.

[2] Vuillaume D. Molecular-scale electronics. C. R. Physique, 2008, 9(1): 78–94.

[3] Mann B, Kuhn H. Tunneling through fatty acid salt monolayers. J. Appl. Phys., 1971, 42(11): 4398–4405.

[4] Aviram A, Ratner M A. Molecular rectifyers. Chem. Phys. Lett., 1974, 29: 277–283.

[5] Binnig G, Rohrer H. Scanning tunneling microscopy: from birth to adolescence. Rev. Mod. Phys., 1987, 59(3): 615–625.

[6] 王伟. 分子电子器件的电性能测试方法. 物理, 2007, 36(4): 288–294.

[7] Ulman A. Formation and structure of self-assembled monolayers. Chem. Rev., 1996, 96(4): 1533–1554.

[8] Schreiber F. Self-assembled monolayers: from "simple" model systems to biofunctionalized interfaces. J. Phys. Condens. Matter., 2004, 16(28): 881–900.

[9] Love J C, Estroff L A, Kriebel J K, et al. Self-assembled monolayers of thiolates on metals as a form of nanotechnology. Chem. Rev., 2005, 105(4): 1103–1170.

[10] Bigelow W C, Pickett D I, Zisman W A. Oleophobic monolayers. 1. Films adsorbed from solution in non-polar liquids. J. Colloid. Sci., 1946, 1: 513–538.

[11] Maoz R, Sagiv J. On the formation and structure of self-assembling monolayers. J. Colloid Interf. Sci., 1984, 100(2): 465–496.

[12] 李珩, 温永强, 杨青林, 等. 有机分子修饰硅表面. 化学进展, 2008, 20(12): 1964–1971.

[13] Vuillaume D. Nanometer-scale organic thin film transistors from self-assembled monolayers. J. Nanosci. Nanotechnol., 2002, 2(3/4): 267-279.

[14] Simmons J G. Electric tunnel effect between dissimilar electrodes separated by a thin insulating film. J. Appl. Phys., 1963, 34(9): 2581-2590.

[15] Hesto G, Barbottin P, Vapaille A. The nature of electronic conduction in thin insulating layers. Instabilities in Silicon Devices, North Holland, Netherlands, 1986, 1: 263.

[16] Aswal D K, Lenfant S, Guerin D, et al. A tunnel current in self-assembled monolayers of 3-mercaptopropyltrimethoxysilane. Small, 2005, 1(7): 725-729.

[17] Salomon A, Cahen D, Lindsay S M, et al. Comparison of electronic transport measurements on organic molecules. Adv. Mater., 2003, 15(22): 1881-1890.

[18] Fontaine P, Goguenheim D, Deresmes D, et al. Monolayers as ultrathin gate insulating films in metal-insulator-semiconductor devices. Appl. Phys. Lett., 1993, 62(18): 2256-2258.

[19] Xu B, Tao N J. Measurement of single-molecule resistance by repeated formation of molecular junctions. Science, 2003, 301(5637): 1221-1223.

[20] Zhou J, Uosaki K. Dielectric properties of organic monolayers directly bonded on silicon probed by current sensing atomic force microscope. Appl. Phys. Lett., 2003, 83 (10): 2034-2036.

[21] Green M L, Gusev E P, Degraeve R, et al. Ultrathin (<4 nm) SiO2 and Si—O—N gate dielectric layers for silicon microelectronics: understanding the processing, and physical and electrical limits. J. Appl. Phys., 2001, 90(5): 2057-2121.

[22] Metzger R M. Unimolecular electrical rectifiers. Chem. Rev., 2003, 103(9): 3803-3834.

[23] Lenfant S, Krzeminski C, Delerue C, et al. Molecular rectifying diodes from self-assembly on silicon. Nano. Lett., 2003, 3(6): 741-746.

[24] Budavari S, The Merck Index: An encyclopedia of chemicals, drugs and biologicals (12th ed.). Whitehouse Station: Merck & Co. Inc., 1996.

[25] Guisinger N P, Greene M E, Basu R, et al. Room temperature negative differential resistance through individual organic molecules on silicon surfaces. Nano. Lett., 2004, 4(1): 55-59.

[26] Roth K M, Dontha N, Dabke R B, et al. Molecular approach toward information storage based on the redox properties of porphyrins in self-assembled monolayers. J. Vac. Sci. Technol. B., 2000, 18(5): 2359-2364.

[27] 左国防, 卢小泉. 卟啉自组装膜与分子信息存储. 化学通报, 2009, 72(5): 412-420.

[28] Liu Z, Yasseri A A, Lindsey J S, et al. Molecular memories that survive silicon device processing and real-world operation. Science, 2003, 302(5650): 1543-1545.

[29] Mahapatro A K, Janes D B. Electrical readouts of single and few molecule systems in metal-molecule-metal device structures. J. Nanosci. Nanotechnol., 2007, 7(6): 2134-2138.

[30] Reese C, Bao Z. Organic single-crystal field-effect transistors. Materials today, 2007, 10(3): 20-27.

[31] Briseno A L, Mannsfeld S C B, Jenekhe S A, et al. Introducing organic nanowire transistors. Materials today, 2008, 11(4): 38-47.

[32] Kagan C R, Afzali A, Martel R, et al. Evaluations and considerations for self-assembled monolayer field-effect transistors. Nano. Lett., 2003, 3(2): 119-124.

第五章 自组装膜应用研究

自 20 世纪 80 年代 Sagiv[1]将自组装膜(SAM)用于半导体的钝化以来,自组装膜的应用正不断地扩展,包括制备金属表面防蚀剂、防磨损保护层、非线性光学材料及电化学和生物传感技术等领域[2]。

第一节 具有光功能的自组装膜

光活性是构建分子装置最重要的功能性之一,基于分子与其性质之间的电子/能量传递是通过光诱导产生的,建构高效的分子光装置要求分子在固体表面上以分子精度定向排列。自组装技术是分子化学吸附于固体表面,具有良好的稳定性,已成为有序分子层建构广泛应用的方法。烷基硫醇在金属,尤其在金表面的自组装得到广泛的研究也是因其在许多领域中的潜在应用,如传感器、湿润性控制以及生物分子和分子电子器件等[3-5]。因此,具有功能性的烷基硫醇自组装膜得到了广泛的研究。自 20 世纪 90 年代以来,它已经成为建构高效分子光装置的最佳选择之一[6-10]。本节主要讨论各种 SAM 的光性质;以各种 SAM 修饰电极、SAM 纳米簇修饰电极的光诱导电子传递;通过 SAM 控制的光—电化学性质;通过 SAM 的光致异构化的电子传递控制以及 SAM 光化学的应用,包括传感器以及微纳光形图案等。

1. 自组装单分子膜光诱导电子传递

模拟一个天然光合成系统对于人工光电子转换装置具有重要的现实意义。在天然系统中,各种功能性分子,如光吸收体—叶绿素以及电子给体,如脱镁叶绿素、苯醌等,以分子尺寸有序地进行组织,并伴随着一个微可逆的电子传递,完成了非常有效的光诱导电荷分离以及光诱导电子传递[11,12]。因此,模拟天然系统中的光—电转换机制去实现高效的人工光合成系统具有重要的意义。人工光合成早期的研究是由 Fujihira 等人[13-18]以 L-B 膜有序地排列分子,然而,相比较于天然系统,这种系统的量子效率非常低(0.4%~1.5%),显然,这些系统中分子的排列水平没有达到高效光诱导电子传递的足够高度。

为了精确地排列分子,常常运用烷基硫醇在金表面的自组装。研究者在 SAM 上构建了光诱导电子传递分子系统——含有光吸收体、电子传递中介以及表面活性基团的 SAM 修饰金属电极[19-21],这种烷基硫醇复合分子是卟啉—醌—巯基(PQSH)及卟啉—Fc—巯基(PFcSH),卟啉和—SH 分别扮演了电子吸收体和表面活性基团,醌和 Fc 在 PQSH 和 PFcSH 中作为电子传递中介,在 PFcSH 中,卟啉、Fc 和—SH 通过烷基链分离(图 5-1)。

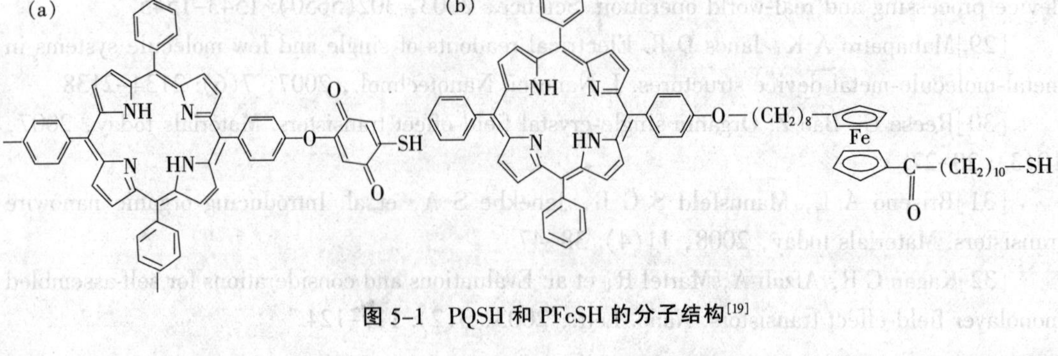

图 5-1 PQSH 和 PFcSH 的分子结构[19]

图 5-2 表示 PQSH 和 PFcSH 自组装膜修饰金电极的 CV 图,一对可逆的氧化还原峰归于醌和 Fc 基团的氧化还原反应(210 mV 和 610 mV,vsAgCl/Ag),从峰的电荷量,估计表面结合的分子的量为 $3.7×10^{13}$ 分子/平方厘米和 $1.4×10^{14}$ 分子/平方厘米,表面覆盖量的不同,归因于 PFcSH 存在长的烷基链。PFcSH 的分子层显然比 PQSH 排列更加有序。当 PQSH 和 PFcSH 修饰金电极浸入含有甲基紫精(作为电子受体)的电解质溶液中,并被光照射时,如果扫描电势比 210 mV 和 610 mV 更负,则有稳定的阴极光电流产生,这与单独修饰于 SAM 上的醌和 Fc 的氧化还原电势相一致。另外,当以白光 Xe 灯照射 PQSH/SAM,以单色光(430 nm)照射 PFcSH/SAM 时,产生稳定的光电流可持续几个小时而没有衰减,延长 SAM 光照时间后,电极附近的溶液颜色变蓝,因为 $MV^{2+}/MV^{·+}$ 的还原电势为 -630 mV,表明 MV^{2+} 被还原为甲基紫精激发态 $MV^{·+}$。研究者通过光的照射,在 PQSH/SAM 和 PFcSH/SAM 上测量的电子传递分别达到 0.8 eV 和 1.2 eV。SAM 修饰金电极的光电流谱与两种化合物的吸收光谱相一致,证明了卟啉基团在两种 SAM 中确实扮演了光活性位点的角色,两种 SAM 的量子产率分别达到了 3% 和 10%。因为量子效率依赖于烷基链长以及电极表面平坦性,研究者通过 ARXPS 谱和 SAM 结构研究表明 PFcSH/SAM 相对高效的原因在于其相对高的空间定位,即在功能基团之间引入了烷基链,致使其电子和能量传递更不可逆[22-24]。

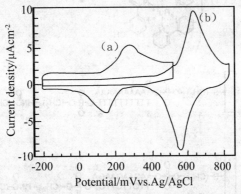

0.1 mol/L Na_2SO_4 和 0.1 mol/L $NaClO_4$,5 mmol/L MV^{2+}
图 5-2 PQSH 和 PFcSH/SAM 修饰金电极的 CV 图[22]

研究者应用烷基硫醇 SAM 也提出了更快的光诱导电子传递[25,26],Imahori 等合成了 FuPorFcSH 复合分子及其衍生物(图 5-3),并将其作用于金和 ITO 电极表面,这种 SAM 修饰电极的量子效率可达到 20%~25%,研究者认为系统中应用 C_{60} 是因其具有小的重组能,即使在金属电极上也可以满足高效光电流产生的要求。研究者也以 FuPorFcSH 和芘—SH 混合 SAM 作为天线分子构建了天然模拟光—电系统,其光俘获效率达到 0.6%~1.5%[27]。另外,研究者也构建了 C_{60}—聚噻吩—SH 复合分子 SAM,同样也观察到了相对高的光电流[28,29]。Ishida 和 Majima[30] 通过表面等离子激发观察到了比常规直接光激发更大的光电流。以 Ru 络合物作为光吸收体,Yamada 等人在金和 ITO 上建构了 $Ru(bpy)_3^{2+}$—紫精衍生物/SAM,并且观察了系统的光电行为[31-35]。另外,C_{60} 作为光吸收体在金表面结合光系统 I 蛋白,并固定于烷基硫醇 SAM 中的光—电转换也被研究[36-38]。

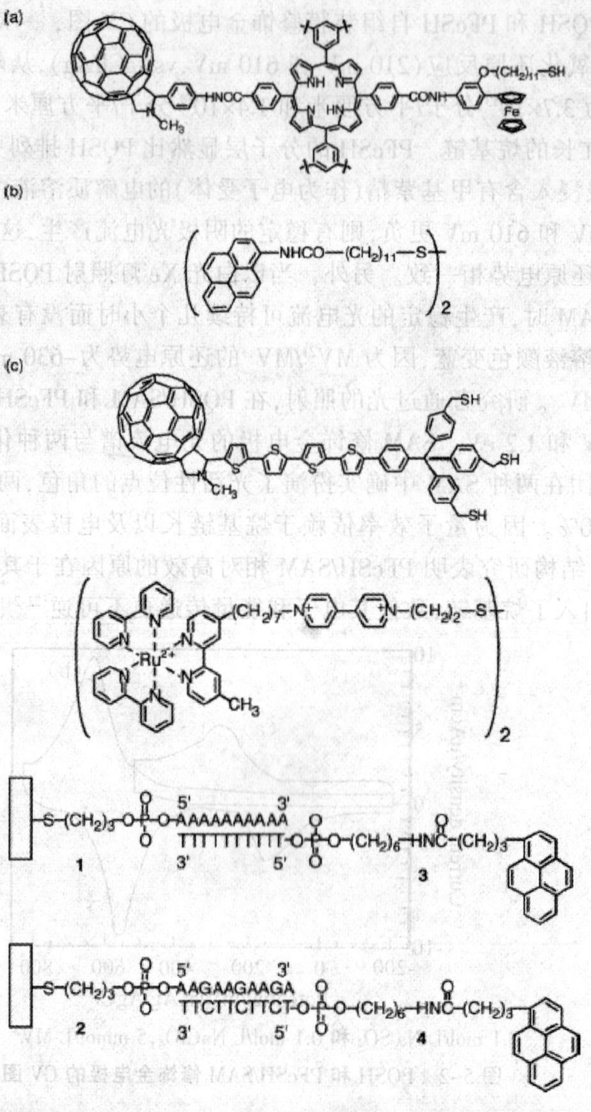

图 5-3 (a)FuPorFcSH 分子，(b)芘—SH 分子，(c)C_{60}—低聚噻吩分子，(d)Ru(bpy)$_3^{2+}$—紫精—SH 分子，(e 和 f)通过低聚核苷酸结合芘至金电极[25]

SAM 作为基底的多层膜上也观察到了光电流的产生。Reese 和 Fox 在金基底 SAM 上构建了—SH 尾端的低聚核苷酸,并且构建了双向—SH 尾端的低聚核苷酸 SAM,在含有 MV^{2+} 作为电子受体的溶液中,观察到了相应的光电流[39]。Thompson 等人[40]以锌阳离子和磷酸阴离子通过静电相互作用构建了卟啉 SAM 的光活性多层膜,并产生了光电流。Shinkai 等人[41]也构建了 C_{60}—高氯酸络合物多层膜,并将阴离子卟啉聚合物以磺化尾端—SH 分子构建于 ITO 电极上,同样也观察到了相对高的光电流。

烷基硫醇在半导体基底也可以形成 SAM[42-44],这种以半导体作为基底的 SAM 具有良好的光电化学性质。相对有序的烷基硫醇单层膜是将半导体基底浸入含有适当烷基硫醇的热溶液中几个小时[45],以此方法,Gu 和 Waldeck 建构了 n-InP 半导体电极,并以几个不同烷基链长的烷基硫醇修饰,系统研究了烷基链长与光电流之间的关系[46,47]。关于半导体基底光电流产生的原理可参阅其他文献[48-50]。

2. 自组装膜覆盖的纳米簇光诱导电子传递

因为烷基硫醇分子在基底上是以单分子层的形式存在，在一个给定面积，功能分子的浓度与三维系统相比要低得多，因此，系统中光电流产生的效率也非常低。尽管在给定面积上功能分子的量可以通过多层膜的形式增加，但通过连续表面反应建构多分子层是非常困难的，因为对于所连接的分子，其表面反应的产率通常比较低。显然，这可以通过金属纳米簇的应用相对容易地构建三维系统。在这个三维系统中，具有几个功能基团的烷基硫醇 SAM 覆盖在金属纳米粒子表面，烷基硫醇 SAM 覆盖的纳米簇非常稳定，且其功能基团很容易引导位置的交换[51-53]。应用硫醇 SAM 修饰的金属或半导体纳米簇已应用于光诱导电子传递，研究表明 SAM 覆盖的半导体纳米簇不仅有助于多层形式的稳定，而且也可以控制半导体的光电化学性质[52-55]。

Yamada 等人[54,55]在 ITO 电极上修饰卟啉—SH 分子金纳米簇观察了光电流；Imahori 等人[56,57]研究了以卟啉复合分子 SAM 覆盖的金纳米簇的光物理性质。他们也观察了以卟啉—SH 和 C_{60}—SH 复合分子覆盖的金纳米簇修饰的 SnO_2 电极上的光电流[58]；Li 等人[59]报道了金电极上以静电沉积卟啉—紫精复合分子的金纳米簇 SAM 的光电流。

半导体纳米簇多层膜可在双—SH/SAM 覆盖的金基底上形成[60-62]，通过选择性地将基底浸入分散的半导体纳米簇溶液来制备。XPS 测量表明双—SH/SAM 的自由尾端基团与半导体纳米簇上的表面原子以共价形式结合，其结果是在 SAM 上形成了一层半导体纳米簇。

各种半导体纳米簇，如 CdS[60]、ZnS[63]、PdS[64]、CdSe[65]，以上述方法在金电极上修饰制备的 SAM，均可以观察到相对较高的光电流。另外，Woo 等人证明了碲纳米簇可以通过电化学方法沉积于金电极上，(β-CD)修饰 SAM 作为分子模板，这可通过 β-CD 在 Au(111) 电极的适当排列获得，并且在电极上发现有阴离子光电流产生[66]。

烷基硫醇 SAM 覆盖的纳米簇多层膜在电极表面上也可以通过静电相互作用很容易地建构[6-8,67]。将金属基底选择性地浸入包含有阴离子电解质、分散的以尾端阳离子硫醇 SAM 覆盖的半导体纳米簇溶液，在半导体纳米簇修饰的电极上可以观察到相对较大和稳定的光电流，并且系统的光电化学性质也可在量子尺度效应的基础上进行讨论[66,67]。

3. 自组装膜上的光致异构化

通过固定在 SAM 上的功能基团的结构变化，可以控制电子传递，应用 SAM 的光致异构化性质，研究者构建了光开关系统。

偶氮基是最常用的光致异构化基团(图 5-4)，经常在 SAM 上引入这种基团去构建光开关系统[68,69]。Mirkin 等人合成了 Fc—偶氮基—SH 分子，并且在 FcAzSH 和 AzSH 分子混合 SAM 修饰的金电极上完成了光生电子传递，这种光子电子传递的电子源是电解质溶液中的亚铁氰化物。当在溶液中加入亚铁氰化物时，在 FcC 的氧化还原电流附近出现的催化电流归因于亚铁氰化物的氧化还原反应，表明混合 SAM 中的 Fc 扮演了电子传递中介催化剂，当光照射后，出现的可逆氧化还原峰归因于固定在混合 SAM 上的 Fc 和亚铁氰化物两者的氧化还原。因此，就完成了光生电子传递。

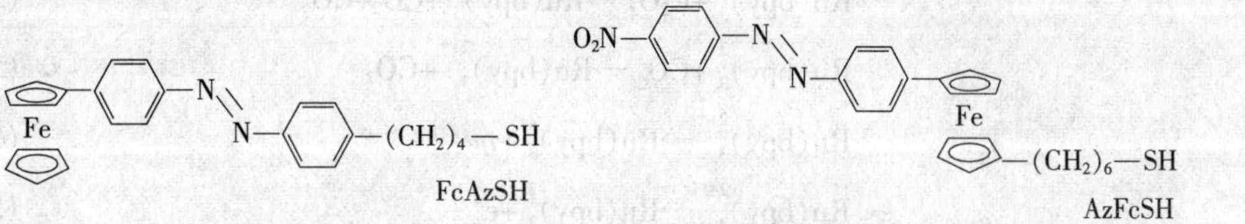

图 5-4 FcAzSH 和 AzFcSH 的分子结构[68]

Fujishima 等人发现了 Az—SH 分子 SAM[68,69]和 Az 衍生物 L-B 膜[70]修饰金电极特殊的电化学性质。cis-Az 至氢化 Az 的还原电位比其cis-形式更正,表明氢化 Az 的还原形式,即使其最初的形式是cis-形式,在+200 mV 附近仅仅电氧化至trans-Az。利用这种现象,研究者在以 Az—Fc—SH 分子 SAM 修饰的金电极上有效控制了系统的电荷传递速率[71]。在以100% trans-形式的AzFcSH修饰金电极上,其CV图仅显示一对氧化还原峰,这归因于在 0~750 mV 范围内存在 Fc 的氧化还原电位,并且其峰形不随电势扫描和 UV 照射而变化。以 20% cis-形式和 80% trans-形式 AzFcSH 修饰的金电极,经 UV 照射后,其 CV 图在 200~750 mV 范围出现唯一的氧化还原峰归因于 Fc,当电势扫描至 0 mV 时,出现的一对峰归因于偶氮基团的氧化还原,然而,在第二次扫描时,这个峰消失了。而当 UV 照射后,氧化还原电势及峰的分离重新回到起始值,后者电极氧化还原电势的变化是可逆的。基于这种结果的在线 FTIR 分析了系统的电化学性质,也表明 Fc 在 SAM 上的氧化还原电势及其电荷传递速率可以通过 SAM 中的 Az 的cis-形式和trans-形式在其电子和光化学结构的转变来可逆地控制。

水杨酸—吡喃/生物质光致异构在烷基硫醇 SAM 上也被应用于开/关电子传递[72]。Willner 等人[73]使用合成的 β-1-[3,3-二甲基-6′-硝基-(indoline-2,2′-2H-苯基吡喃)]-丙酸和一个氨基尾端烷基硫醇 SAM 构建了水杨酸—吡喃/生物质尾端和 4-巯基吡啶 SAM 修饰的金电极。使用这种 SAM,他们研究了生物活性分子的光开关系统。

除了 Az 和水杨酸—吡喃/生物系统的光致异构化,1,2-二苯乙烯和二苯基乙烯也被应用于烷基硫醇 SAM 修饰电极的光致异构化[74,75]。

在偶氮基团控制分子重构的光致异构化中,Matsui 等人[76]使用多肽纳米管成功地控制了分子重构,这种多肽纳米管包含的—NH_2 功能基团可以经由氢键形式去结合诸如蛋白、纳米晶、卟啉等。

4. 自组装膜的化学发光

光活性 SAM 的化学发光性质也得到了广泛研究。Fox 和 Wooten[77]构建了蒽—SH 分子 SAM,测量了 SAM 的 FTIR 光谱和发光强度,并且研究了这种 SAM 中的蒽的二联体形式。Guo 等人[78]在金属卟啉—SH 分子修饰的金电极上,通过在溶液中与相应的金属卟啉的脱辅基红蛋白重构研究了活性肌球素蛋白质的光活性和电化学活性。在开环条件下可以观察 Fc—SH 衍生物和 Zn—四卟啉—SH 衍生物混合 SAM 的荧光性,Roth 等人[79]以此定量地检测了 SAM 系统中光子储存的电荷量。Bohn 等人[80]建构了蛋白联系的 SAM,并以聚苯乙烯纳米球掺杂作为 SAM 中的荧光标记物,测量了其荧光强度。

在以 $Ru(bpy)_3^{2+}$—SH 分子 SAM 修饰的金和 ITO 电极系统的电致发光现象也有报道[81,82],$Ru(bpy)_3^{2+}$+SAM 修饰电极在含有 $C_2O_4^{2-}$ 的溶液中可以观察到依赖电势的发光光谱及其发射强度。下述反应被认为是在正的电势区域观察到的发射反应机理[81]:

$$Ru(bpy)_3^{2+} \rightarrow Ru(bpy)_3^{3+} + e^- \tag{5-1}$$

$$Ru(bpy)_3^{3+} + C_2O_4^{2-} \rightarrow Ru(bpy)_3^{2+} + CO_2 + CO_2^{\cdot -} \tag{5-2}$$

$$Ru(bpy)_3^{3+} + CO_2^{\cdot -} \rightarrow Ru(bpy)_3^{2+*} + CO_2 \tag{5-3}$$

$$Ru(bpy)_3^{2+*} \rightarrow Ru(bpy)_3^{2+} + \eta \nu \tag{5-4}$$

$$Ru(bpy)_3^{2+*} \rightarrow Ru(bpy)_3^{3+} + e^- \tag{5-5}$$

在发射过程中，$Ru(bpy)_3^{3+}$ 氧化 $C_2O_4^{2-}$ 形成 CO_2 和 $C_2O_4^{·-}$，其被还原为 $Ru(bpy)_3^{2+}$，即给出一个电子到电极表面，即式(5-1)。因此，在此 SAM 中，$Ru(bpy)_3^{2+/3+}$ 的头基扮演了氧化还原的中介。可以观察到阴极电流的线性增长。当电势变得更正时，$CO_2^{·-}$ 还原 $Ru(bpy)_3^{3+}$ 为 $Ru(bpy)_3^{2+*}$，其光发射效率由反应(5-5)的电子传递速率所控制，这依赖于 SAM 上 $Ru(bpy)_3^{2+}$ 头基和电极之间的距离。Bard 等人使用三聚丙胺的氧化反应及上述反应(5-1)和(5-4)，在 SAM 修饰的 Au 和 Pt 电极上，通过 $Ru(bpy)_3^{2+*}$ 的产生也观察到了 ECL[83]。

$$TPrA \rightarrow TPrA^{·+} + e^- \tag{5-6}$$

$$Ru(bpy)_3^{3+} + TPrA \rightarrow Ru(bpy)_3^{2+} + TPrA^{·+} \tag{5-7}$$

$$TPrA^{·+} \rightarrow TPrA^{·} + H^+ \tag{5-8}$$

$$Ru(bpy)_3^{3+} + TPrA^{·} \rightarrow Ru(bpy)_3^{2+*} + 产物 \tag{5-9}$$

通过 $Ru(bpy)_3^{2+}$/TPrA 系统的 ECL 行为，研究者对疏水表面的 DNA 及蛋白的固定进行了研究[84]。$Ru(bpy)_3^{2+}$ 修饰电极除了上述描述的 ECL 行为外，也被应用去构建一个光电化学微指令排列[85]。

SAM 在电化学发光装置中的应用也有报道，Yamashita 等人[86]从三角架形 π 共轭—SH 和二硫醇 SAM 修饰的金电极上，研究了电子中空注射(图 5-5)。他们在经过真空蒸发的 SAM 上，构建了一层 TPD，Alq3 以及 Mg—Ag 合金作为中空通道，发射并传递电子。例如，沉积在裸金和裸 ITO 基底上的结构分子层，可以比较这些装置的 EL 性质。移向高电压装置的 I—V 以及 L—V 曲线的顺序为，SAM(b)<SAM(a) ≈ 裸 ITO<裸金<SAM(c)。显然，与裸金装置相比，SAM(b)装置显示了对 EL 有意义的促进行为，产生了更大的还原工作电势，发出了更强的光，有更高的电流及更好的稳定性；相反，SAM(c)装置较低的 EL 性质归因于其较低的 SAM 修饰密度。通过 SAM 在金/TPP 界面产生真空水平的移动，并以这些 SAM 修饰金电极的 CV 结果表明了三角架形 SAM(b)更加密集，SAM(a)不太密集，二硫化物 SAM(c)最弱，这在金/TPP 界面影响了孔的能垒注射高度，因此也影响了装置的 EL 性质。

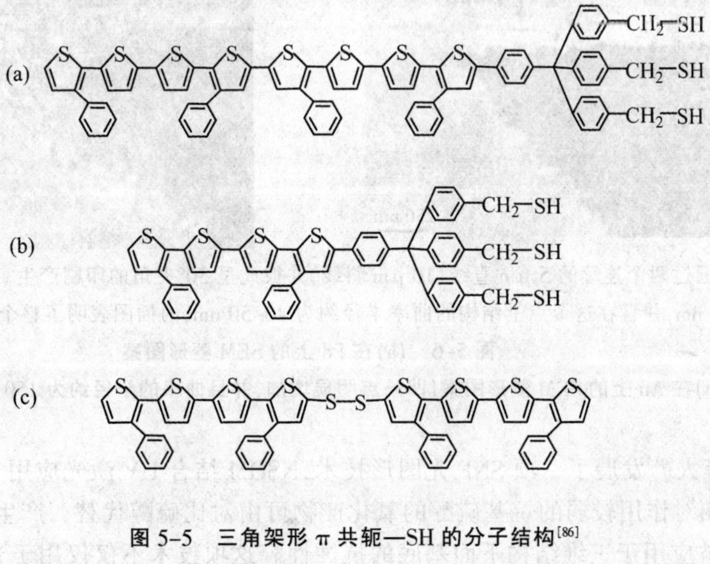

图 5-5 三角架形 π 共轭—SH 的分子结构[86]

应用光活性 SAM，通过检测生物物质的荧光性质而建立的生物传感装置是非常重要的。Sato 等人[87]将发光物质/H_2O_2 系统修饰于 Fc—SAM 金电极上，研究了系统的电氧化化学发光。当发光物质和 H_2O_2 在电解质溶液中时，出现的一对氧化还原峰归因于 Fc 的氧化还原反应，L 的催化氧化电流也被测量，并且同时观察到了光的发射，ECL 的强度依赖于溶液的 pH 值，应用这种系统，研究者在葡萄糖氧化酶存在下检测了葡萄糖。

将卟啉结合于 SAM 中研究其发光现象也有相关报道，Reich 等人[88]建构了 Au, Au/Ni 纳米线，其表面被卟啉—SH 复合分子修饰，荧光显微镜应用于 Au/Ni 纳米线功能性的优化，并用于纳米线—分子结合体的选择和稳定。聚苯酚基团结合于烷基硫醇也用于发光研究[89]，对光反应变色的 SAM 覆盖的金纳米颗粒的发光性研究也有报道[90]。

5. 自组装膜光形图案化

微电子装置的组合水平在逐年增强，对图形精度的要求也越来越高。自从 Whitesides[91] 首次提出烷基硫醇 SAM 微触印刷的概念之后，烷基硫醇 SAM 在光形图案领域出现了大量研究[92]。

Crooks 等人[93]应用以下过程，报道了具有高决定性的过程：首先，将 TEM 微柱作为光掩膜，置于 HOOC—$(CH_2)_{10}$—C≡C—C≡C—$(CH_2)_{10}$—SH—Au/Cr/Si 密集的 SAM 上面，然后，将此系统暴露于 UV 灯下，在 SAM 未标记区域，产生了化学聚合，再通过电化学脱附反应，在未聚合区域选择性地脱附，而聚合区域通过多重 Au/S 和范德华力相互作用强烈地键合于基底表面，使其电势偏移。微米级的硫醇 SAM 的移去所暴露的区域，在 O_2 饱和的 1 mol/L KOH+10 mol/L KCN 溶液中，可以通过腐蚀格子图像渗入金表面。图 5-6 是一个 400 目小格应用于金表面的图形的显微图，在溶解水平上，观察到了良好的保护性质的重现性。然而，接近检查揭示了六边形横向区域比其原始保护要少得多，这可以增加保护角的衍射；即从调制转变功能，在接近保护边缘处趋势还原(减少)光流动性。

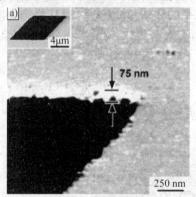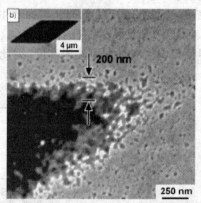

这种图案通过两个连续的 5 μm 直线(10 μm 斜线)并彼此呈 30°夹角的印刷产生。边界的粗糙性约为 30~70 nm，并且在这个点上结构的曲率半径约为 40~50 nm，内插图表明了整个菱形图案。

图 5-6 (a)在 Pd 上的 SEM 菱形图案
(b)在 Au 上的 SEM 菱形图案(凹陷点明显增加，并且曲率的半径约为 150 nm)[93]

Legget 等人[94]发展了一项 SNF 光图形技术，NSOM 结合 UV 激光应用于硫醇 SAM 的选择性氧化，相对作用较弱的烷基硫醇的氧化产物可由对比硫醇代替，产生的图案尺寸小于 20 nm，这也被应用于三维结构下面基底的抗蚀性。这项技术不仅仅用于光形图案的制备，也可以用于研究 SAM 光氧化的动力学过程。

通常,金基底的硫醇 SAM 在腐蚀期间容易留下缺陷和表面凹陷,因此在硫醇 SAM 图案研究中,金不作为没有任何缺陷的图案表面构建的基底。Whitesides 等人[95]使用 Pd 作为基底材料构建了没有缺陷的图案。图 5-6 为通过微触印刷术在 Pd 和金上腐蚀相结合生成的图案。比较而言,Pd 膜的结构具有更好的边角定义,比其在金膜上约少 85%~90%的蚀坑。

烷基硫醇在金属表面 SAM 的光性质得到了大量研究,研究的方向有非线性光学性质、SPR 以及单分子光谱等,我们期待在光活性 SAM 的研究中进一步拓展各个科学领域,并且在未来纳米技术中发挥更重要的作用。

第二节 自组装膜在金属防护中的应用

自组装膜技术应用于金属的防护研究尽管只有十几年的历史,但是由于自组装膜技术具有许多独特的优点而引起人们的重视。首先,自组装膜制备工艺简单,只需基体与活性分子接触,通过化学吸附即可自发生成自组装膜。自组装膜结构致密、均匀性好,不受基体表面形状的影响。由于自组装膜的厚度约为 2 nm,金属表面成膜后并不影响其外观和其他性能。这对于 Au、Ag、Cu 等制成的文物、工艺品和纪念品的防护具有特别重要的意义。此外,通过改变膜的化学组成和端基可以获得不同的厚度及功能。同时,自组装膜技术有助于在分子水平上研究膜结构、膜厚、致密度和结晶度等对金属防护性能的影响。

1. 自组装膜在金防护中的应用

Au 作为贵金属,在一般环境中具有较高的耐蚀性。但是在特定的化学和电化学条件下,溶液中含有络合剂如 Cl^-、Br^-、I^-、CN^-和 $CS(NH_2)_2$ 等时,Au 被溶解。Francis 和 Richard[96]采用原位电化学扫描隧道显微镜观察分析了 Au(111)表面用十六硫醇改性前后,在 10 mmol/L KCN 溶液中的腐蚀溶解行为。Au 表面改性后,腐蚀电位提高数百毫伏,腐蚀速率明显减小。腐蚀开始发生的位置由改性前的晶界边缘处转变为单分子膜缺陷的晶阶处。随后,研究者又探讨了 Au 在含 Br^- 的溶液中烷基硫醇的链长和端基对提高 Au 耐蚀性的影响[97],烷基($C_{11}CH_3$、$C_{15}CH_3$、$C_{11}OH$、$C_{10}COOH$)硫醇能有效抑制 Au 的腐蚀溶解。长链的烷基($C_{15}CH_3$)硫醇防护性能优于短链($C_{11}CH_3$)硫醇,相同链长的硫醇对 Au 的防护性能由强到弱依次为:$C_{11}OH>C_{10}COOH>C_{11}CH_3$。可见,自组装膜的厚度、组成和端基对 Au 防护性能的影响相当复杂。

近年来,有学者发现 Au 在自组装过程中出现腐蚀溶解现象[98,99]。在 O_2 饱和的硫醇溶液中,硫醇的诱导和氧化作用导致 Au 腐蚀。此时,硫醇的腐蚀效应强于吸附效应。溶解的最大和最小速率分别为 9.2×10^{11} 金原子/(平方厘米·秒)和 5.4×10^{10} 金原子/(平方厘米·秒)。硫醇和溶解氧的浓度是 Au 溶解的决定因素。

2. 自组装膜在银防护中的应用

Ag 具有优异的物理性能,在诸如珠宝、艺术、摄影材料和电子工业等领域得到了广泛的应用。但是 Ag 的腐蚀问题在工业应用和银质文物的保护方面非常突出,自组装膜作为一种超薄有机膜,完全能满足要求。目前,自组装膜成为 Ag 制品防腐蚀变色的一个重要研究方向。Magali 等[100]研究了十六烷硫醇自组装膜对 Ag 的抗变色能力的影响。银片经三次有机溶剂除油,稀 H_2SO_4 活化,然后在 30 ℃的 0.15 mol/L 正十六硫醇的异丙醇溶液中浸泡 1 h,表面形成自组装膜。10 mmol/L Na₂S 点滴试验表明,其保护效率达到 90%。在大气中存放 6 个月后,防护效果未受影响。Burleigh 等[101]进一步优化了十六烷硫醇自组装膜工艺,即清洗除油、10% H_2SO_4 酸蚀活化、92%硫醇溶液中浸泡 2 min 或 2%硫醇溶液中浸泡 1 h,再分别用三氯乙烯、去离子水清洗后,在 100 ℃烘干 30 min。Ag 强化变色实验表明,十六烷硫醇自组装膜的防护性能优于表面沉积聚胺基三唑膜[102]。

尽管硫醇在 Ag 防护方面性能优异,但是散发恶臭气味,使其应用受到一定的影响。全氟硫醇沸点比硫醇低,巯基气味较淡。Burleigh 等[103]将 Ag 浸在质量分数分别为 0.01%、0.1% 和 1%全氟烷基乙硫醇的正丙醇溶液中制备自组装膜。在含有空气、水蒸气、1 mg/L H₂S 的密闭容器中进行变色加速试验。结果表明,全氟烷基硫醇自组装膜的抗变色能力优于十六硫醇自组装膜。硅烷自组装膜无异味,在 Ag 抗变色应用上有一定的潜力。研究者[104]用 3-巯基丙基三甲氧基硅烷(3-MPS)在 Ag 表面制备自组装膜,得到 3-MPS 的最佳浓度为 0.1 mmol/L,最佳组装时间为 6 h。在 0.1 mol/L NaOH 中 Ag 的腐蚀电位为 −97 mV,腐蚀电流密度为 0.22 μA/cm²,极化电阻 R_p 为 362.1 kΩ·cm²,缓蚀效率达到 82%。

3. 自组装膜在铜防护中的应用

Cu 以优异的导热、导电和力学性能,在化学和微电子工业中大量应用。但 Cu 腐蚀在应用中常常发生,在环境中存在氧化剂时,腐蚀更为严重。因此,为了有效地抑制腐蚀,在 Cu 表面覆着保护层成为一种重要的防护措施。自组装膜对 Cu 表面 O_2、H_2O 和溶液中离子的扩散传递具有一定的抑制作用,从而起到防护作用。

烷基硫醇类自组装膜是 Cu 表面防护常用的自组装膜体系。相对 Au 和 Ag 而言,Cu 难免在制备自组装膜过程中被氧化。Cu 表面的氧化层将影响自组装膜的结合力和膜的缺陷。Hutt 等[105]采用一系列浸蚀方法除去 Cu 箔表面的自然氧化层,在十八烷基硫醇自组装膜中未检出氧。Whelan 等[106]采用低浓度的十二烷基硫醇溶液作为处理 Cu 表面氧化的还原剂,CuO 可完全被去除或转化,随后形成有钝化作用的硫醇自组装膜。通过比较机械抛光、硝酸浸蚀、盐酸浸蚀、阴极还原等四种 Cu 表面预处理工艺,得到优化的十二烷基硫醇自组装膜制备工艺[107]。Ma 等[108]在经 HNO₃ 浸蚀的 Cu 表面,采用十八硫醇、十二硫醇和己硫醇等 3 种正硫醇形成自组装膜,在 0.2 mol/L NaCl 溶液中,3 种烷基硫醇的缓蚀率均达到 90%以上。研究者[109]也在 3% NaCl 溶液中研究 Cu 表面吸附十二烷基硫醇后的腐蚀行为。结果表明,随着十二烷基硫醇吸附时间的增加,缓蚀效果增强。室温下吸附时间达到 15 min 以上时,即可达到较为理想的缓蚀效果。

烷基硫醇的碳链长度对 Cu 的防护性能有一定的影响[110,111]。交流阻抗测试表明,当烷基硫醇的碳原子数 $n \geq 16$ 时,膜电阻随链增加呈线性增加,较碳原子数 $n \leq 12$ 的膜电阻增大几十倍。在 100 kPa 的 O_2 和 100% RH 下的暴露试验表明,烷基硫醇的防护性能与烷基链碳原子数呈指数关系。长碳链的烷基硫醇更易保持结构的稳定性,这是由于长碳链的膜分子间存在较大的范德华力。烷基碳链中引入功能基团,将影响自组装膜的防护性能。与相同膜厚的烷基硫醇相比,长链的烷基硫醇自组装膜对 Cu 的防护性能,与链长和醚键的位置均有关。醚键离 Cu 表面越远,其初始膜电阻值越接近烷基硫醇;醚键离 Cu 表面越近,其初始膜电阻值与相应的烷基硫醇相比越小。与传统的 Cu 有机缓蚀剂相比,烷基硫醇对 Cu 的防护效果具有优势。Tan 等[112]考察了 Cu 表面分别吸附十二硫醇(DT)、巯基苯并噻唑(MBT)、苯并三唑(BTA)、咪唑(IMD)和苯并噻唑(BT)等形成有机膜后,在 0.5 mol/L H_2SO_4 溶液中的抗蚀性能。

有机膜通过阻止酸溶液中溶解氧与 Cu 表面的接触可降低腐蚀速率,它们对 Cu 的氧化抑制效果依次为 DT>MBT>BT>BTA>IMD。为减少硫醇类自组装膜的缺陷,改善膜的耐蚀性能,许多学者探讨了自组装膜的改性技术。Wang 等[113-115]首先在 Cu 表面吸附苯基硫脲形成自组装膜,用十二硫醇进行改性,然后加载交流电进一步改性。交流阻抗测得复合膜层电荷转移电阻较高,表明膜层可抑制 Cu 表面发生腐蚀电化学反应。计算得出表面膜的覆盖率>99.0%,在 0.5 mol/L NaCl 溶液中的缓蚀效率>97.2%。Aramaki[116]在 Cu 表面形成烷基硫醇自组装膜后,采用烷基三氯硅烷进行改性,得到超薄膜,可明显提高抗水溶液和大气腐蚀的性能。

席夫碱类化合物含有特殊的功能基团—C═N,一些芳香族席夫碱的苯环上还带有—OH,它们使给电子基团与苯环形成大的共轭 π 键,而 Cu^{2+} 则提供空轨道。因此,Cu^{2+} 与席夫碱易形成螯合物后覆盖在 Cu 表面,起到保护作用。Quan 等[117]在 Cu 表面制备席夫碱自组装膜,采用 HNO_3 浸蚀预处理并在自组装过程中施加电势,提高耐蚀性能。Chen 等[118]合成了具有特殊结构的席夫碱,并在 Cu 表面形成自组装膜。实验表明,在 1 mmol/L 席夫碱溶液中成膜 2 h 后,缓蚀效率最佳。此外,带有巯基的席夫碱对 Cu 有良好的防护能力[119]。

随着人们对自组装膜认识和有机合成技术的提高,Cu 表面出现了一些新的自组装膜体系,如烷基硫代硫酸钠[120]($Me(CH_2)_{n-1}S_2O_3^-Na^+$)、烷基羟氨基酸[121,122]($CH_3(CH_2)_nCONHOH$)和咔唑[123](Carbazole)等。

4. 自组装膜在钢铁防护中的应用

钢铁作为工业大量使用的金属,在大气中极易被氧化,腐蚀现象相当严重。采用自组装膜可以有效地阻止钢铁的氧化,减轻腐蚀。烷基膦酸与钢铁基体具有较好的结合力。Felhosi 等[124]用 1-烷基膦酸衍生物在钢铁表面形成自组装膜得到的膜致密连续,能有效阻止钢基体活性溶解。烷基二膦酸在钢铁表面也可形成自组装膜,Felhosi 等[125]研究发现,含有奇数个亚甲基的二膦酸盐防护性能比含有偶数个碳原子的烷基碳链好。1,7-二磷酸庚烷的缓蚀效果最佳。由于钢铁表面易产生氧化物,形成的自组装膜存在大量的针孔和塌陷等缺陷,因此,多层膜对于减少缺陷、提高防护性能更为有利。Miecznikowski 等[126]通过多层顺序吸附金属阳离子和高铁氰酸根阴离子,在不锈钢表面沉积得到稳定的双层自组装膜。Aramaki 等[127]用硅烷耦合剂对烷基硫醇和烷基胺类自组装膜进行改性,在钢铁表面制备二维自组装膜,有效地保护钢铁免遭腐蚀。据报道,有机硅类的自组装膜复合膜也具有优异的耐蚀性能[128]。

5. 自组装膜在其他材料防护中的应用

烷基硫醇自组装膜的基底除了金、银、铜、铁几种金属以外,还有铂、镍、不锈钢等。2003 年,Li 等[129]用接触角法、椭圆光度法、红外反射吸收光谱法等方法研究了不同链长的烷基硫醇 $CH_3(CH_2)_mSH$ (m=5,7,9,11,13,15,17,19,21)在铂和氧化铂表面的自组装膜,发现铂表面的自组装膜比氧化铂表面的自组装膜更致密、有序,随烷基链长的增加,自组装膜的有序性提高,并且由红外光谱中亚甲基和甲基的振动强度估测出烷基链在铂表面的倾斜角度小于 15°。在此基础上,2006 年,Petrovykh 等[130]用 X 射线光电子能谱、椭圆偏振光谱等方法研究了铂表面的烷基硫醇自组装膜在空气中的稳定性。结果表明,在一周时间左右,自组装膜无变化,但约一个月后,自组装膜大部分被氧化,这是由于铂表面的部分硫醇盐被氧化所致。

1997 年,Delhalle 等[131]在多晶镍电极上组装了正癸烷基硫醇和正十二烷基硫醇,发现这层膜能抑制 SO_2 及空气的侵蚀作用。他们同时指出,电极未经过预处理也能得到较好的组装膜，但经表面脱氧处理的镍电极上得到的自组装膜明显具有更好的化学稳定性和更高的分子有序度等特点。这说明在金属和氧化物上自组装膜是有差异的。

2002年，Ruan等[132]用直链烷基硫醇和烷基胺在不锈钢表面制备了双层膜，研究了不同碳链长度的自组装膜的性能。结果也表明，对直链的烷基硫醇和烷基胺两种自组装膜而言，碳链越长其自组装膜的有序度越高。

目前，自组装膜的作用机理还并不十分清晰。在金属腐蚀与防护方面，还存在着化学稳定性、重现性差等问题，为了进一步拓宽自组装膜的应用范围，必须深刻了解分子结构与功能的关系。自组装膜技术在Au、Ag、Cu等金属防护上的应用已趋于成熟，尤其在Ag、Cu制品抗变色方面成为一种重要的手段。同时，自组装膜对Au、Ag和Cu质文物的防护，具有不可替代的优势。在钢铁等工业应用广泛的金属上，如果具有缓蚀功能自组装膜的研究取得较大进展，能够从分子设计和自组装过程优化等角度解决自组装膜与活泼性基体结合力差的问题，自组装膜将逐步取代目前在金属防护领域中广泛应用的液相和气相缓蚀剂，从而给金属的防护技术带来巨大的变革。由于自组装膜的尺度为分子水平，可以通过设计分子结构单元来赋予膜特定的功能，这为在分子水平上研究金属腐蚀和防护机理开拓了新途径。可见，无论是基础理论研究方面还是应用研究方面，自组装膜技术都将推动金属的腐蚀与防护科学的进步，并将产生巨大的社会效益和经济效益。

第三节 自组装膜在电分析化学中的应用

1. 自组装膜在电分析化学中的应用

自组装膜具有分离检测的功能，如Kondo等[133]将含有甲基紫精的溶液作为电极接收器，在硫醇自组装膜上吸附卟啉和二茂铁，然后根据光电流受碳链影响这一特点，可将卟啉、二茂铁分离。Peng等[134]用2-硫基-3-辛基噻吩在金电极上形成自组膜来实现中性有机物多巴胺和$[H_2TMPyP]_4^+$离子的分离。引入特定基团或化合物的自组装膜修饰电极对金属离子有选择性的响应，由此可制成离子选择性电极。Turyan[135]研究了ω-(巯基-甲基)吡啶自组装膜电极对Cr(VI)的选择性测定，表明Cr(III)的存在对测定不产生干扰。Whitesides等[136]利用自组装膜技术制成了对酸度有响应的pH微传感器，氢醌被键合在烷烃硫醇自组装膜上，它在膜上的氧化还原取决于溶液的pH值，且峰电位变化为59 mV/pH。自组装膜在电催化中起两种作用：首先为电催化剂或酶的固定提供一种灵活的方法；其次自组装膜阻滞靶向氧化还原电对接近电极表面。有关多电子氧化还原体系的电催化主要有钴卟啉自组装膜催化O_2的两电子还原[137]；含吡啶铬自组装膜催化NO还原[138]；铁卟啉、血红蛋白自组装膜电极催化H_2O_2还原[139,140]等。此外，利用在铂电极上4-吡啶氢醌自组装膜对肼类化合物表现出很高的电催化活性，可用于肼类化合物的分析检测[141]。研究者[142]也研究了抗坏血酸在硫堇衍生化自组装膜修饰金电极上的催化氧化行为，发现硫堇衍生自组装膜修饰电极对抗坏血酸的氧化产生明显的电催化作用。

将生物大分子如蛋白质、DNA等直接固定在自组装膜上，研究它们的电化学行为也引起了人们的广泛兴趣。例如，汪尔康等研究了细胞溶素(PL)直接固定在巯基丙酸(MPA)自组装膜上的界面特征[143]，发现PL/MPA膜对溶液中的氧化还原电对的扩散有部分的抑制作用，且电极表面具有微电极阵列特征。Valdermaras[144]研究了过氧化物酶(MP-11)键合在

4,4-二硫联吡啶(DTDP)自组装膜上的直接电化学行为,并用椭圆偏振光法研究了膜的形成过程,同样发现该修饰电极具有微电极性质。

2. 自组装膜与生物传感技术

长链硫醇在基底上形成的紧密有序自组装膜层,提供了一个与生物膜相似、能够固定有机或生物分子的微环境。被固定的生物分子(如核酸、抗体、酶甚至整个生物细胞)具有生物活性和高选择识别性能,再通过结合电化学技术将信号转化并输出,可以作为高效灵敏生物传感器[145,146]。利用这一原理,研究者已制备出了识别铜离子的传感器、检测pH的电化学传感器、抗生蛋白和生物碱性酶传感器等多种传感器[147,148]。

近年来,氧化还原蛋白质和酶的直接电化学的研究引起了越来越多研究者的兴趣。这些研究不仅可以获得其内在热力学和动力学性质的重要信息,而且对于了解生命体内的能量转换和物质代谢、生物分子的结构和各种物化性质,探索其在生命体内的生理作用及其作用机制,开发新型的生物传感器等均具有重要的意义。结构有序、活性良好的生物分子有序膜是实现新型生物传感器的要素之一,因此构建含生物活性分子有机薄膜的研究很有意义。

借助生物分子之间和生物分子内部进行有效的、有选择性的电子转移来研究生物反应,有助于了解生命体的物质代谢和能量转换过程。自组装(自组装膜)技术是一种固定酶的有效方法,分子通过化学键相互作用自发吸附在固液界面,形成热力学稳定的能量最低的有序膜。其中以烷基硫醇在金电极表面的自组装最典型,并得到广泛应用。自组装膜的主要特征是具有组织有序、定向、密集和完好的单分子层,而且十分稳定。它具有明晰的微结构,在研究界面电子转移、催化和分子识别及构建第三代生物传感器方面具有开拓性意义。

近几年自组装膜技术在生物传感器制备方面的应用得到了更加快速的发展。如Gooding等[149]对硫醇自组装膜在酶电极和传感器方面的应用进行了总结,Vijayamohanan等人[146]对最新的自组装单层膜生物传感器进行了概括。除了与电化学结合的传感技术之外,与压电技术相结合的传感器制造技术也有了快速的发展。该技术使用石英晶体微天平(QMB)通过检测振荡频率的改变对微小质量的变化(如吸附)进行监控。压电晶体传感技术对分子识别过程的监控,可用于对单层膜吸附、酶固定及选择性吸附有机和生物分子等过程的研究。Kepley等[150]利用该技术对含二价铜离子的自组装膜对有机膦酸盐的选择性识别进行了研究。Richert等[151]将QMB应用于两种不同的自组装单层膜体系,分别制备了间苯二酚自组装膜的气体传感器和缩氨酸抗原自组装膜的液体传感器。Zhou等[152]在QMB的金电极上组装了带有两种不同ω-功能基团的硫醇,考察了QMB传感器与有机分子蒸气间的相互作用机理。研究发现,通过氢键和偶极—诱导偶极的相互作用,该传感器可以对醋酸和乙醇等有机分子蒸气产生快速的感应。Crooks等[153]也通过此法实现了超分子高分子膜层及树枝状聚酰胺膜层对小分子有机物的识别。

此外,Sung[154]研制了以Pt电极为基体材料的聚烷基硅烷自组装膜葡萄糖传感器。Ruan等[155]通过共价键合硫堇与过氧化物氧化酶在胱胺自组装膜上制成过氧化氢传感器。Willner[156]以胱氨酸或半胱氨酸的自组装单分子膜为基础,通过缩合反应键合的媒介体(如TCNQ、二茂铁、醌类等)或生物酶是生物光催化和电催化反应活性组装体,依此研制出了葡萄糖氧化酶、谷胱甘肽、胆红素氧化酶和苹果酸酶等生物传感器。Saitoh等[157]以不同分子结构和原子团的硫醇化合物为敏感膜材料,利用分子自组装方法将敏感膜固定在沉积有Au膜涂层的晶片上,制作成检测气体(如乙醇、丙醇、苯、己烷等)的化学传感器。孙灏等[158]研制出了电流式葡萄糖传感器,这种传感器是固定葡萄糖氧化酶于聚合物复合膜中,并且这种复合膜形成于金电极上的硫醇自组装膜之上。此外,利用自组装技术还可制备 DNA 修饰电极,从而为电化学 DNA 传感器的研究开辟了一条道路。如 Bard 等[159]通过巯丁基磷酸在 Au 电极上的有序吸附和反应步骤,制出一层有序的链烷磷酸铝膜,膜中含有金属中心离子 Al^{3+},能与带负电的 DNA 链作用,将 DNA 固定于电极表面得到 DNA 修饰电极,并将电致化学发光与 ssDNA 修饰电极相结合进行基因传感器研究。

自组装膜有较高的稳定性和较宽的基底选择范围,因此自组装膜特别适合于固定生物分子。将自组装膜作为平台连接生物分子制成的生物传感器容易形成有序、孔隙大小随意可调且稳定的单分子层。此外,可灵活设计的自组装膜功能端基,能实现亲水或憎水基表面的不同需要。由于自组装膜提供的膜表面类似于细胞的微环境并具有适当的稳定性,因此是生物传感器理想的选择基底。根据识别机理,自组装膜生物传感器可分为电化学传感器、光学传感器、热传感器和质量传感器[160]。

2.1 电化学传感器

自组装膜生物传感器的主要优点是它具有稳定性,这与电活性的 SAM 作为电子通道有关,因此适合用作检测生物分子的基底。如葡萄糖氧化酶在胱胺和二茂铁氧化还原介质单分子层作用下,可制得葡萄糖传感器[161,162]。由于金表面硫醇自组装膜的优良性能,人们选取双官能团的含硫化合物先在金表面形成自组装膜,再利用其另一端的官能团可将蛋白质、酶等固定在金表面获得多层有序膜,此类膜可望用于研究膜中生物分子之间定向的电子转移、能量传递以及生物化学传感器的设计与制造[163]。以巯基化合物 L-半胱氨酸(cysteine,cys)为电子转移促进剂的研究报道非常多,因为 L-半胱氨酸分子小,很容易与金电极形成金—硫键,自组装效果非常好,在金电极与酶分子之间形成了一座桥梁,极大地缩短了酶分子与电极之间的距离,促进了电子的转移。干宁等[164]将自组装技术和气相沉积 TiO_2 凝胶技术结合,制备了巯基丁二酸铜(II)(CuL_2SH)/TiO_2 凝胶/HRP 复合电极,考察了该传感器的电化学性质及对 H_2O_2 的催化效果,并用于实际样品测定。研究者利用层层自组装法在铂(Pt)电极表面构建了聚丙烯胺葡萄糖氧化酶膜,研究了自组装薄膜的表面微观形貌和电化学性质[165]。葡萄糖响应线性范围为 $5.0\times10^{-4}\sim2.1\times10^{-2}$ mol/L;检出限为 1.0×10^{-4} mol/L(S/N=3),响应时间为 3.80 s,表观米氏常数为 17.79 mmol,对抗坏血酸等具有较强的抗干扰能力。

2.2 表面等离子体共振(SPR)传感器

表面等离子体共振传感器属光学型传感器,它能检测覆盖在金属表面单分子膜厚度的变化,还可用于监测 SAM 原位生长和溶液中分子主体—客体作用,这种方法一般检测大的生物分子。如环糊精和杯状芳烃修饰的长链 SAM 表面,由于其孔穴可螯合有机小分子和相关的大生物分子,因此具有较高的选择性。

2.3 石英晶体微天平

石英晶体微天平(QCM)是另一种重要的生物传感器,它根据共振频率变化引起的原位质量变化进行分析检测。这种质量敏感技术在研究单分子层信息、酶的固定、小分子有机物以及大的生物分子选择性响应等方面具有独到之处。

虽然自组装膜生物传感器具备很多优点,但目前在实际应用中仍然存在很多问题[166]。如固着的酶对 pH 值、离子强度和温度非常敏感,这些参数中的一个发生微小变化,有时就能改变其生物活性;有些自组装膜的化学稳定性不好,实验过程中单分子膜的化学氧化、电诱导和热解吸等都会对生物传感器的应用产生不利影响；由于极高的表面能，憎水基自组装膜表面积累的一些杂质还会导致自组装膜生物传感器识别中心的吸附或阻塞等。但是随着自组装膜技术的不断完善和发展，其在生物传感器领域中必将发挥更大优势，具有更多的实际应用价值和良好的发展空间。

第四节 自组装膜的其他应用

1. 自组装单分子膜的摩擦学行为研究

自组装膜的摩擦学行为与其组成和结构密切相关。末端基团的性质、分子链长、堆积密度以及组装分子同基底的结合方式等均可能对自组装膜表面的物理、化学性能及摩擦学行为产生重要影响[167]。到目前为止，硫化物在金属表面形成的自组装单分子膜的摩擦学行为是研究得最为深入的一类。如金基底表面硫醇自组装膜的摩擦学行为优于硅基底表面形成相同的自组装膜，这是由于硅烷自组装膜的有序性比硫醇自组装膜的低[168]。另外，也可以通过控制自组装单分子膜的化学组成和空间结构来改善薄膜的摩擦行为。如 Zhang[169] 报道认为烷基硫醇在金基底上组装成膜时存在一定的倾斜角，所以其摩擦力是各向异性的。

相同官能团长碳链的硫醇自组装膜显示出略高的摩擦系数。这是因为长碳链间具有更高的交叉度和更强的横向黏着力。Vegte[170] 发现，不对称烷基硫化物 $CH_3(CH_2)_nS(CH_2)_9$—CH_3(n=9,11,13,15,17)链长增加时，摩擦力和黏着力都增加；而 Brewer[171] 则指出，以 CH_3 为末端基团的硫醇自组装膜，短碳链的摩擦系数远比长碳链的大，极性末端基团，如—OH、—COOH 的摩擦系数随碳链的增加变化不大，这是由于其分子间存在氢键。一般认为，自组装膜的摩擦学行为与分子的堆积密度密切相关。Lee[172] 运用极化调节红外反射吸收光谱仪及 AFM 比较了三种同源 C_{17}—SH 的堆积密度与摩擦力的关系，发现堆积密度高者摩擦力较低，他们认为，这是由于在松堆积密度膜上滑动，分子变形产生的激发要消耗更多的能量，因此就产生了更大的摩擦力响应。

自组装膜末端基团的化学性质决定其表面能的大小，并进一步影响其表面的润湿性、黏着力和摩擦力。Ahn[173] 研究了不同末端基团烷基硫醇自组装单分子膜的摩擦学行为，发现其动摩擦系数大小顺序为：—COOH>—OH>—CH_3，这与表面能的趋势相同。Brewer[171] 证明了这种趋势，指出不同末端基团的烷基单分子膜的摩擦力大小顺序为：—COOH—Si>—COOCH$_3$—Si>—CH_3—Si≈—CH_3—Au，表明随着表面特性变化从亲水到疏水，摩擦系数降低，且 CH_3—自组装膜在不同的基底上摩擦学行为基本相似；Lee[174] 以苯为末端基团的硫醇($C_6H_5(CH_2)_nSH$, n=12~15)自组装膜为研究对象，发现摩擦力大小为 C_{60}—自组装膜>Ph—自组装膜>CH_3—自组装膜，这说明由于苯环基团的空间尺寸较大，它可以有效地保护下面的基团，从而消除了奇—偶效应。

2. 微触点技术

制备具有复杂几何图形的表面微结构是半导体工业、纳米加工、电子工业中的关键技术，具有巨大的应用前景。微触点印制技术是其中最为方便和有效的方法之一，如通过使用高分子弹性"印章"和自组装单层膜技术可以在基片膜上印刷微米和纳米量级图纹[175]。图 5-7 说明了微触点印刻技术的原理及制备过程，其中 PDMS 是经过光刻处理而具有精细图纹的聚二甲基硅烷弹性印章。以烷基硫醇为"墨水"，用 PDMS 印章在镀金的基底表面上"盖印",精细图纹就通过自组装膜从印章传递到金基底表面[176]。

Aizenberg 等[177]利用印制自组装膜为模板实现了成核密度、晶粒大小和晶格取向可控的碳酸钙晶体生长,在金、钯、银表面制备了排列有序的方解石二维阵列。该技术除了用于晶体模板外,还可用于微电极制作,其原理是利用单分子膜层对化学腐蚀液的阻隔作用,用蚀刻剂进行表面腐蚀,从而在基片上得到与原蚀刻图案完全一样的精细电极图纹。另外,该技术还可作为细胞生物学模板进行生物、微观化学反应等方面的研究[175]。

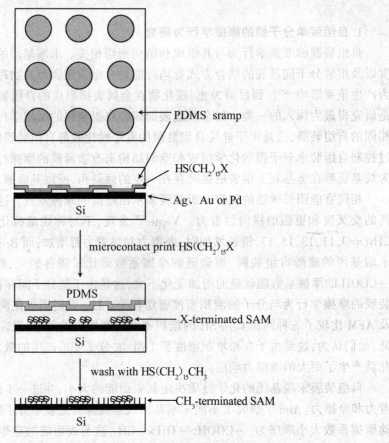

图 5-7 微触点印刻技术的原理及制备过程[176]

3.电寻址固定

实验发现,硫醇在金属电极上的吸附组装和脱附可通过一定的电极电势加以控制。利用这一特点可以通过外加正电势来实现自组装单层的快速沉积[178]。如果外加一定的负电势,还会出现自组装膜层脱附的现象,即通过电势控制 S—Au 键的吸附与脱附产生一定的化学记忆效应,实现了自组装膜在一系列电极上的自动寻址固定。Sullivan 等人[179]利用该原理使一系列不同硫醇按照一定的顺序自动寻址沉积于多达 64 个电极排列上。这一技术的优点在于整个制备过程不需要任何微观机械设备的参与,即可实现微观结构的制备和控制(最新的研究表明:电极排列可达到的最小间隔为 6 μm[180])。同时,利用电势控制沉积这一性质,Wang 等[181]还实现了对含有硫醇的 DNA 自组装膜层在金电极上的电化学控制脱附,并提出可利用此方法向人体内输送疾病的治疗基因。

4.分布阻隔技术

分布阻隔技术主要是利用了分子印记(Molecular imprinting)原则,在二维表面上组装具有一定纳米结构的硫醇单分子膜层[182]。首先,在金基底表面吸附少量具有一定分子结构的硫醇分子作为模板,在未吸附的空隙表面再次组装上另一种长链硫醇,由于两种硫醇链长不同,就在表面形成了具有形似纳米孔结构的自组装膜。如果所形成的纳米孔结构(即作为模板的硫醇分子的结构)具有与被分析物形似的结构,那么就可以实现化学识别的作用。利用这一原理可以制造化学传感器,Yang 等人[183]的研究表明,用带有硫醇基团的胆固醇分子作为模板制备的纳米孔结构自组装膜对吩噻嗪具有很好的选择性。

5.自组装膜在生物分子器件中的应用

微电子学和信息技术的发展,要求大规模集成电路的尺寸微型化。以硅材料为基础的集成电路的加工已达到物理极限(0.2 μm),要想进一步提高集成度必须另辟蹊径。自组装体系实现了在分子尺寸范围对电子的控制,从而为将分子聚集体构筑成有特殊功能的器件——分子器件创造了条件[184]。细胞色素 c 是一种阳离子型电子转移蛋白,具有氧化和还原两种状态,其导电率相差 1000 倍,可做记忆元件[185]。Kimizuka 等[186]研究的 cyt-c/TiO$_2$ 自组装膜,为实现分子开关和分子记忆元件提供了一种思路。细菌视红紫质(bR)是一种嗜盐菌紫膜内的蛋白质,是具有能量转换和活性离子输送功能的膜缔合蛋白,具有光驱泵的功能,是一种光驱动开关的原型,由光辐射启动的质子泵在膜两边形成的电位,经离子灵敏场效应放大后,可给出较好的开关信息。沈家骢等[187]和 Li 等[188]研究的 bR 自组装膜,为进一步制备光电器件提供了新的途径。DNA 分子以核苷酸碱基编码方式存储遗传信息,是一种存储器的分子模型[186]。DNA 分子组装膜的研究,为分子存储器的研制提供了合理的方向。

参考文献

[1] Sagiv J. Formation and structure of oleophobic mixed monolayers on solid surfaces. J. Am. Chem. Soc., 1980, 102: 92.

[2] 杨长江,梁成浩.自组装膜在金属防护中的应用.中国腐蚀与防护学报,2007, 27(5): 315-320.

[3] Kondo T, Uosaki K. Self-assembled monolayers (SAMs) with photo-functionalities. Journal of Photochemistry and Photobiology C: Photochemistry Reviews, 2007, 8: 1-17.

[4] Ulman A. An Introduction to Ultrathin Organic Films from L - B to Self-assembly, New York: Academic Press, 1991.

[5] Finklea H O. Electroanal Chemistry, vol. 19. New York: Marcel Dekker, 1996.

[6] Kondo T, Okamura M, Uosaki K. Layer-by-layer assembly of gold nanoclusters modified with self-assembled monolayers. Chem. Lett., 2001: 930-931.

[7] Song W, Okamura M, Kondo T, et al. Electron and ion transport through multilayers of au nanoclusters covered by self-assembled monolayers. J. Electroanal. Chem., 2003, (554-555): 385-393.

[8] Wenbo S, Okamura M, Kondo T, et al. Construction and electrochemical characteristics of multilayer assemblies of Au nanoclusters protected by mixed self-assembled monolayers on tin-doped indium oxide. Chem. Phys., 2003, 5: 5279-5284.

[9] Sumi T, Wano H, Uosaki K. Electrochemical oxidative adsorption and reductive

desorption of a self-assembled monolayer of decanethiol on the Au (111) surface in KOH+ethanol solution. J. Electroanal. Chem., 2003, (550 – 551): 321–325.

[10] Wano H, Uosaki K. In Situ, Real-time monitoring of the reductive desorption process of self-assembled monolayers of hexanethiol on Au (111) surfaces in acidic and alkaline aqueous solutions by scanning tunneling microscopy. Langmuir, 2001, 17: 8224–8228.

[11] Stryer L. Biochemistry. 3rd ed.. New York: Freeman, 1988.

[12] Deisenhofer J, Epp O, Miki K, et al. Resonance raman spectrum of distorted porphyrin radical cation reveals orbital mixing. J. Mol. Biol., 1984, 180: 385–398.

[13] Sakata Y, Tatemitsu H, Bienvenue E, et al. Evidences for the formation of bisbenzamidine-heme complexes in cell-free systems. Chem. Lett., 1988, 1625–1628.

[14] Kondo T, Okamura M, Uosaki K. Anion effect on the electrochemical characteristics of a gold electrode modified with a self-assembled monolayer of ferrocenylhexanethiol in aqueous and dichloromethane solutions. J. Organometall. Chem., 2001, (637 – 639): 841–844.

[15] Fujihira M. Photoelectric conversion by monolayer assemblies. Mol. Cryst. Li q. Cryst., 1990, 183: 59–69.

[16] Fujihira M, Nishiyama K, Yamada H. Photoelectrochemical responses of optically transparent electrodes modified with Langmuir–Blodgett films consisting of surfactant derivatives of electron donor, acceptor and sensitizer molecules. Thin Solid Films, 1985, 132: 77–82.

[17] Fujihira M, Yamada H. Molecular photodiodes consisting of unidirectionally oriented amphipathic acceptor-sensitizer-donor triads. Thin Solid Films, 1988, 160: 125–132.

[18] Sakomura M, Fujihira M. Direction of photocurrents of molecular photodiodes with oriented triad molecules. Thin Solid Films, 1994, 243: 616–619.

[19] Kondo T, Ito T, Nomura S, et al. Photoelectrochemical characteristics of a self-assembled monolayer of porphyrin-mercaptoquinone coupling molecules. Thin Solid Films, 1996, (284 – 285): 652–655.

[20] Kondo T, Yanagida M, Nomura S, et al. pH–dependent photoinduced electron transfer at the gold electrode modified with a self-assembled monolayer of a porphyrin-mercaptoquinone coupling molecule. J. Electroanal. Chem., 1997, 438: 121–126.

[21] Uosaki K, Kondo T, Zhang X–Q, et al. Very efficient visible-light-induced uphill electron transfer at a self-assembled monolayer with a porphyrin ferrocene thiol linked molecule. J. Am. Chem. Soc., 1997, 119: 8367–8368.

[22] Yanagida M, Kanai T, Zhang X-Q, et al. Angle-resolved X-ray photoelectron spectroscopic study on a self-assembled monolayer of a porphyrin-ferrocene-thiol linked molecule on gold: Evidence for a highly ordered arrangement for efficient photoinduced electron transfer. Bull. Chem. Soc. Jpn., 1998, 71: 2555–2559.

[23] Kondo T, Kanai T, Iso-o K, et al. Photochemical properties of gold nanoparticles modified with self-assembled monolayers containing porphyrin and ferrocene groups. Phys. Chem., 1991, 212: 23–30.

[24] Kondo T, Yanagida M, Zhang X-Q, et al. Light-controlled conductance using molecular switches. Chem. Lett., 2000, 964–965.

[25] Yamada H, Imahori H, Nishimura Y, et al. Photovoltaic properties of self-assembled monolayers of porphyrins and porphyrin-fullerene dyads on ITO and gold surfaces. J. Am. Chem.

Soc., 2003, 125: 9129–9139.

[26] Hasobe T, Imahori H, Ohkubo K, et al. Structure and photoelectrochemical properties of ITO electrodes modified with self-assembled monolayers of meso,meso-linked porphyrin oligomers. J. Porphyrins, Phthalocyanines, 2003, 7: 296–312.

[27] Imahori H, Norieda H, Yamada H, et al. Light-harvesting and photocurrent generation by gold electrodes modified with mixed self-assembled monolayers of boron-dipyrrin and ferrocene-porphyrin-fullerene triad. J. Am. Chem. Soc., 2001, 123: 100–110.

[28] Hirayama D, Takimiya K, Aso Y, et al. Large photocurrent generation of gold eelectrodes modified with [60]fullerene-liked oligothiophenes bearing a tripodal rigid anchor. J. Am. Chem. Soc., 2002, 124: 532–533.

[29] Hirayama D, Yamashiro T, Takimiya K, et al. Preparation and photoelectrochemical properties of gold electrodes modified with [60]fullerene-linked oligothiophenes. Chem. Lett., 2000, 570–571.

[30] Ishida A, Majima T. Surface plasmon excitation of porphyrin self-assembly monolayers on an Au surface. Nanotechnology, 1999, 10: 308–314.

[31] Akiyama T, Nitahara S, Inoue S, et al. Bi-directional photocurrent generation dependent on the wavelength of irradiation of a mixed monolayer assembly. Photochem. Photobiol. Sci., 2004, 3: 26–28.

[32] Terasaki N, Akiyama T, Yamada S. Structural characterization and photoelectrochemical properties of the self-assembled monolayers of tris (2,2'-bipyridine)ruthenium (ii) viologen linked compounds formed on the gold surface. Langmuir, 2002, 18: 8666–8671.

[33] Akiyama T, Inoue M, Kuwahara Y, et al. Novel photoelectrochemical cell using a self-assembled monolayer of a ruthenium (ii) tris(2,2'-bipyridine) thiol derivative. Jpn. J. Appl. Phys., 2002, 41: 4737–4378.

[34] Urasaki K, Tokunaga K, Sekine Y, et al. Hydrogen production by steam reforming of ethanol using cobalt and nickel catalysts supported on strontium titanate. Chem. Lett., 2005, 5: 668–669.

[35] Nitahara S, Akiyama T, Inoue S, et al. A photoelectronic switching device using a mixed self-assembled monolayer. J. Phys. Chem. B., 2005, 109: 3944–3948.

[36] Enger O, Nuesch F, Fibbioli M, et al. Photocurrent generation at a fullerene self-as sembledmonolayer-modified gold electrode cast with a polyurethane membrane. J. Mater. Chem., 2000, 10: 2231–2233.

[37] Ko B S, Babcock B, Jennings G K, et al. Effect of surface composition on the adsorption of photosystem I onto alkanethiolate self-assembled monolayers on gold. Langmuir, 2004, 20: 4033–4038.

[38] Kincaid H A, Niedringhaus T, Ciobanu M, et al. Entrapment of photosystem I within self-assembled films. Langmuir, 2006, 22: 8114–8120.

[39] Reese R S, Fox M A. Non-covalent assembly of multilayer thin film supramolecular structures. Can. J. Chem., 1999, 77: 1077–1084.

[40] Abdelrazzaq F B, Kwong R C, Thompson M E. Photocurrent generation in multilayer organic-inorganic thin films with cascade energy architectures. J. Am. Chem. Soc., 2002, 124: 4796–4803.

[41] Ikeda A, Hatano T, Shinkai S, et al. Efficient photocurrent generation in novel self-assembled multilayers comprised of [60]fullerene-cationichomooxacalixarene inclusion complex and anionic porphyrin polymer. J. Am. Chem. Soc., 2001, 123: 4855–4856.

[42] Sheen C W, Shi J-X, Martensson J, et al. Structural and spectroscopic characterization of chiral ferric tris-catecholamides: unraveling the design of enterobactin. J. Am. Chem. Soc., 1992, 114: 1512–1514.

[43] Noble-Luginbuhl A R, Nuzzo R G. The assembly and characterization of SAMs formed by the adsorption of alkanethiols on zinc selenide substrates. Langmuir, 2001, 17: 3937–3944.

[44] Lim H, Carraro C, Maboudian R. Chemical and thermal behavior of alkanethiol and sulfur passivated InP(100). Langmuir, 2004, 20: 743–747.

[45] Baum T, Ye S, Uosaki K. Formation of self-assembled monolayers of alkanethiols on GaAs surface with in situ surface activation by ammonium hydroxide. Langmuir, 1999, 15: 8577–8579.

[46] Gu Y, Waldeck D H. SEM images of a polymer pen array (A) with and (B) without a glass support. J. Phys. Chem. B., 1998, 102: 9015–9028.

[47] Gu Y, Kumar K, Lin A, et al. Studies of the antenna effect in polymer molecules 29. Isomerization of provitamin D3 photosensitized by polymers containing pendant naphthalene groups. J. Photochem. Photobiol. A., 1997, 105: 189–194.

[48] Gerischer H. NATO Advanced Study Institutes Series, Series B: Physics, Photovolt. Photoelectrochem. Sol. Energy Convers., 1981, B69: 199.

[49] Bockris O' M J, Uosaki K. The theory of the light-induced evolution of hydrogen at semiconductor electrodes. J. Electrochem. Soc., 1978, 125: 223–227.

[50] Fujishima A, Honda K. Electrochemical photolysis of water at a semiconductor electrode. Nature, 1972, 238: 37–38.

[51] Brust M, Walker M, Bethell D, et al. Synthesis of thiol-derivatised gold nanoparticles in a two-phase Liquid-Liquid system. J. Chem. Soc. Chem. Commun., 1994, 801–802.

[52] Brust M, Fink J, Bethell D, et al. Synthesis and reactions of functionalised gold nanoparticles. J. Chem. Soc. Chem. Commun., 1995, 1655–1656.

[53] Hostetler M J, Templeton A C, Murray R W. Dynamics of place-exchange reactions on monolayer-protected gold cluster molecules. Langmuir, 1999, 15: 3782–3789.

[54] Terasaki N, Otsuka K, Akiyama T, et al. Fabrication of a photoelectrochemical cell using a self-assembled monolayer of tris(2,2'-bipyrisine)ruthenium(ii)-violgen linked thiol on multistructured gold nanoparticles. Jpn. J. Appl. Phys., 2004, 43: 2372–2375.

[55] Yamada S, Tasaki T, Akiyama T, et al. Gold nanoparticle-porphyrin self-assembled multistructures for photoelectric conversion. Thin Solid Films, 2003, (438–439): 70–74.

[56] Imahori H, Kashiwagi Y, Endo Y, et al. Structure and photophysical properties of porphyrin-modified metal nanoclusters with different chain length. Langmuir, 2004, 20: 73–81.

[57] Imahori H, Fukuzumi S. Porphyrin monolayer-modified gold clusters as photoactive materials. Adv. Mater., 2001, 13: 1197–1199.

[58] Hasobe T, Imahori H, Kamat P V, et al. Photovoltaic cells using composite nanoclusters of porphyrins and fullerenes with gold nanoparticles. J. Am. Chem. Soc., 2005,

127: 1216-1228.

[59] Li G, Fudickar W, Skupin M, et al. Starre lipidmembranen und nanometerlücken-motive zur gestaltung molekularer landschaften. Angew. Chem. Int., 2002, 41: 1828-1852.

[60] Nakanishi T, Ohtani B, Uosaki K. Construction of semiconductor nanopartile layers on gold by self-assembly technique. Jpn. J. Appl. Phys., 1997, 36: 4053-4056.

[61] Radhakrishnan C, Lo M K F, Warrier M V, et al. Photocatalytic reduction of an azide-terminated self-assembled monolayer using CdS quantum dots. Langmuir, 2006, 22: 5018-5024.

[62] Ogawa S, Hu K, Fan F-R F, Bard A J. Photoelectrochemistry of films of quantum size lead sulfide particles incorporated in self-assembled monolayers on gold. J. Phys. Chem. B., 1997, 101: 5707-5711.

[63] Nakanishi T, Ohtani B, Uosaki K. Structure and photoelectrochemical properties of laminated monoparticle layers of CdS and ZnS on gold. Jpn. J. Appl. Phys., 1999, 38: 518-521.

[64] Miyake M, Torimoto T, Nishizawa M, et al. Effect of surface charges and surface states of chemically modified cadmium sulfide nanoparticles immobilized to gold electrode substrate on photoinduced charge transfers. Langmuir, 1999, 15: 2714-2718.

[65] Bakkers E P A M, Roest A L, Marsman A W, et al. Characterization of photoinduced electron tunneling in gold/SAM/Q-CdSe systems by time-resolved photoelectrochemistry. J. Phys. Chem. B., 2000, 104: 7266-7272.

[66] Woo D-H, Choi S-J, Han D-H, et al. Nanosized tellurium clusters grown at thiolated b-cyclodextrin molecular template modi?ed electrodes : Electrochemical deposition and STM characterization. Phys. Chem. Chem. Phys., 2001, 3: 3382-3386.

[67] Miyake M, Torimoto T, Sakata T, et al. photoelectrochemical characterization of nearly monodisperse CdS nanoparticles-immobilized gold electrodes. Langmuir, 1999, 15: 1503.

[68] Walter D G, Campbell D J, Mirkin C A. Photon-gated electron transfer in two-component self-assembled monolayers. J. Phys. Chem. B., 1999, 103: 402-405.

[69] Wang R, Iyoda T, Tryk D A, et al. Electrochemical modulation of molecular conversion in an azobenzene-terminated self-assembled monolayer film: an in situ UV-visible and infrared study. Langmuir, 1997, 13: 4644-4651.

[70] Wang R, Jiang L, Iyoda T, et al. Investigation of the surface-morphology and photoisomerization of an azobenzene-containing ultrathin-film. Langmuir, 1996, 12: 2052-2057.

[71] Kondo T, Kanai T, Uosaki K. Control of the charge-transfer rate at a gold electrode modified with a self-assembled monolayer containing ferrocene and azobenzene by electro-and photochemical structural conversion of cis and trans forms of the azobenzene moiety. Langmuir, 2001, 17: 6317-6324.

[72] Katz E, Lion-Dagan M, Willner I. Control of electrochemical processes by photoisomerizable spiropyran monolayers immobilized onto Au-electrodes: Amperometric transduction of optical signals. J. Electroanal. Chem., 1995, 382: 25-31.

[73] Shipway A N, Willner I. Electronically transduced molecular mechanical and information functions on surfaces. Acc. Chem. Res., 2001, 34: 421-432.

[74] Wang Z, Cook M J, Nygard A-M, et al. Metal ion chelation and sensing using a self-

assembled molecular photoswitch. Langmuir, 2003, 19: 3779-3784.

[75] Nakashima N, Nakanishi T, Nakatani A, et al. Photoswitching of the vectorial electron transfer reaction at a diarylethene-modified electrode. Chem. Lett., 1997, 591-592.

[76] Banerjee I A, Yu L, Matsui H. Application of host-guest chemistry in nanotube-based device fabrication: photochemically controlled immobilization of azobenzene nanotubes on patterned α-CD monolayer/Au substrates via molecular recognition. J. Am. Chem. Soc., 2003, 125: 9542-9543.

[77] Fox M A, Wooten M D. Characterization, adsorption, and photochemistry of self-assembled monolayers of 10-thiodecyl 2-anthryl ether on gold. Langmuir, 1997, 13: 7099-7105.

[78] Guo L-H, McLendon G, Razafitrimo H, et al. Photo-active and electro-active protein films prepared by reconstitution with metalloporphyrins self-assembled on gold. J. Mater. Chem., 1996, 6: 369-374.

[79] Roth K M, Lindsey J S, Bocian D F, et al. Characterization of charge storage in redox-active self-assembled monolayers. Langmuir, 2002, 18: 4030-4040.

[80] Plummer S T, Bohn P W. Spatial dispersion in electrochemically generated surface composition gradients visualized with covalently bound fluorescent nanospheres. Langmuir, 2002, 18: 4142-4149.

[81] Sato Y, Uosaki K. Electrochemical and electrogenerated chemiluminescence properties of tris (2,2'-bipyridine)ruthenium (II)-tridecanethiol derivative on ITO and gold electrodes. J. Electroanal. Chem., 1995, 384: 57-66.

[82] Obeng Y S, Bard A J. Electrogenerated chemiluminescence. 53. Electrochemistry and emission from adsorbed monolayers of a tris(bipyridyl)ruthenium(II)-based surfactant on gold and tin oxide electrodes. Langmuir, 1991, 7: 195-201.

[83] Miao W, Bard A J. Electrogenerated Chemiluminescence. 72. Determination of Immobilized DNA and C-Reactive Protein on Au(111) Electrodes Using Tris(2,2'-bipyridyl)ruthenium(II) Labels. Anal. Chem., 2003, 75: 5825-5834.

[84] Naoi K, Ohko Y, Tatsuma T. Switchable Rewritability of Ag-TiO$_2$ Nanocomposite Films with Multicolor Photochromism. Chem. Commun. 2005, 1288-1290.

[85] Szunerits S, Walt D R. Fabrication of an optoelectrochemical microring array. Anal. Chem. 2002, 74: 1718-1723.

[86] Zhu L, Tang H, Harima Y, et al. Enhanced hole injection in organic light-emitting diodes consisting of self-assembled monolayer of tripod-shaped π-conjugated thiols. J. Mater. Chem., 2002, 12: 2250-2254.

[87] Sato Y, Sawaguchi T, Mizutani F. Potential-dependent chemiluminescence of luminol on the self-assembled monolayer. Electrochem. Commun., 2001, 3: 131-135.

[88] Reich D H, Tanase M, Hultgren A, et al. Biological applications of multifunctional magnetic nanowires (invited). J. Appl. Phys., 2003, 93: 7275-7279.

[89] Wang L, Chen W, Huang C, et al. Ultra-fast electron transfer from oligo (p-phenylene-ethynylene)thoil to gold. J. Phys. Chem. B., 2006, 110: 674-676.

[90] Matsuda K, Ikeda M, Irie M. Photochromism of diarylethene-capped gold nanoparticles. Chem. Lett., 2004, 456-457.

[91] Whitesides G M, Mathias J P, Seto C T. Molecular self-assembly and nanochemistry:

a chemical strategy for the synthesis of nanostructures. Science, 1991, 254: 1312-1319.

[92] Brolo A G, Irish D E, Smith B D. Applications of surface enhanced Raman scattering to the study of metal-adsorbate interactions. Journal of Molecular Structure, 1997, 405(1): 29-34.

[93] Chan K C, Kim T, Schoer J K, et al. Polymeric self-assembled monolayers. 3. pattern transfer by use of photolithography, electrochemical methods, and an ultrathin self-assembled diacetylenic resist. J. Am. Chem. Soc., 1995, 117: 5875-5876.

[94] Chong K S L, Sun S, Leggett G J. Measurement of the kinetics of photo-oxidation of self-assembled monolayers using friction force microscopy. Langmuir, 2005, 21: 3903-3909.

[95] Love J C, Wolfe D B, Chabinyc M L, et al. Self-assembled monolayers of alkanethiolates on palladium are good etch resists. J. Am. Chem. Soc., 2002, 124: 1576-1577.

[96] Francis P Z, Richard M C. In-situ electrochemical scanning tunneling microscopy study of cyanide-induced corrosion of naked and hexadecyl mercaptan-passivated Au(111). Langmuir, 1997, 13(2): 122-126.

[97] Francis P Z, Richard M C. Corrosion passivation of gold by nalkanethiol self-assembled monolayer: effect of chain length and end group. Langmuir, 1998, 14(12): 3279-3286.

[98] Cao Z, Gu N. Investigation on gold corrosion by in situ quartz crystal microbalance and atomic force microscopy in self-assembled processes of alkanethiol monolayers. Mater. Lett., 2005, 59(28): 3687-3693.

[99] Cao Z, Xiao Z L, Gu N, et al. Corrosion behaviors on polycrystalline gold substrates in self-assembled processes of alkanethiol monolayers. Anal. Lett., 2005, 38(8): 289-304.

[100] Magali E, Michel K, Hisasi T. The formation of self-assembling membrane of hexadecane-thiol on silver to prevent the tarnishing. Electrochi. Acta, 2004, 49: 1937-1943.

[101] Burleigh T D, Gu Y, Donahey G, et al. Tarnish protection of silver using a hexadecanethiol self-assembled monolayer and descriptions of accelerated tarnish tests. Corrosion, 2001, 57(12): 1066-1074.

[102] Bernard M C, Dauvergane E, Evesque M, et al. Reduction of silver tarnishing and protection against subsequent corrosion. Corros. Sci., 2005, 47: 663-679.

[103] Burleigh T D, Shi C, Kilic S, et al. Self-assembled monolayers of perfluoroalkyl amideethanethiols, fluoroalkylthiols, and alkylthiols for the prevention of silver tarnish. Corrosion, 2002, 58(1): 49-56.

[104] 刘金红, 王怡红, 郭志睿, 等. 银表面分子自组装膜的防腐性能. 化工学报, 2004, 55(10): 1674-1677.

[105] Hutt D A, Liu C. Oxidation protection of copper surfaces using self-assembled monolayers of octadecanethiol. Appl. Surf. Sci., 2005, 252(2): 400-411.

[106] Whelan C M, Kinsella M, Ho H M, et al. In-situ cleaning and passivation of oxidized Cu surfaces by alkanethiols and its application to wire bonding. J. Electro. Mater., 2004, 33(9): 1005-1011.

[107] Feng Y, Teo W K, Siow K S, et al. Corrosion protection of copper by a self-assembled monolayer of alkanethiol. J. Electrochem. Soc., 1997, 144(1): 55-64.

[108] Ma H Y, Yang C, Yin B S, et al. Electrochemical characterization of copper surface modified by n-alkanethiols in chloride-containing solution. Appl. Surf. Sci., 2003, 218: 143-

[109] 闻荻江, 冯芳. 正十二烷基硫醇对铜在酸性介质中的缓蚀行为. 化工学报, 2005, 56(7): 1363-1367.

[110] Jennings G K, Munro J C, Yong T H, et al. Effect of chain length on the protection of copper by n-alkanethiols. Langmuir, 1998, 14(21): 6130-6139.

[111] Jennings G K, Yong T H, Munro J C, et al. Structural effects on the barrier properties of self-assembled monolayers formed long chain ω-alkoxy-n-alkanethiols on copper. J. Am. Chem. Soc., 2003, 125(10): 2950-2957.

[112] Tan Y S, Srinivasan M P, Pehkonen S O, et al. Self-assembled organic thin films on electroplated copper for prevention of corrosion. J. Vac. Sci. Technol. A., 2004, 22(4): 1917-1925.

[113] Wang C, Chen S, Zhao S. Inhibition effect of as-treated, mixed self-assembled film of phenylthiourea and 1-dodecanethiol on copper corrosion. J. Electrochem. Soc., 2004, 151(1): B11-B15.

[114] Wang C T, Chen S H. A new approach for preparing effective inhibition film on copper based on self-assembled process. Chin. Chem. Lett., 2003, 14(3): 308-311.

[115] 王春涛, 陈慎豪, 赵世勇, 等. 烯丙基硫脲和十二烷基硫醇对铜的缓蚀作用. 化学学报, 2003, 61(2): 151-155.

[116] Aramaki K. Preparation of protective films on copper by chemical modification of alkanethiol self-assembled monolayer. Zairyo to Kankyo, 2002, 51(4): 136-143.

[117] Quan Z, Chen S, Cui X, et al. Factors affecting the quality and corrosion inhibition ability of self-assembled monolayers of a Schiff base. Corrosion, 2002, 58(3): 248-256.

[118] 陈玉红, 王庆飞, 丁克强, 等. B2-et-2B 自组装膜对铜的缓蚀作用. 河北师范大学学报: 自然科学版, 2003, 27(4): 377-379.

[119] 崔维真, 丁克强, 王庆飞, 等. 席夫碱自组装膜对铜的腐蚀保护. 河北师范大学学报: 自然科学版, 2002, 26(4): 381-383.

[120] Lusk A T, Jennings G K. Characterization of self-assembled monolayers formed from sodium S-alkyl thiosulfates on copper. Langmuir, 2001, 17(25): 7830-7836.

[121] Telegdi J, Rigó T, Kálmán E. Molecular layers of hydroxamic acids in copper corrosion inhibition. J. Electroanal. Chem., 2005, 582(1-2): 191-201.

[122] Telegdi J, Rigó T, Kálmán E. Nanolayer barriers for inhibition of copper corrosion. Corros. Eng. Sci. Technol., 2004, 39(1): 65-70.

[123] Wang D Q, Chen S H, Ma H Y, et al. Protection of copper corrosion by carbazole and N-vinylcarbazole self-assembled films in NaCl solution. J. Appl. Electrochem., 2003, 33(2): 179-186.

[124] Felhosi I, Telegdi J, Palinkas G, et al. Kinetics of self-assembled layer formation on iron. Electrochi. Acta, 2002, 47(13-14): 2335-2340.

[125] Felhosi I, Kalman E, Poczik P. Corrosion protection by self-assembly. Russ. J. Electrochem., 2002, 38(3): 230-237.

[126] Miecznikowski K, Chojak M, Steplowska W, et al. Two-layer structures of ultra-thin metal hexacyanoferrate films: Charge trapping and possibility of application to corrosion protection. Pol. J. Chem., 2004, 78(9): 1183-1193.

[127] Aramaki K. Preparation of protective films on iron by chemical modification of self-assembled monolayers. Zairyo to Kankyo, 2003, 52(7): 332-339.

[128] Ohkubo H, Itoh J, Nagano H, et al. Corrosion prevention of pure iron using self-assembled monolayer coating. Zairyo to Kankyo, 2003, 52(6): 316-318.

[129] Li Z Y, Chang S C, Williams R S. Self-assembly of alkanethiol molecules onto platinum and platinum oxide surfaces. Langmuir, 2003, 19 (17): 6744-6749.

[130] Petrovykh D Y, Kimura-Suda H, Opdahl A, et al. Alkanethiols on platinum: multicomponent self-assembled monolayers. Langmuir, 2006, 22 (6): 2578-2587.

[131] Mekhalif Z, Riga J, Pireaux J J, et al. Self-assembled monolayers of n-dodecanethiol on electrochemically modified polycrystalline nickel surfaces. Langmuir, 1997, 13 (8): 2285-2290.

[132] Ruan C M, Bayer T, Meth S, et al. Creation and characterization of n-alkylthiol and n-alkylamine self-assembled monolayers on 316l stainless steel. Thin Solid Films, 2002, 419 (1/2): 95-104.

[133] Toshihiro K Kana. Effects of alkyl chain length on the efficiency of photoinduced electron transfer at gold electrodes modified with SAMs of molecules containing porphyrin, ferrocene, and thiol separated each other by alkyl chains. Phys Chem., 1999, 212(1): 23-27.

[134] Peng Z, Dong S. Formation of a self-assembled monolayer of 2-mercapto-3-n-octylthiophene on gold. Langmuir, 2001, 17: 4904-4909.

[135] Turyan I, Mandler D. Selective determination of Cr(vi) by a self-assembled monolayer-based electrode. Anal. Chem., 1997, 69: 894-897.

[136] Henke C, Steinem D, Janshoff A, et al. Self-assembled monolayers of monofunctionalized cyclodextrins on to gold: a mass spectrometric characterization and impedance analysis of host-guest interaction. Anal. Chem., 1996, 68: 3158-3165.

[137] Hutchison J E, Postlethwaite T A, Chen C, et al. Electrocatalytic activity of an immobilized cofacial diporphyrin depends on the electrode material. Langmuir, 1997, 13: 2143-2148.

[138] Maskus M, Abruna H D. Synthesis and characterization of redox-active metal complexes sequentially self-assembled onto gold electrodes via a new thiol-terpyridine ligand. Langmuir, 1996, 12: 4455-4462.

[139] Mabouk P A. Direct electrochemistry for the imidazole complex of microperoxidase-11 in dimethyl sulfoxide solution at naked electrode substrates including glassy carbon, gold, and platinum. Anal. Chem., 1996, 6: 189-191.

[140] Diab N, Schuhmann W. Electropolymerized manganese porphyrin/polypyrrole films as catalytic surfaces for oxidation of nitric oxide. Electrochim. Acta., 2001, 47: 265-273.

[141] You T, Niu L, Gui J Y, et al. Detection of hydrazine, methylhydrazine and isoniazid by capillary electrophoresis with a 4-pyridyl hydroquinone self-assembled microdisk platinum electrode. J. Pharm. Biomed. Anal., 1999, 19(1-2): 231-237.

[142] 徐静娟, 方惠群, 陈洪渊. 硫堇衍生化自组装膜修饰金电极的电化学性质及其对抗坏血酸的电催化氧化. 高等学校化学学报, 1997, 5: 706-710.

[143] Li J H, Lin D, Wang E K, et al. Interfacial characteristics of the self-assembly system of poly-L-lysine 3-mercaptopropionic acid gold electrode. J. Electroanal. Chem., 1997, 431: 227-

230.

[144] Valdemaras R, Thomas A. Direct electrochemistry of microperoxidase-11 at gold electrodes modified by self-assembled monolayers of 4,4'-dithiodipyridine and 1-octadecanethiol. J. Electroanal. Chem., 1997, 427: 1-5.

[145] 翟怡,张金利,王一平,等. 自组装单层膜的制备与应用. 化学进展. 2004, 16(4): 477-484.

[146] Vijayamohanan K, Aslam M. Applications of self-assembled monolayers for biomolecular electronics. Appl. Biochem. Biotech., 2001, 96: 25-39.

[147] Zhang Q, L Archer A. Boundary lubrication and surface mobility of mixed alkylsilane self-assembled monolayers. J. Phys. Chem. B, 2003, 107: 13123-13132.

[148] Nirmalya K, Chaki K, Vijayamohanan K. Self-assembled monolayers as a tunable platform for biosensor applications. Biosensors & Bioelectronics, 2002, 17: 1-12.

[149] Gooding J J, Hibbert D B. The application of alkanethiol self-assembled monolayers to enzyme electrodes. Trends in Analytical Chemistry, 1999, 18: 525-533.

[150] Kepley L J, Crooks R M, Ricco A J. A selective saw-based organophosphonate chemical sensor employing a self assembled, composite monolayer: a new paradigm for sensor design. Anal. Chem., 1992, 64: 3191-3198.

[151] Richert J, Weiss T, Gopel W. Self-assembled monolayers for chemical sensors: molecular recognition by immobilized supramolecular structure. Sensors and Actuators B, 1996, 31: 45-50.

[152] Zhou X C, Zhong L, Li S F Y, et al. Crystal microbalance coated with self-assembled monolayers. Sensors and Actuators B, 1997, 42: 59-65.

[153] Crooks R M, Ricco A J. New organic materials suitable for use in chemical sensor arrays. Acc. Chem. Res., 1998, 31: 219-227.

[154] Sung K J, Wilson G S. Polymeric mercap tosilane-modified platinum electrodes for elimination of interferants in glucose biosensors. Anal. Chem, 1996, 68: 591-596.

[155] Ruan C, Yang R, Chen X H, et al. A reagentless amperometric hydrogen peroxide biosensor based on covalently binding horseradish peroxidase and thionine using thiol-modified gold electrode. J. Electroanal. Chem, 1998, 455: 121-125.

[156] Andrew N, Shipway, Willner I. Nanoparticles as structural and functional units in surface-confined architectures. Chem. Commun., 2001, 2035-2045.

[157] Saitoh A, Munoz S, Moriizu M T, et al. Quartz crystal microbalance odor sensor coated with mixed-thiol-compound sensing film. Japanese Journal of App lied Physics, 1998, 37 (5B): 2849-2852.

[158] 孙灏,王洪恩. 硫醇/金自组装膜上组装葡萄糖氧化酶及其应用. 济宁医学院学报, 1999, 22 (3): 4-7.

[159] Xu X H, Bard A J. Immobilization and hybridization of DNA on an aluminum(iii) alkanebisphosphonate thin film with electrogenerated chemiluminescent detection. J. Am. Chem. Soc., 1995, 117: 2627-2631.

[160] 秦玉华,张术勇,庞琳,等. 自组装单分子膜在生物传感器中的应用. 东北电力学院学报, 2004, 24(1): 27-30.

[161] Andrew N, Willner I. Nanoparticles as structural and functional units in surface 2

confined architectures. Chem. Commun., 2001, (20): 2035.

[162]Alfonta L, Willner I. Electronic transduction of biocatalytic transformation on nucleic acid-functionalized surfaces. Chem. Commun., 2001, (16): 1492.

[163]王海雄,吴侯,翁新楚,等.活性炭吸附法固定猪胰脂酶的初步研究.上海大学学报:自然科学版, 2003, (5): 428-432.

[164]千宁,余杨锋,王志颖,等.辣根过氧化物酶在巯基丁二酸铜(II)自组装单分子层修饰电极表面的固定及对H_2O_2传感研究.传感技术学报, 2007, 20(2): 262-266.

[165]尹峰,赵紫霞,吴宝艳,等.于多壁碳纳米管和聚丙烯胺层层自组装的葡萄糖生物传感器.分析化学研究简报, 2007, 35(7): 1021-1024.

[166]李兰扣,董江庆,徐晓燕.自组装技术及其在生物传感中的应用研究.河北化工, 2009, 32(7): 52-54.

[167]杨广彬,张平余,吴志申,等.分子有序膜及其摩擦学行为研究进展.河南大学学报:自然科学版. 2005, 35(2): 25-32.

[168]李景虹.自组装膜电化学.北京:高等教育出版社, 2002.

[169]Zhang L Z, Leng Y S, Jiang S Y. Tip-based hybrid simulation study of frictional properties of self-assembled monolayers: Effects of chain length, terminal group, scan direction, and scan velocity. Langmuir, 2003, 19: 9742-9747.

[170]van der Vegte E W, Subbotin A, Hadziioannou G. Nanotribological properties of unsymmetrical n-dialkyl sulfide monolayers on gold: Effect of chain length on adhesion, friction, and imaging. Langmuir, 2000, 16: 3249-3256.

[171]Brewer N J, Beake B D, Legett G. J. Friction force microscopy of self-assembled monolayers: Influence of adsorbate alkyl chain length, terminal group chemistry, and scan velocity. Langmuir, 2001, 17: 1970-1974.

[172]Lee S H, Shon Y S, Colorado R, et al. The influence of packing densities and surface order on the frictional properties of alkanethiol self-assembled monolayers (SAMs) on gold: A comparison of SAMs derived from normal and spiroalkanedithiols. Langmuir, 2000, 16: 2220-2224.

[173]Ahn H S, Cuong P D, Park S K, et al. Effect of molecular structure of self assembled monolayers on their tribological behaviors in nano-and microscales. Wear, 2003, 255: 819-825.

[174]Lee S H, Puck A, Graupe M, et al. Structure, wettability, and frictional properties of phenyl-terminated self-assembled monolayers on gold. Langmuir, 2001, 17: 7364-7370.

[175]潘力佳,何平笙.纳米器件制备的新方法——微接触印刷术.化学通报, 2000, 12: 12-17.

[176]Kumar A, Abbott N A, Kim E, et al. Patterned self-assembled monolayers and meso-scale phenomena. Acc. Chem. Res., 1995, 28: 219-226.

[177]Aizenberg J, Andrew J B, Whitesides G M. Control of crystal nucleation by patterned self-assembled monolayers. Nature, 1999, 398: 495.

[178]Ma F, Lennox R B. Potential-assisted deposition of alkanethiols on Au: Controlled preparation of single-and mixed-component SAMs. Langmuir, 2000, 16: 6188-6190.

[179]Sullivan M G, Utomo H, Fagan P J, et al. Automated electrochemical analysis with combinatorial electrode arrays. Anal. Chem., 1999, 71: 4369-4375.

[180] Lee S W, Laibinis P E. Protein-resistant coatings for glass and metal oxide surfaces derived from oligo (ethylene glycol)-terminated alkytrichlorosilanes. Biomaterials, 1998, 19: 1669–1675.

[181] Wang J, Rivas G, Jiang M, et al. Electrochemically induced release of DNA from gold ultramicroelectrodes. Langmuir, 1999, 15: 6541–6545.

[182] 钱林茂, 雒建斌, 温诗铸. 二氧化硅及其硅烷自组装膜微观摩擦力与粘着力的研究(Ⅰ)摩擦力的实验与分析. 物理学报, 2000, 49(11): 2240–2246.

[183] Yang Z P, Kauffmann J M, Valenzuela M I A, et al. Electroanalytical behaviour of a nanoarray self-assembled thiocholesterol gold electrode. Mikrochim. Acta, 1999, 131: 85–90.

[184] 李扬眉, 陈志春, 何琳, 等. 生物大分子自组装膜及其应用研究进展. 化学进展, 2002, 14(3): 212–216.

[185] 孙小强. 超分子化学导论. 北京: 中国石化出版社, 1996. 125.

[186] Kimizuka N, Tanaka M, Kunitake T. Spatially controlled synthesis of protein/inorganic nano-assembly: alternate molecular layers of Cyt c and TiO2 nanoparticles. Chem. Letter, 1999, 15: 1333–1334.

[187] 李正强, 王力彦, 张希, 等. 菌紫质分子沉积膜的制备. 高等学校化学学报, 1995, 16: 958–960.

[188] Li M, Li B F, et al. The fast photovoltaic response from multilayer by alternate layer-by-layer assembly of polycation and bacteriorhodopsin. Chem. Letter, 2000: 266–267.

第六章 卟啉及金属卟啉的合成及其应用概述

卟啉类化合物已被广泛地应用于光电转换、催化、导电、铁磁体、载氧体、分子电子器件和医学等方面。在卟啉环周边进行化学修饰,引入特定官能团,使之与底物产生多重相互作用,可以达到对分子的精确识别[1,2]。在仿生化学的应用研究中,还可以将视野拓宽到过氧化物酶以外的酶体系,增加目标分析物的种类。近年来,利用卟啉独特的电子结构和光电性能,设计和合成光电功能材料及光电器件的研制已成为国内外十分活跃的研究领域。特别是卟啉化合物的合成、光物理和光化学性质的研究及其应用已成为现代卟啉化学研究的重要领域[3-16]。通过认识卟啉在光合作用中的地位和卟啉给—受体分子的光诱导电子转移和电荷分离的能力,还可以开发其光学、电学等方面的应用[17-21]。功能卟啉超分子跨越化学、生物和医学等多个学科,展现了卟啉类化合物广阔的开发和应用前景。

第一节 卟啉的合成

在由非卟啉前体合成卟啉的方法中,按照缩合方式的不同大致可以分为两种:经典方法——由等量的芳醛和吡咯缩合合成卟啉;模块法——由两分子二吡咯甲烷与两分子醛或一分子胆色素与一分子吡咯缩合而得卟啉。

1. Adler-Longo 法

Adler 和 Longo 于 1967 年改进了 Rothemund 的合成方法[22],在相对温和的反应条件下使许多带有其他取代基的芳醛可以与吡咯反应生成卟啉(图 6-1),产率大大提高,可达 20% 左右。这就使克级规模合成卟啉类化合物成为可能。应用 Adler-Longo 法,研究者已经获得了大量的取代苯基卟啉。当选取适当比例的两种醛合成卟啉时,还可以获得含有不对称官能团的卟啉。后来,研究者[23]对 Adler-Longo 法又作了进一步的调整,他们采用有机酸和极性溶剂代替丙酸介质,反应过程中产生的杂质明显减少,四苯基卟啉(TPP)的产率最高达到 50%。

总的来说,Adler-Longo 法的优点是操作比较简单,实验条件也不是很苛刻,反应路线比较成熟,反应产率相对较高,到目前为止在卟啉合成中仍有广泛应用[24-26]。然而,Adler-Longo 法也有一些缺点,如由于反应在酸性体系中进行,不能选用对酸敏感的醛类化合物作为反应物;另外在高温会产生大量焦油状的副产物,这使得分离纯化变得十分困难。

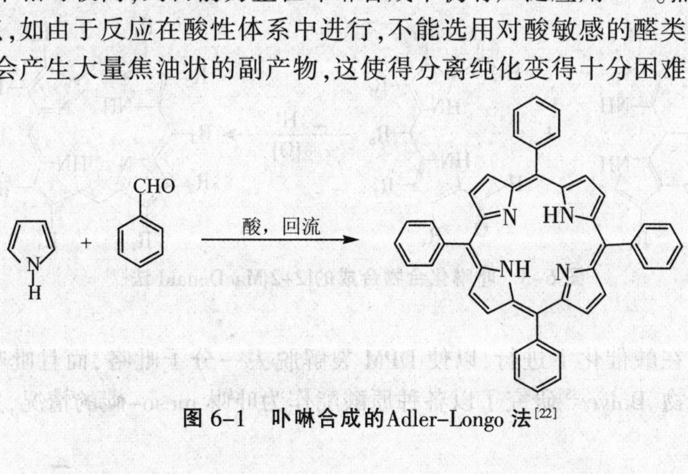

图 6-1 卟啉合成的 Adler-Longo 法[22]

2. Lindsey 法

1986 年 Lindsey 等[27,28]提出了一种新的合成方法。整个反应分两步进行,先得到合成卟啉的中间体——卟啉原(Porphyrinogen)(Dolphin[29]曾证实了这一反应中间体的存在),再加入氧化剂将其氧化成卟啉,从而使得反应可以在常温下进行(图 6-2)。

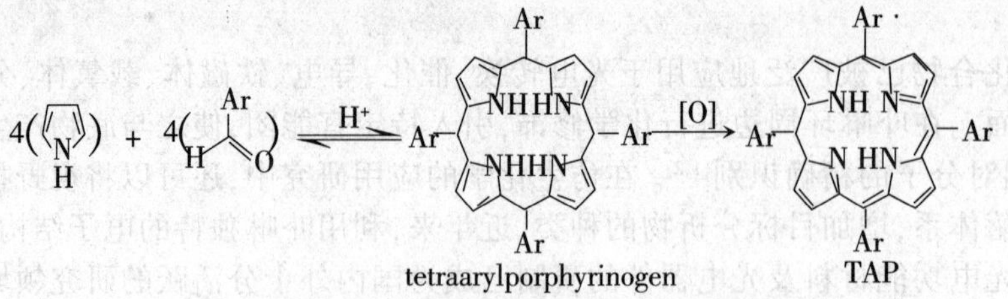

图 6-2 卟啉化合物合成的 Lindsey 法[27]

由于 Lindsey 法的反应温度较低,一般不会产生焦油状副产物,目标产物的分离提纯就显得比较容易;同时,温和的反应条件也允许反应物先经过化学修饰,连接上一些敏感基团。Lindsey 等用此方法合成了 30 多种卟啉,平均产率在 30%~40%。由于 Lindsey 法的成功,人工合成的卟啉类化合物大大扩充,其功能也越来越多,是卟啉合成的重大进步。但是,Lindsey 法的实验条件要求反应过程中无水,其所用的试剂都比较昂贵,而且要求浓度较低(10^{-2} mol/L),所以不适用于卟啉的大规模合成。

Lindsey 法经过改良,可得到立体位阻较大的苯基 2,6-位有取代的四芳基卟啉[30],其非共平面的构型表现出不同的光学特性[31]。如四-(2,4,6-三甲基苯基)卟啉[32]就是这样一个例子,合成时在 BF_3 中加入乙醇作为协同催化剂,产率达 30%。Llama 等人[33]合成 TPP 时加入了过渡金属盐作为氧化剂,这一改良使产率高达 68%,而且比传统的 Lindsey 法的反应浓度更高。

3. 卟啉合成的模块法

3.1 [2+2]法

所谓[2+2]法,即两分子二吡咯甲烷(DPM)缩合来生成卟啉母核。较早由 MacDonald[34]等人合成,所以[2+2]合成法也称为 MacDonald 法,其基本合成路线见图 6-3。最初为了模拟自然界卟啉的合成过程,MacDonald 把物质的量比为 1:1 的 DPM 和 α,α′-二甲酰基二吡咯甲烷在酸催化下缩合产生尿卟啉。α 和 α′-位的甲酰基成为卟啉的 meso-碳。

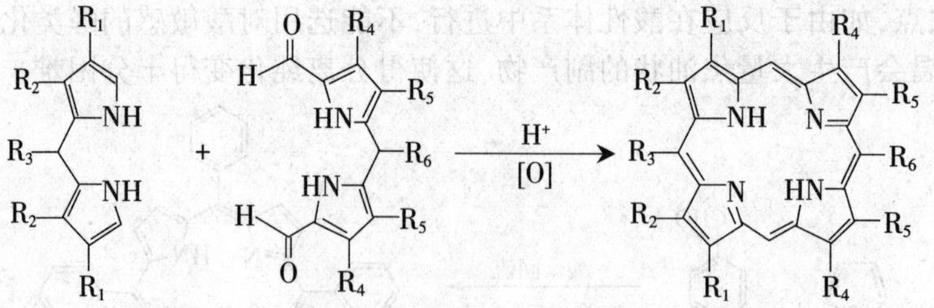

图 6-3 卟啉化合物合成的[2+2]MacDonald 法[34]

此类反应要在酸催化下进行,以使 DPM 裂解脱去一分子吡咯,而且吡咯自缩合会产生难以分离的副产物。Bolye[35]研究了以各种原酸酯作为卟啉 meso-碳的情况,发现原酸酯的取代基对卟啉合成影响很大;取代基为强的吸电子基团、立体位阻较大均对反应不利。Smith 等[36,37]也对[2+2]法作了有意义的改进,从吡咯出发合成α-位带羟基的DPM,相同的两分子 DPM缩合,可以得到结构对称的卟啉,而且可以抑制上述的裂解反应。

[2+2]法能够方便地控制卟啉大环的取代基,利用这一优点,可以先合成各种取代的 DPM,然后合成各种β-位和meso-位取代的卟啉。相对于 Adler—Longo 法和 Lindsey 法,[2+2] 法灵活性和区域选择性更大,并且在合成低聚卟啉上运用得也很多,因而在近年来十分流行[38-41], 产生了许多基于[2+2]法的其他合成卟啉的方法。使用[2+2]法得到的卟啉大致归为四类:

(1)β-位和meso-位都有取代基;
(2)全部β-位和两个meso-位有取代基;
(3)β-位没有取代基,两个meso-位有取代基;
(4)β-位没有取代基,四个meso-位有取代基。

3.2 [3+1]法

与[2+2]法相似,[3+1]法是将一分子有饱和碳相连的胆色素和一分子α,α′-二甲酰基 吡咯环合。受到[2+2]法的启发,Grigg等[42]用一分子呋喃(或噻吩、吡咯)和一分子胆色素合 成了类卟啉化合物,如21-氧代、硫代卟啉。与[2+2]法一样,由吡咯的α-位甲酰基形成了卟 啉的meso-碳,但是Grigg等得到的是两个异构体。Momenreau 等[43]对此法进行了改进,先合 成对称的吡咯和胆色素,那么缩合的产物就是唯一的。Senge[44]为得到5,10-二取代卟啉,把 一分子胆色素、一分子吡咯和两分子芳醛(或脂肪醛)缩合,得到了预期的卟啉(图6-4),但 产率很低。而Lash[45]用吡咯二醛和胆色素在酸催化下得到卟啉原,再经DDQ(四氯苯醌)氧 化获得了八烷基卟啉,收率达到60%。

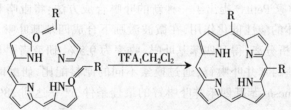

图6-4　[3+1]卟啉合成法[44]

[3+1]法需要在成功合成胆色素的前提下进行,一般总产率较低。但此方法能合成一 些结构复杂的卟啉,还可用来合成芳香环扩展的卟啉和类卟啉化合物。

4. 由2-位取代吡咯合成卟啉

此种方法是以"首尾相接(head-to-tail)"的方式,在酸性条件下环合四分子吡咯,形成卟 啉化合物。对反应物的要求是:

(1)吡咯的2-位或5-位含有一个取代甲基,取代甲基有利于在酸中形成高度亲电的氮 杂戊二烯;

(2)吡咯剩余的2-位或5-位可以是未被取代的,也可以是被一个在酸性条件下很容易 消除的基团(如羧基)所取代。用这种方法合成β位卟啉是很有用的。这种方法存在的缺点 是:在合成卟啉前需要花费很多步骤制备出相应的取代吡咯,并且反应时头尾相连的方式使 此法不能用于合成大多数天然卟啉。Ogoshi等人[46]利用这种方法合成了一种电子缺陷(不稳 定)卟啉分子,该卟啉中含有4个三氟甲烷基团(图6-5)。

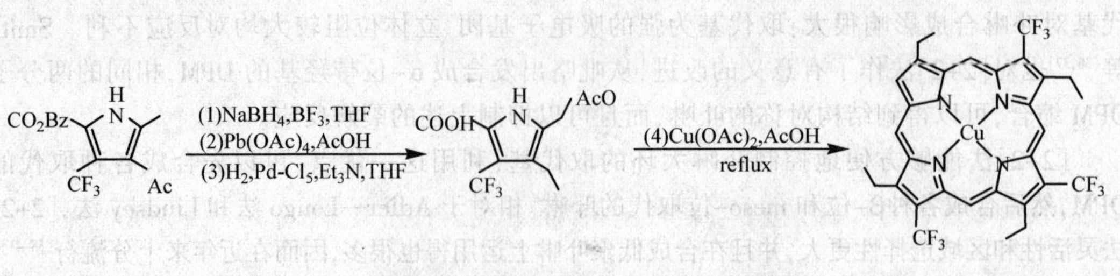

图 6-5　2-取代吡咯合成卟啉示意图[46]

5. 由线形四吡咯出发合成卟啉

将线形连四吡咯火胆色烷(bilane)成环也可得到卟啉(图 6-6)。应用这一合成方法可以获得不对称卟啉和那些在 β 位有多种取代基团的卟啉[47]。

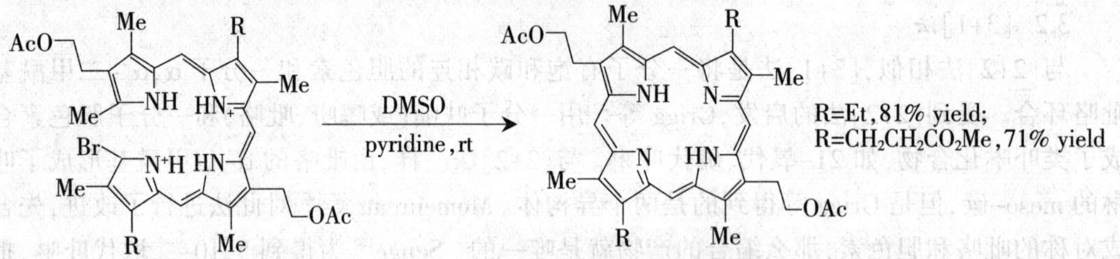

图 6-6　由线形四吡咯出发合成卟啉[47]

6. 微波激励法

1992 年法国化学家 Petit[48]提出了一种新的卟啉合成方法：将吡咯和苯甲醛吸附于无机载体硅胶上，利用载体的酸性催化作用，在微波激励下合成四苯基卟啉。反应 10 min 后，直接加入层析柱进行层析分离，得到四苯基卟啉，收率为 9.5%。研究者[49]也采用无溶剂微波固相法，合成 meso-苯基四苯并卟啉锌，通过摸索不同的原料配比、研磨时间和微波辐射时间等条件，确定了形成 meso-苯基四苯并卟啉锌的最佳条件，产率达到 28%。

第二节　金属卟啉的合成

1. 金属卟啉合成的基本反应过程

卟啉配体分子(H_2P)的四个氮原子，可以与金属配位化合，形成卟啉金属衍生物，这一过程称为卟啉金属化。最简单的金属卟啉配合物为单中心金属离子配合物(MP)，其反应按下式进行：

$$H_2P + MX_m \rightleftharpoons MP + mHX$$

反应式中正方向为金属化，而逆方向为去金属化。

金属卟啉配合物从结构简单的 1:1 型配合物，到结构复杂的金属卟啉分子聚合物，具有多种多样的分子空间构型，因而，它们的制备方法也不尽相同。但都具备以下五个基本反应过程：

(1) 氢离子电离平衡

卟啉配体分子的 N_4 环中，存在两个氢，由于卟啉共轭作用，这两个氢与环中四个氮原子的成键相同。以中性四苯基卟啉分子为例，受介质影响，这四个 N 原子既能接受质子，形成 H_4PP^{2+} 和 H_3PP^+ 离子，又能电离出质子，形成 HPP^- 和 PP^{2-}。

在酸性环境中，卟啉以 H_4PP^{2+} 和 H_3PP^+ 形式存在，不利于形成金属卟啉配合物；相反，在碱性环境中，得到 PP^{2-} 离子多，则有利于配合物的形成。考虑到金属盐的水解，卟啉金属化反应多数在弱酸性环境下进行。

(2) 金属盐的电离平衡

金属盐（例如醋酸盐）的反应活性低，必须完全电离，以离子形式存在，或由惰性的金属盐转化成反应活性高的溶剂化分子，才能与卟啉分子进行配位反应（如 $Zr(OPh)_4$ 的反应活性远远高于其醋酸盐[50]）。

此外，鉴于大多数卟啉分子只能在有机溶剂中溶解，即配位反应必须在极性较小的介质中进行，所以反应中多采用在有机溶剂中溶解度较好的金属醋酸盐。

(3) 金属进入 N_4 环的配位平衡

在具备了以上条件后，金属离子才能与卟啉分子的四个氮原子形成配位键。像 Ni(II) 离子，本身为二价，且采用 d^2sp 的平面四边形杂化轨道，正好符合卟啉分子中 N_4 平面环的空间结构和电价要求，至此，已经全部完成配位过程。而其他金属，尚需以下两个步骤。

(4) 电荷平衡

金属卟啉轴向配位的动力之一为化合物的电中性需求。当金属离子所带电荷高于 +2 价时，则金属卟啉带正电，需要体系中的阴离子，在金属卟啉分子平面的轴向接近中心离子，与中心金属离子配位，抵消卟啉环所带电荷，使其呈电中性。

(5) 完全配位平衡

多数金属离子在形成配合物时，采用 d^2sp^3 或 sp^3d^2 的六配位八面体形状的杂化轨道。N_4 平面环中的四个氮原子与单一轴向电荷平衡阴离子的配位，并未达到金属离子饱和配位的要求，因此需要反应体系中电中性分子（一般多为溶剂分子）中所含有的电负性较大的原子提供孤对电子，在卟啉环的另一轴向，与中心金属离子配位，达到饱和配位的要求。

另外，鉴于金属卟啉配合物的结构和性质各不相同，卟啉与金属离子的金属化过程，需选择不同的反应体系。金属离子是否可进入卟啉配体的环中间，形成平面配合物，取决于金属离子直径的尺寸。直径在 0.6~0.65 Å 之间的金属离子，可以进入 N_4 环中，形成平面配合物；卟啉环可通过 S_4 型畸变，缩小环的孔径，再与小直径离子配位；对于直径略大的离子，卟啉分子则可与其进行变形（拱形）配位，通过电子云的重叠，来扩大孔径，形成平面配合物。大部分金属卟啉均以平面结构存在；而前过渡系的 Ti、Zr、Hf、V、Nb、Ta、Cr、Mo 和 W 等大直径金属离子，不能进入卟啉环内，只能在 N_4 环平面外形成配合物。

2. 金属卟啉合成的主要方法

2.1 醋酸盐法

在金属化过程中，卟啉环上的两个质子解离下来，与醋酸盐作用，形成弱酸——醋酸，使化学平衡向形成金属离子方向移动，有利于金属卟啉配合物的生成。反应结束后，在反应体系中加入甲醇或水，充分冷却，就可将产物从溶液中结晶出来。此外，可以用 $CHCl_3$/MeOH 混合溶液代替冰醋酸进行反应，$CHCl_3$ 和 MeOH 的作用分别是使卟啉和醋酸盐充分溶解。除了在醋酸中不稳定的金属离子外，所有的二价金属的金属卟啉配合物都可用醋酸盐法合成。固体醋酸盐的加入，可进一步缓冲反应溶液，增强卟啉配体分子中氢的解离[51]。

2.2 吡咯法

若卟啉配体分子在醋酸介质中不稳定,反应溶剂应选用碱性溶剂——吡咯[52]。吡咯对卟啉配体分子和二价金属盐有较好的溶解性能,而且具有良好的配位性能。通过吡咯分子对金属卟啉的配位作用,可将产物从反应体系中沉淀出来,简化分离提纯。但也正是由于吡咯的这一配位作用,可与金属离子生成配合物阻碍了金属卟啉配合物的形成。因此,以吡咯为反应溶剂,需要较长的反应时间,一些反应甚至根本无法进行。

2.3 乙酰丙酮金属盐法

使用乙酰丙酮金属盐合成金属卟啉的优点在于,这种盐本身在有机溶剂中溶解度大,易处理,且弱酸盐的电离平衡,有利于卟啉分子中氢的解离。同吡咯一样,其唯一缺陷在于,可与半径较小、价态较高的金属离子形成稳定的螯合物,阻碍卟啉与金属离子之间的配位。所以,经常采用加入苯酚或咪唑的方法,与金属离子生成加合物,增加金属离子的反应活性。离子半径大的ⅢA和ⅢB族金属、镧系金属和前过渡金属的卟啉配合物,均以此方法合成。

2.4 苯酚法

一些前过渡金属并不存在乙酰丙酮化合物,或金属化合物本身反应活性低,这些元素的金属卟啉配合物的合成,可采用苯酚法。在苯酚沸腾温度下(295℃),金属以 $M(OPh)_n$ 活性中间体存在,便于卟啉分子金属化。较为典型的化合物为 $Zr(OPh)_4$ 和 $Al(OPh)_3$。Ta、Mo、W 和 Re 的卟啉配合物,也用本法合成。由于反应温度高,易挥发物质将不断逸出,因而反应要在敞开体系中进行。

2.5 苯甲腈法

苯甲腈可较好地溶解卟啉配体分子,其配位能力弱,属弱碱性溶剂,沸点高,可增强惰性无水氯化物的溶解性和反应活性。形成金属卟啉过程中所产生的 HCl 气体,可通过氮气除去。此方法为 Pd、Pt[53]和 Nb[54]等金属卟啉的特效制备方法。反应中,先将 $PtCl_2$ 加入沸腾的 PhCN 中,再加入卟啉配体分子,以防卟啉配体高温分解。

2.6 DMF 法

Adler 等人[55]在 DMF(N,N-二甲基甲酰胺)溶剂中,研究了 68 种金属卟啉的合成。这种溶剂的特点为:沸点高,对卟啉配体和金属氯化物(特别是水解强、难以处理的金属氯化物)的溶解性好。由于沸点高,反应中产生的 HCl 气体被蒸出体系,使平衡向生成金属卟啉配合物方向移动。还可通过再次加入过量金属盐,迫使反应生成金属卟啉配合物。

2.7 有机金属化合物法

烷基金属化合物本身为 Lewis 酸,有助于活化卟啉的氮原子与金属离子配位;同时,配位后所产生的烷基负离子是很强的碱,因而反应活性极高,需要在低温下进行反应。但该法要求过程中无氧无水,条件较为苛刻。

2.8 金属羰基化合物法

金属羰基化合物中 CO 的离去,生成具有 Lewis 酸作用的配位非饱和化合物,易于进攻卟啉 N 的孤对电子,形成配位键。此法适用于制备ⅣB 至ⅦB 族的金属卟啉配合物。以 $Re_2(CO)_{10}$ 为例,反应中金属铼从 $Re(0)$ 氧化成 $Re(Ⅰ)$,卟啉上的 H 被还原成 H_2。

2.9 烷氧基金属化合物法

一些对酸敏感的金属卟啉配合物,包括 $Ca(Ⅱ)$、$Sr(Ⅱ)$、$Ba(Ⅱ)$ 和碱金属的卟啉配合物,要通过烷氧基金属化合物方法来合成。具体为醇解 Grinard 试剂,所得到的烷氧基金属化合物,直接在该醇溶液中与卟啉配体完成配位反应。

第三节 卟啉及金属卟啉应用概述

卟啉由于具有独特的结构及性能,近年来在材料化学、医学、生物化学、分析化学、能源化学和地球化学等领域都有着广泛的应用。卟啉化学的研究也得到迅速的发展。以下就目前卟啉化合物及其金属配合物在不同领域的应用分别加以简要阐述。

1. 卟啉在材料化学中的应用

金属卟啉配合物作为一种功能性客体,在光电材料方面用途广泛,可用于磁性材料和非线性光学材料等。卟啉具有很强的发光特性,尤其是与过渡金属配位后发出很强的磷光,而这种磷光又可被 O_2 猝灭,所以被用作光敏器件的磷光材料[56]。

卟啉类化合物因为具有光敏性好、性能稳定和易于修饰等优点成为分子器件研究的理想模型化合物。分子器件是指在分子水平上由光子、电子或离子(总称为介体 mediator)操纵的器件[57]。选择卟啉作为分子器件的原因是:

(1)卟啉化合物具有大 π 共轭结构,从而使卟啉分子的 HOMO 与 LUMO 之间的能量差降低,使卟啉在可见光区有发射,且发光量子效率高。

(2)卟啉结构易于修饰,通过改变卟啉化合物的周边取代基可以改变卟啉分子的荧光发射波长。

(3)卟啉易于形成金属配合物[58]。

在分子器件中,卟啉可以作为卟啉分子导线、卟啉分子开关[59-61]、卟啉超分子天线[62-66]、卟啉光能转化器[67]、光控制输入输出器[68]、光电子闸门[69]、癌治疗荧光探针[70]和生物传感器[71]等。如 Wasielewski 等[72]设计了一个电子给体—受体—给体卟啉分子开关(图 6-7),这是一个光强度依赖性的光开关。

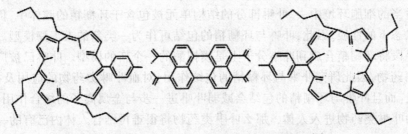

图 6-7 Wasielewski 等设计的分子开关[72]

2. 卟啉在医学中的应用

卟啉因其结构具有特殊的性能——与癌细胞有特殊的亲和力,在医学研究中可作为检测癌症的光敏剂和抗癌药物。随着人们对卟啉类物质研究的不断深入,以卟啉类光敏剂为核心的光动力疗法也逐渐成为继手术、放疗和化疗之外的第四种成熟的癌症治疗方法。光动力疗法(photodynami therapy,PDT)的基础是先给予肿瘤组织有选择性摄入作用的光敏剂,然后用一定波长的光照射肿瘤组织,由光敏剂诱发光动力效应,产生单线态氧($\cdot O_2$)活性物质,它们与生物分子中的氧化敏感基团作用,导致其氧化失活,最终引起肿瘤细胞死亡而显示治疗作用。目前应用在人类临床试验和已批准使用的所有光敏剂全部是卟啉类化合物[73,74],主要有如下几种：血卟啉[75-77]、苯并卟啉衍生物[78,79]、短肽链修饰的卟啉衍生物[80-83]、糖基卟啉衍生物[84]、糖肽卟啉衍生物[85,86]、核苷卟啉衍生物[87]、类固醇卟啉衍生物[88]、硼烷卟啉衍生物、meso-四(间-羟基苯基)二氢卟吩(m-THPC)[89,90]和酞菁类[91]等。

3. 卟啉在生物化学中的应用

金属卟啉是血红素、细胞色素和叶绿素的生物大分子核心部分(图6-8)。利用人工合成的金属卟啉配合物对血红蛋白、血红素和细胞色素等生物大分子功能进行模拟一直是人们研究的热点。

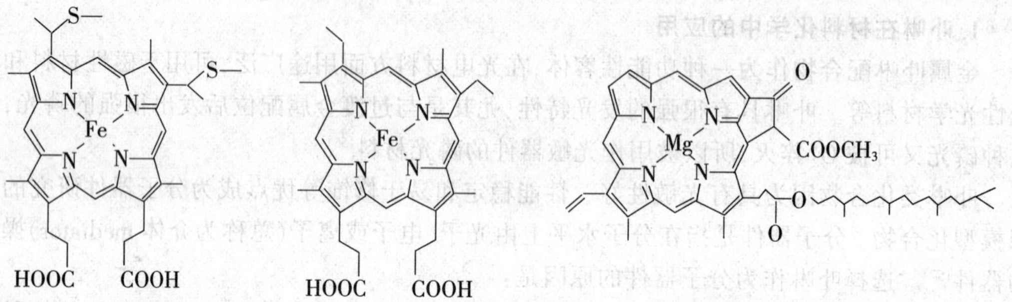

图6-8 细胞色素、血红素和叶绿素的结构

卟啉与环糊精体系的研究与生命科学密切相关。卟啉或金属卟啉与环糊精包结作用的研究在卟啉化学以及人工酶模拟中占有重要的地位。

1992年Dick用紫外—可见光谱和二维核磁研究了7种卟啉与Me—β—CD的包结作用。Dick认为环糊精类似于血红素外面的蛋白附属物，血红蛋白中的蛋白质部分除了能给血红素的活动中心提供一个疏水性的环境外，还能作为一个载体，将血红素转移到适当的细胞环境中。同样，环糊精与卟啉的包结作用中，环糊精的作用类似于蛋白质附属物，包结卟啉后，给卟啉提供一个疏水性的环境，阻碍了卟啉进一步形成聚集体，而且作为一个载体，将卟啉运送到适当的细胞环境中去。卟啉部分的结构单元被包含于环糊精的腔体中，仍然显示了它在自然状态下的性质。因此，卟啉与环糊精的包结可作为一类新的人工酶模型。此外，用有机合成的手段将环糊精合成到卟啉分子上是模拟酶的一个新的模型。卟啉已被广泛地认可为一种抗癌药物，因此研究卟啉与环糊精的包结作用，对血卟啉类药物的应用及控制、释放有重要意义。而且卟啉与环糊精的包结会减弱卟啉进一步与金属离子的络合作用，若以环糊精为载体的卟啉类药物进入人体，那么卟啉类药物将很难再结合人体内已有的一些有益的金属离子，如Fe^{2+}，这一结果为卟啉药物在医学上的合理应用提供了依据。

天然生物分子大多是手性的，因此手性分子识别是许多生物化学现象的本质。研究主体化合物对客体分子的手性分子识别机理，对从分子水平上模拟生物功能，研究生物体内的各种生物现象具有重要意义，它的应用也为发展特异性、专一性、选择性高的生物活性分子分析方法提供了广阔的前景[92]。

卟啉及其金属配合物在生物传感器方面的应用也比较广泛。文献[93]评述了固定于电聚合薄膜中的金属卟啉配合物在仿生催化和生物传感器方面的应用进展。

4. 卟啉在分析化学中的应用

卟啉类试剂早在20世纪50年代就作为贵金属银的光度分析显色剂而得到应用。到70年代由于卟啉化合物与金属离子配合物Soret谱带的选用及表面活性剂的引入，试剂可溶性及其测定金属离子的灵敏度大大改善,卟啉试剂被广泛应用于光度法测定许多过渡金属离子。金属卟啉配合物可制备 ISE 膜,并可被电聚到硅氧烷、石墨、银电极上。Sun 等[94]阐述了应用 TPPS$_4$ 电位法测定食用海草中的碘,测定结果与光度法结果十分一致。

作为一类重要的分析试剂,卟啉也是超分子化学中极为活跃的一类环状化合物。卟啉及其金属配合物、类似物的超分子功能已应用于生物相关的物质分析,并展示了更加诱人的前景[95]。卟啉及其金属配合物种类多,分子具有刚性结构,周边官能团的位置和方向可以加以控制,且分子有较大表面,其轴向配体周围空间大小和相互作用方向的控制余地较大,故作为受体有显著优点,可以进行分子大小和形状、官能团和手性异构体识别[96]。利用卟啉周边官能团的氢键作用,可以识别糖分子[97]。卟啉环上连接有冠醚、二酰氨基吡啶等部位的受体,能对底物的大小和形状进行专一识别[98-104],卟啉和金属卟啉的分子识别作用在色谱分离中已得到广泛的应用[105]。将卟啉或金属卟啉固定在硅胶上,作为液相色谱固定相分离稠环芳烃[106]和富勒烯[107,108]时对溶质有很高的选择性,这是溶质与卟啉间强烈的 π—π 相互作用所致。也可利用中心金属离子与阴离子的选择性配位,将金属卟啉硅胶柱用于包括芳香羧酸和磺酸等多种阴离子的分离。利用芳香氨基酸与固定化卟啉间的 π—π 相互作用以及金属离子与氨基酸的配位作用,可实现肽的选择性分离[109-113]。

5. 卟啉在能源方面的应用

随着化石资源的枯竭以及环境污染的日趋严重,可持续性绿色能源及其利用转换技术的开发迫在眉睫。燃料电池和金属空气电池是高效、清洁的能源转换系统,它们在电动车上的发展潜力引起了国际上的广泛重视。燃料电池和金属空气电池的发展关键在于氧电极催化剂的不断更新。催化剂材料的研究一直备受关注。燃料电池常用的电催化剂均以铂金属为主,铂在低温燃料电池中是一种很好的电催化剂,但它价格昂贵,储量少,易被 CO 毒化(燃料气中存在的 CO 或直接甲醇阳极氧化时的中间产物),限制了铂作为电催化剂的应用。为了解决这一问题,世界各国科研工作者都在努力寻找非贵金属化合物来代替铂,例如:金属氧化物、合金和过渡金属大环化合物等。在酸性电解液中,金属卟啉是空气电极最有希望的催化剂,它能有效促进 H_2O_2 分解,使电池的工作电压提高,增加放电容量[114]。金属酞菁和金属卟啉配合物具有高的共轭结构和化学稳定性,对分子氧还原表现出良好的电催化活性,近几年来成为氧还原电催化剂的研究热点。

此外,在利用太阳能方面,卟啉的应用也很重要。人们研究光能转换主要是模拟自然界绿色植物的光合作用。目前研究较多的是卟啉和醌类体系,最成功的一个体系是 Moore 等人[115]把胡萝卜素分子、卟啉分子和苯醌分子连在一起,合成了一类化合物,在光照后可实现有效的电荷分离。在这个体系中,胡萝卜素分子具有天线功能,能吸收太阳光,成为单重激发态,单重态发生能量转移,生成卟啉单重激发态。此外,胡萝卜素分子对卟啉分子还具有保护作用,由于三重态卟啉对单重态氧具有较强的光敏作用,而胡萝卜素分子还可通过猝灭三重态卟啉,避免其与单重态氧作用,这个体系是目前最接近活体系统的人工光合反应模型。

6. 卟啉在地球化学中的应用

在富含有机质的沉积物如原油、油页岩中,存在复杂的金属卟啉混合物,已经分离鉴定出的具有不同碳骨架的卟啉被统称为"石油卟啉"[116-118]。石油卟啉由叶绿素、血红素降解得到,并基本完整地保存在古代沉积物中,因此又被称为生物标记物。石油卟啉多以镍、钒化合物的形式存在,比较稳定,因而成为地球化学的分子化石。按含侧链不同,石油卟啉可分为初卟啉(ETIP)和脱氧初卟啉(DPEP)。在石油成熟过程中,ETIP 和 DPEP 的比例发生变化,平均碳数减少。因此,分析石油卟啉的组成、类型、含量等,对获得地球化学信息有重要意义[119]。

除此之外,卟啉及金属卟啉化合物在主客体化学[120,121]、应用开环聚合催化[122]以及有机合成等领域也得到了广泛的研究和应用,而且随着对卟啉化合物研究的进一步深入,其应用必将越来越广泛。

参考文献

[1]Fujimoto K, Toyoshi T, Doi Y, et al. Synthesis and molecular recognition properties of a self-assembling molecule consisted of a porphyrin core and two hydrogen-bonding moieties. Materials Science and Engineering C, 2007, 27: 142-147.

[2]Purello R, Raudino A, Scolaro L M, et al. Ternary porphyrin aggregates and their chiral memory. J. Phys. Chem. B, 2001, 105: 2474-2474.

[3]Gust D, Moore T A, Moore A L, et al. Long-lived photoinitiated charge separation in carotene-diporphyrin triad molecules. J. Am. Chem. Soc., 1991, 113: 3638-3649.

[4]Shimidzu T, Segawa H. Porphyrin arrays connected with molecular wire. Thin Solid Films, 1996, 273: 14-19.

[5]Mizusekia H, Belosludova R V, Farajian A A, et al. Molecular orbital analysis of frontier orbitals for molecular electronics: a case study of unimolecular rectifier and photovoltaic cell. Science and Technology of Advanced Materials, 2003, 4: 377-382.

[6]Hasobe T, Kamat P V, Troiani V, et al. Enhancement of light-energy conversion efficiency by multi-porphyrin arrays of porphyrin-peptide oligomers with fullerene clusters. J. Phys. Chem. B, 2005, 109(1): 19.

[7]Marvaud V, Launay J P. Control of intramolecular electron transfer by protonation: oligomers of ruthenium porphyrins bridged by 4,4'-azopyridine. Inorg. Chem., 1993, 32: 1376-1382.

[8]Hasobe T, Kashiwagi Y, Absalom M A, et al. Supramolecular photovoltaic cells using porphyrin dendrimers and fullerene. Adv. Mater., 2004, 16(12): 975-979.

[9]Crossley M J, Burn P L. An approach to porphyrin-based molecular wires: synthesis of a bis(porphyrin)tetraone and its conversion to a linearly conjugated tetrakisporphyrin system. J. Chem. Soc. Chem. Commun., 1991, 1569-1571.

[10]Kawao M, Ozawa H, Tanaka H, et al. Synthesis and self-assembly of novel porphyrin molecular wires. Thin Solid Films, 2006, 499: 23-28.

[11]Gregg B A, Fox M A, Bard A J. 2, 3, 7, 8, 12, 13, 17, 18 - Octakis (beta-hydroxyethyl)porphyrin (octaethanolporphyrin) and its liquid crystalline derivatives:synthesis and characterization. J. Am. Chem. Soc., 1989, 111: 3024-3029.

[12]Bruce D W. Calamitic nematic liquid crystal phases from ZnIIcomplexes of 5,15-disubstituted porphyrins. J. Chem. Soc. Chem. Commun., 1994, 18: 2089-2090.

[13]Shackletle L W, Elsenbacemer R L, Chance R R, et al. Electrochemical doping of poly-(p-phenylene) with application to organic batteries. J. Chem. Soc. Chem. Commun., 1982, 6: 361-362.

[14]H Anderson L, Martin S J, Bradly D D C. Synthesis and third-order nonlinear optical

properties of a conjugated porphyrin polymer. Angew. Chem. Int. Ed. Engl., 1994, 33: 655–657.

[15]Tagami K, Tsukada M. Electronic transport through tape-porphyrin molecular bridges. Thin Solid Films, 2004, 464–465: 429–432.

[16]Pandey R K, Shiau F Y, Medforth C J, et al. Efficient synthesis of porphyrin dimers with carbon-carbon linkages. Tetrahedron Lett., 1990, 31: 789–792.

[17]Angelini N, Micali N, Mineo P, et al. Uncharged water-soluble Co(II)-porphyrin:a receptor for aromaticα-amino acids. J. Phys. Chem. B, 2005, 109: 18645–18651.

[18]Fukuzumi S, Ohkubo K, Wenbo E, et al. Metal-centered photoinduced electron transfer reduction of a gold (III) porphyrin cation linked with a zinc porphyrin to produce a long-lived charge-separated state in nonpolar solvents. J. Am. Chem. Soc., 2003, 125(49): 14984–14985.

[19]Aiello I, Donna L Di, Ghedini M, et al. Charge-transfer matrixes as a tool to desorb intact labile molecules by matrix-assisted laser desorption/ionization.use of 2,7-dimethoxynaphthalene in the ionization of polymetallic porphyrins. Anal. Chem., 2004, 76: 5985–5989.

[20]Ohta N, Iwaki Y, Ito T, et al. Photoinduced charge transfer along a meso, meso-linked porphyrin Array. J. Phys. Chem. B, 1999, 103: 11242–11245.

[21]Yang S I, Prathapan S, Miller M A, et al. Synthesis and excited-state photodynamics in perylene-porphyrin dyads 2. effects of porphyrin metalation state on the energy-transfer, charge-transfer, and deactivation channels. J. Phys. Chem. B, 2001, 105: 8249–8258.

[22]Adler A D, Sklar L, Longo F R. A mechanistic study of synthesis of meso-tetraphenylporphin. J. Heterocycl Chem., 1968, 5: 669–670.

[23]潘继刚,何明威,刘轻轻.四苯基卟啉及其衍生物的合成.有机化学,1993,13: 533–536.

[24]朱宝库,徐志康,徐又一.四(4-硝基苯基)卟啉和四(4-氨基苯基)卟啉的合成.应用化学,1999,16(1): 58–60.

[25]郑维忠,曾庆平,王先元.高分子键联金属卟啉的合成及催化性能的研究.有机化学,1995,15: 520–524.

[26]Lavallee D K, Xu Z, Pina R. Synthesis and properties of new cationic-periphery porphyrins, tetrakis (p-(aminomethyl)phenyl) porphyrin and N-methyltetrakis (p-(aminomethyl) phenyl) porphyrin. J. Org. Chem., 1993, 58: 6000.

[27]Lindsey J S, Schreiman I C, Hsu H C, et al. Rothemund and Adler-Longo reactions revisited:synthesis of tetraphenylporphyrins under equilibrium conditions. J. Org. Chem., 1987, 52: 827–836.

[28]Montanari F, Casella L. Metallopophyrins Catalyzed Oxidations. Kluwer Academic Publishers, 1994.

[29]Dolphin D. Porphyrinogens and porphodimethenes, intermediates in the synthesis of meso-tetra-phenylporphins from pyrroles and benzaldehyde. J. Heterocycl. Chem., 1970, 7: 275–283.

[30]Lindsey J S, Wagner R W. Investigation of the synthesis of ortho-substituted tetraphenylporphyrins. J. Org. Chem., 1989, 54: 828–836.

[31]Barkigia K M, Chantranupong L, Smith K M, et al. Structural and theoretical models

of photosynthetic chromophores. Implications for redox, light-absorption properties and vectorial electron flow. J. Am. Chem. Soc., 1988, 110: 7566-7567.

[32] Groves J T, Nemo T E. Aliphatic hydroxylation catalyzed by iron porphyrin complexes. J. Am. Chem. Soc., 1983, 105: 6243-6248.

[33] Gradillas A, Del Campo C, Sinisterra J V, et al. Novel synthesis of 5,10,15,20-tetraarylporphyrins using high-valent transition metal salts. J. Chem. Soc., Perkin Trans., 1995, 2611-2613.

[34] Arsenault G P, Bullock E, MacDonald S F. Pyrromethanes and porphyrins therefrom. J. Am. Chem. Soc., 1960, 82: 4384-4379.

[35] Fox S, Hudson R, Boyle R W. Use of orthoesters in the synthesis of meso-substituted porphyrins. Tetrahed. Lett., 2003, 44: 1183-1185.

[36] Wallace D M, Leung S H, Senge M O, et al. Rational tetraarylporphyrin syntheses: tetraarylporphyrins from the MacDonald route. J. Org. Chem., 1993, 58: 7245.

[37] Wallace D M, Smith K M. Stepwise syntheses of unsymmetrical tetra-arylporphyrins. Adaptation of the macdonald dipyrrole self-condensation methodology. Tetrahed. Lett., 1990, 31: 7265-7268.

[38] Abdalmuhdi I, Chang C K. A novel synthesis of triple-deckered triporphyrin. J. Org. Chem., 1985, 50: 411-413.

[39] Clarke O J, Boyle R W. Selective synthesis of asymmetrically substituted 5,15-diphenylporphyrins. Tetrahed. Lett., 1998, 39: 7167-7168.

[40] Ravikanth M, Ptrachan S J, Li F, et al. Trans-substituted porphyrin building blocks bearing iodo and ethynyl groups for applications in bioorganic and materials chemistry. Tetrahedron, 1998, 54: 7721-7734.

[41] 李广年, 邹胜德, 滕有为. 5,15-二(对-取代苯基)八烷基卟啉及其金属配合物的合成. 有机化学, 1985, 5: 300-305.

[42] Broadhurst M J, Grigg R, Johnson A W. Synthesis of porphin analogues containing furan and/or thiophen rings. J. Chem. Soc.(C), 1971, 3681-3690.

[43] Boudif A, Momenteau M. A new convergent method for porphyrin synthesis based on a "3+1" condensation. J. Chem. Soc., Perkin Trans., 1996, 1, 1235-1242.

[44] Senge M O. Synthetic access to 5,10-disubstituted porphyrins. Tetrahedron Letters, 2003, 44: 157-160.

[45] Lash T D. Porphyrins with exocyclic rings. Part 9. Synthesis of porphyrins by the "3+1" approach. J. Porphyrins Phthalocyanines, 1997, 1: 29-44.

[46] Aoyagi K, Toi H, Aoyama Y, et al. Facile syntheses of perfluoroalkylporphyrins-electron deficient porphyrins. Chem. Lett., 1988, 2, 1891-1894.

[47] Hin P Y, Wijesekern T, Dolphin D. An efficient route to vinylporphyrins. Can. J. Chem., 1990, 68: 1867-1875.

[48] Petit A, Loupy A, Maiuardb P, et al. Microwave irradiation in dry media a new and easy method for synthesis of tetrapyrrolic compounds. Synthetic Communications, 1992, 22(8): 1137-1142.

[49] 谭迪, 杨克儿, 佟珊玲, 等. 无溶剂微波法合成meso-苯基四苯并卟啉锌. 化学与生物工程, 2006, 23(7): 51-53.

[50] Buchler J W, Folz M, Habets H, et al. Nucleation theory and dynamics of first-order phase transitions near a critical point. Chem. Ber., 1976, 109: 1477-1485.

[51] Fuhrhop J H, Smith K M, in Porphyrins and Metallporphyrins. Amsterdam: Elsevier, 1975.

[52] Fischer H, Neumann W. Neumann, über einige Derivate vona Atioporphyrin I. Justus Liebigs. Ann. Chem., 1932, 494: 225-245.

[53] Buchler J W, Puppe L. Metallkomplexe mit tetrapyrrol-liganden, x 1). synthese und konfiguration weiterer. Justus Liebigs Ann Chem., 1974, 1: 1046-1062.

[54] Tsutsui M, Hrung C P, Ostfeld D, et al. An iron-bridged diborane (6) derivative. Preparation of $K^+[-Fe(CO)_4B_2H_5]^-$. J. Am. Chem. Soc., 1975, 75: 3953-3954.

[55] Adler A D, Longo F R, Kampas F, et al. On the preparation of metalloporphyrins. J. Inorg. Nucl. Chem., 1970, 32: 2443-2445.

[56] 汪进, 白永平, 魏月贞, 等. 八乙基卟啉合成方法的改进. 化学世界, 2000, (01): 27-30.

[57] Crossley M J, Burn P L. An approach to porphyrin-based molecular wires:synthesis of a bis(porphyrin) tetraone and its conversion to a linearly conjugated tetrakisporphyrin system. J. Chem. Soc., Chem. Commun., 1991, 21: 1569-1571.

[58] 金志平, 彭孝军, 孙立成. 卟啉超分子化合物在分子器件中的应用. 化学通报, 2003, (07): 464-473.

[59] Parthenopoulos D A, Rentzepis P M. Three-dimensional optical strorage memory. Science, 1989, 245, 843-845.

[60] Lieberman K, Harush S, Lewis A, et al. A light source smaller than the optical wavelength. Science, 1990, 247: 59-61.

[61] Feldstein M J, Vorhringer P, Wang W, et al. Femtosecond optical spectroscopy and scanning probe microscopy. J. Phys. Chem., 1996, 100: 4739-4748.

[62] Yang S I, Seth J, Balasubramanian T, et al. Interplay of orbital tuning and linker lo cation in controlling electronic communication in porphyrin-based nanostructures. J. Am. Chem. Soc., 1999, 121: 4008-4018.

[63] Li F, Yang S I, Seth J, et al. Design, synthesis, and photodynamics of light-harvesting arrays comprised of a porphyrin and one, two or eight boron-dipyrromethene accessory pigments. J. Am. Chem. Soc., 1998, 120: 10001-10017.

[64] Yang S I, Lammi R K, Riggs J A, et al. Excited-state energy transfer and ground-state hole/electron hopping in p-phenylene-linked porphyrin dimers. J. Phys. Chem. B, 1998, 102: 9426-9436.

[65] Gust D, Moore T A, Moore A L. Mimicking photosynthetic solar energy transduction. Acc. Chem. Res., 2001, 34: 40-48.

[66] Chio M S, Aida T, Yamazaki T, et al. A Large dendritic multiporphyrin array as a mimic of the bacterial light-harvesting antenna complex: molecular design of an efficient energy funnel for visible photons. Angew. Chem. Int. Ed., 2001, 40(17): 3194-3198.

[67] Gust D, Moore T A. Molecular mimicry of photosynthetic energy and electron transfer. Acc. Chem. Res., 1993, 26: 198-205.

[68] Birge R. Nanotchnology research and perspectives. Cambridge, MA: MIT Press,

1992.

[69] Wagner R W, Lindsey J S, Seth J, et al. Molecular optoelectronic gates. J. Am. Chem. Soc., 1996, 118: 3996-3997.

[70] Yang S I, Seth J, Balasubramanian T, et al. Templated synthesis, excited-state photodynamics, and studies of electronic communication in a hexameric wheel of porphyrins. J. Am. Chem. Soc., 1999, 121, 4008-4018.

[71] de Silva A P, Gunaratne H Q N, Gunnlaugsson T, et al. Signaling recognition events with fluorescent sensors and switches. Chem. Rev., 1997, 97: 1515-1566.

[72] O'Nelil M P, Niemczky M P, Wasielewski M R, et al. Complexation of a tris-bipyridine cryptand with americium(III). Science, 1993, 25: 1115-1117.

[73] Ali H, van Lier J E. Metal complexes as photo-and radiosensitizers. Chem. Rev., 1999, 99: 2379-2450.

[74] Bonnett R. Photosensitizers of the porphyrin and phthalocyanine series for photodynamic therapy. Chem. Soc. Rew., 1995, 24: 19-33.

[75] Schwartz S, Absolon K, Vermound H. Some relationship of porphyrins, X-rays and tumors. Univ. Minnesota Med. Bull., 1995, 27: 1-37.

[76] Lipson R L, Baldes A M, Olsen. The use of a derivative of hematoporphyrin in tumor detection. J. Natl. Cancer. Inst., 1961, 26: 1-11.

[77] Mang T, Dougherty T. Characterization of intra-tumoral porphyrin following injection of hematoporphyrin derivative or its purified. Photochem. Photobiol., 1987, 46: 67-70.

[78] 马金石. 卟啉类第二代光敏剂的发展. 感光科学与光化学, 2002, 20(2): 131-148.

[79] 余建鑫, 许德余. 苯并卟啉羟基和醚类衍生物的简便合成. 有机化学, 1999, 19(5): 475-478.

[80] Choma C T, Kaestle K, Akerfeldt K S, et al. A general method for coupling unprotected peptides to bromoacetamido porphyrin templates. Tetra. Lett., 1994, 35(34): 6191-6194.

[81] Gibson S L, Nguyen M L, Havens J J, et al. δ-Aminolaevulinic acid-induced photodynamic therapy inhibits protoporphyrin IX biosynthesis and reduces subsequent treatment efficacy in vitro. Biochem. Biophys. Res. Commun., 1999, 265(2): 315-321.

[82] Barinaga M. Peptide-guided cancer drugs show promise in mice. Science, 1998, 279 (5349): 323-324.

[83] Arap W, Pasqualini R, Ruoslahti E. Cancer treatment by targeted drug delivery to tumor vasculature in a mouse model. Science, 1998, 279(5349): 377-380.

[84] 金晓敏, 吴健. 卟啉类光敏药物的研究进展. 中国药物化学杂志, 2002, 12(1): 52-56.

[85] Sol V, Blais J C, Bolbach G, et al. Toward glycosylated peptidic porphyrins: a new strategy for PDT? Tetra. Lett., 1997, 38(36): 6391-6394.

[86] Sol V, Blais J C, Carre V, et al. Synthesis, spectroscopy and photocytotoxycity of glycosylated amino acid porphyrin derivatives as promising molecules for cancer therapy. J. Org. Chem., 1999, 64(12): 4431-4444.

[87] Cornia M, Menozzi M, Ragg E, et al. Synthesis and utility of novel c-meso-glycosylated metalloporphyrins. Tetrahedron, 2000, 56(24): 3977-3983.

[88] Matile S, Berova N, Nakanishi K, et al. Structural Studies by Exciton Coupled

Circular Dichroism over a Large Distance: Porphyrin Derivatives of Steroids, Dimeric Steroids, and Brevetoxin B. J. Am. Chem. Soc., 1996, 118: 5198-5206.

[89]Bonnett R. Photosensitizers of the porphyrin and phthalocyanine series for photodynamic therapy. Chem. Soc. Rev., 1995, 24: 19-33.

[90]Bonnett R. Photodynamic therapy in historical perspective. Rev. Contemp. Pharmacother, 1999, 10: 1-17.

[91]Phillips D. The photochemistry of sensitisers for photodynamic therapy. Pure. Appl. Chem., 1995, 67: 117-126.

[92]邢其毅,徐瑞秋,周政,等. 基础有机化学. 北京: 高等教育出版社, 1994.

[93]Bedioui F, Devynck J, Bied-Charreton C. Immobilization of metalloporphyrins in electropolymerized films: design and applications. Acc. Chem. Res., 1995, 28: 30-36.

[94]Sun C, Zhao J, Xu H, et.al. Fabrication of a multilayer film electrode containing porphyrin and its application as a potentiometric sensor of iodide. Talanta, 1998, 46: 15-21.

[95]郭忠先,沈含熙. 卟啉及其类似物超分子功能的分析应用. 分析化学, 1998, 26(02): 226-233.

[96]刘海洋,黄锦汪,彭斌,等. 金属卟啉配合物的分子识别研究进展. 无机化学学报, 1997, 13(01): 1-10.

[97]Mizutani T, Murakami T, Matsumi N, et al. Molecular recognition of carbohydrates by functionalized zinc porphyrins. J. Chem. Soc., Chem. Commun., 1995, 1257-1258.

[98]Bonar-Law R P, Sanders J K M. Polyol recognition by a steroid-capped porphyrin. enhancement and modulation of misfit guest binding by added water or methanol. J. Am. Chem. Soc., 1995, 117: 259-271.

[99]Nagasaki T, Fujishima H, Takeuchi M, et al. Design and synthesis of a c4-symmetrical hard-soft ditopic metal receptor by calixarene-porphyrin coupling. J. Chem. Soc., Perkin Trans. 1995, 1: 1883-1888.

[100]Konishi K, Yahara K, Toshishige H, et al. A novel anion-binding chiral receptor based on a metalloporphyrin with molecular asymmetry. highly enantioselective recognition of amino acid derivatives. J. Am. Chem. Soc., 1994, 116: 1337-1344.

[101]Konishi K, Kimata S, Yoshida K, et al. 5-Exo or 6-endo exploring transition state geometries of 4-penten-1-oxyl radical ring closures. Angew. Chem. Int. Ed. Engl., 1996, 35: 2820-2823.

[102]Rudkevich D M, Verboon W, Reinhoudt D N. Capped biscalixarene-Zn-porphyrin: metalloreceptor with a rigid cavity. J. Org. Chem., 1995, 60, 6585-6587.

[103]Liang Y, Chang C K, Peng S M. Molecular recognition with C-clamp porphyrins: synthesis, structural, and complexation studies. J. Mol. Recognit., 1996, 9: 149-157.

[104]Alessio E, Macchi M, Heath S, et al. Synthesis and reactivity of Cyanotetrathiaful valenes: X-ray crystal structure of Cyanotetrathiafulvalene. Chem. Commun., 1996, 1411-1412.

[105]Xiao J, Kibbey C E, Coutant D E, et al. ChemInform abstract: immobilized porphyrins as versatile stationary phases in liquid chromatography. J. Liq. Chromatogr. Relat. Technol., 1996, 19: 2901-2932.

[106]Kibbey C E, Meyerhoff M E. Preparation and characterization of covalently bound tetraphenylporphyrin silica-gel stationary phases for reversed phase and anion-exchange

chromatography. Anal. Chem., 1993, 65: 2189-2196.

[107] Evans D R, Fackler N L P, Xie Z, et al. π-Arene/cation structure and bonding. solvation versus ligand binding in iron (ⅲ) tetraphenylporphyrin complexes of benzene, toluene, p-xylene, and [60]fullerene. J. Am. Chem. Soc., 1999, 121(37): 8466-8474.

[108] Sun D, Tham F S, Reed C A. Supramolecular fullerene-porphyrin chemistry. fullerene complexation by metalated "jaws porphyrin" hosts. J. Am. Chem. Soc., 2002, 124(23): 6604-6612.

[109] Borovkov V V, Lintuluoto J M, Sugeta H, et al. Supramolecular chirogenesis in zinc porphyrins: equilibria, binding properties, and thermodynamics. J. Am. Chem. Soc., 2002, 124(12): 2993-3006.

[110] Huang X, Fujioka N, Pescitelli G, et al. Absolute configurational assignments of secondary amines by CD sensitive dimeric zinc porphyrin host. J. Am. Chem. Soc., 2002, 124(35): 10320-10335.

[111] Sirish M, Chertkov V A, Schneider H J. Porphyrin-based peptide receptors: synthesis and NMR analysis. Chem. Eur. J., 2002, 8(5): 1181.

[112] Imai H, Munakata H, Uemori Y, et al. Chiral recognition of amino acids and dipeptides by a water-soluble zinc porphyrin. Inorg. Chem., 2004, 43(4): 1211-1213.

[113] 罗国添,刘海洋,黄锦汪,等.手性氨基酸尾式卟啉锌配合物对氨基酸酯的手性分子识别.中山大学学报,1997,36(4): 125-126.

[114] 黄庆华,李振亚,王为.电池用氧电极催化剂的研究现状.电源技术,2003,27: 241-244.

[115] Moore T A. Photodriven change separation in a carotenporphyrin-quinonetriad. Nature, 1984, 307: 630.

[116] Baker E W, Palmer S E. The Porphyrins. New York: Acedemic Press, 1978.

[117] Callot H J. The Chlorophylls. Boca Raton, FL: CRC Press, 1991.

[118] Filby R H, Van Berkel G J. Metal complexes in fossil fuels geochemistry, characterization, and processing. Washington DC: ACS, 1987.

[119] 陈培榕,王志杰,郭建林,等.薄层层析—干柱层析技术在分离制备石油生物标志物——卟啉中的应用.分析实验室,1997,16(2): 39-41.

[120] 刘彦钦,韩士田,杨秋清,等.尾式卟啉-吡啶(三乙铵)季铵盐的合成.合成化学,2000,8(2): 8-10.

[121] 王景桃,丁西明,许惠君.不对称卟啉化合物的合成及其与氨基酸甲酯的作用.感光科学与光化学,1998,16(4): 331-337.

[122] Gollapalli R D, Francis D S. Molecular recognition directed porphyrin chemosensor for selective detection of nicotine and cotinine. Chem. Commun., 2000, 1915-1916.

第七章 卟啉非线性光学材料研究

在激光发明之前,理论认为电磁波辐射场在介质中传播时,只与麦克斯韦方程中场强的一次项有关,并用该理论成功地解释了光的折射、散射、双折射等线性光学现象。但激光出现以后,人们发现与普通的光源相比,激光是一种高强度、单色性和相干性好的光源,介质在强激光场的作用下,产生的极化强度与入射场强之间不再是简单的线性关系,而与场强的二次、三次或更高次项有关。

1961 年,Franken 等在石英晶体中观察到倍频效应;1962 年,Bass 等人在三甘氨酸硫酸盐晶体中观察到倍频效应;Woodbury 和 Ng 在甲苯中发现了后来被证实的受激拉曼散射。这是早期发现的三个基本的非线性光学效应。同年,Bloembergen 建立了以介质极化和耦合波方程为基础的非线性光学理论,并于 1965 年出版了《Nonlinear Optical Phenomena》一书。非线性光学在激光发明以后迅速发展起来,它所揭示的大量新现象极大地丰富了非线性物理学的内容。1965—1985 年是非线性光学发展成熟的阶段。这一时期,半导体非线性光学得到发展,并通过实验发现了光学双稳、光学混沌及光学压缩态,促进了光计算和量子光学的发展。1984 年,为总结这个阶段的研究成果,非线性光学专家沈元壤出版了《The Principles of Nonlinear Optics》一书。1985 年至 20 世纪末是非线性光学的初步应用阶段。新型非线性光学晶体的提出,推动了非线性光子晶体理论与器件的研究;有机激发态非线性光学材料、光纤放大器及 DWDM 光通信技术的发展,对光限幅器、波长转换器、拉曼放大器、光开关等非线性光学器件提出了需求;光纤放大器的发明及光孤子通信和远程量子信息传输实验的完成,促使光纤通信、孤子通信及量子通信技术得到了进一步发展[1]。迄今为止,非线性光学的理论研究已经比较成熟,但在光学与光子技术上的应用还有待进一步研究发展。

卟啉类化合物是得到广泛研究的典型大环 π 共轭体系有机非线性光限幅材料之一。卟啉类物质由于取代基团的不同而有许多种衍生物,当吡咯质子被金属离子取代后即为金属卟啉。由于卟吩分子是平面结构,环上的碳、氮之间采用 sp^2 杂化,剩余的一个 p 轨道被单电子或孤对电子占有,碳和吡咯型的氮之间都提供单个 π 电子,同氢连接共轭的氮原子提供 2 个 π 电子。因此卟啉分子是具有大 π 键的大分子,有较强的方向性。其电子离域程度大,三阶非线性响应特性非常显著,而高的三阶非线性响应特性为其作为光限幅材料提供了基础。

研究表明,卟啉的光限幅效应不仅体现在可见光区,而且通过对其结构的调整可到达近红外区,可用于对宽波段可调谐激光武器的防护;另外,它还具有限幅效果明显、响应速度快的优点,尤其适合对高能量密度激光的防护。因此,卟啉类材料已成为非线性光学领域研究的热点,而且最近研究发现卟啉中心金属的存在可极大提高三阶非线性光学性质。

第一节 非线性光学理论

非线性光学是非线性物理学的一个分支。非线性物理学是研究在物质间宏观强相互作用下普遍存在着的非线性现象,也就是作用和响应之间的关系是非线性的现象。非线性物理现象包含在物理学的各个领域,形成了非线性力学、非线性声学、非线性热学、非线性电子学以及非线性光学等学科。

1. 非线性光学原理

非线性光学是描述强光与物质发生相互作用的规律,即激光与光学介质之间的非线性作用产生了非线性光学效应。表7-1给出了非线性光学与线性光学的主要区别。非线性光学的基本原理是指强光(如激光)照射介质时,介质分子的极化不再与外加电场呈线性关系,而是按泰勒级数展开:

$$p=\alpha E+\beta EE+\gamma EEE+\cdots\cdots \tag{7-1}$$

相应介质的电极化表示为:

$$P=\chi^{(1)}E+\chi^{(2)}EE+\chi^{(3)}EEE+\cdots\cdots \tag{7-2}$$

p、P分别表示分子、介质的电极化强度,α和$\chi^{(1)}$分别是分子、介质的线性光学系数,β、γ分别为分子二、三阶非线性光学系数或三、四阶张量,$\chi^{(n)}(n\geq 2)$为介质非线性极化率,或$n+1$阶张量,E为电场强度。$\chi^{(2)}$为三阶张量,只有当材料呈非中心对称时才不全为零,即材料在宏观上要求非中心对称。三阶非线性光学效应有四个光电场相互作用,因而比二阶非线性效应丰富得多,除了谐波的产生、和频、差频等混合外,还有其他相互作用过程:

(1)二阶效应中一般产生的信号光频率异于入射光频率,而在三阶效应中产生的信号光频率可以等于某一入射光频率。

(2)不同种类的三阶效应反映了不同的三阶非线性极化率,可以根据需要选择共振和非共振效应,从而在实际中得到了更广泛的应用。

(3)三阶效应可以发生在只有一个入射光电场频率的情况下,产生的效应也只对应于该入射光电场的频率,这种效应可使介质的折射率发生变化,也可以产生一个具有独特相位特性的新的出射光。

(4)只有缺乏中心对称的介质才存在二阶非线性光学效应,但三阶非线性效应可以发生在所有介质中。

表7-1 非线性光学与线性光学的主要区别

线性光学	非线性光学
光在介质中传播,通过干涉、衍射、折射可以改变光的空间能量分布和传播方向,但与介质不发生能量交换,不改变光的频率	一定频率的入射光,可以通过与介质的相互作用而转变成其他频率的光(倍频等),还可以产生一系列在光谱上周期分布的不同频率和光强的光(受激拉曼散射等)
多光束在介质中传播,不发生能量相互交换,不改变各自的频率	多光束在介质中交叉传播,可能发生能量相互转移,改变各自的频率或产生新的频率(三波或四波混频)
光与介质相互作用,不改变介质的物理参量,这些物理参量只是光频的函数,与光场强度变化无关	光与介质相互作用,介质的物理参量如极化率、吸收系数、折射率等是光场强度的函数(非线性吸收和色散、光克尔效应、自聚焦)
光束通过光学系统,入射光强与透射光强之间一般呈线性关系	光束通过光学系统,入射光强与透射光强之间呈非线性关系,从而实现光开关(光限幅、光学双稳、各种干涉仪开关)
多光束在介质中交叉传播,各光束的相位信息彼此不能相互传递	光束之间可以相互传递相位信息,而且两束光的相位可以互相共轭(光学相位共轭)

按照激光与介质的相互作用,可以把非线性光学效应分为以下两类:

一类是被动非线性光学效应。其特点是光与介质间无能量交换,而不同频率的光波间能够发生能量交换,例如倍频、三波混频、参量过程、四波混频、相位共轭等。

另一类是主动非线性光学效应。其特点是光与介质间会发生能量转换,介质的物理参量与光场强度有关,例如非线性吸收(饱和吸收、反饱和吸收、双光子吸收等)、非线性折射(光克尔效应、自聚焦与自散焦、折射率饱和与反饱和等)、非线性散射(受激拉曼散射、受激布里渊散射等)、光学双稳性、光限幅等。

2. 非线性光限幅效应与反饱和吸收

非线性光限幅效应是指激光入射到介质时,当入射光强达到一定值(阈值)后,输出光缓慢增加或不再增加。即当激光激发介质时,在低光强下,介质具有较高的线性透射率;而在高光强下,由于介质的非线性光学效应使透射率下降,简称为光限幅。对于理想的光限幅器,当入射光强超过阈值后,其输出光强将保持为常数。就非线性光学材料而言,对同一波长处的强光可以实现限幅作用,同时对弱光具有较高透明度。

自 1964 年首次报道了光限幅现象以来,基于非线性吸收、非线性折射、非线性散射、非线性反射等非线性光学过程的光限幅效应得到了广泛的研究,其中反饱和吸收光限幅材料是目前各种非线性光限幅材料中研究最多和最深入的一种[2]。

光吸收是光与物质相互作用的一种最基本的方式,光辐射、光散射和光致折射率变化过程都与光吸收有密切关系。一束单色光通过厚度为 L 的介质,其透射率:

$$T = I/I_0 = \exp(-aL) \tag{7-3}$$

其中 a 是吸收系数。弱光作用下,a 为一常数,称为线性吸收。单光子过程有两种相反的效应:吸收系数随入射光强增加而减小(即透射率增加),称为饱和吸收(SA);吸收系数随入射光强增加而增加(即透射率减小),称为反饱和吸收(RSA)。饱和吸收和反饱和吸收都是由于激发态吸收不同于基态吸收而引起的。饱和吸收与基态非线性吸收过程有关。当物质体系吸收入射激光时,它会跃迁到激发态。如果从基态跃迁到激发态的粒子数相当多,而激发态又不对光产生吸收,会出现饱和吸收效应。对于二能级系统,当入射光强增大达到饱和吸收时,从基态跃迁到第一激发态的粒子数和从第一激发态跃迁到基态的粒子数相等。有些分子存在单重态和三重态两个激发态能级。基态能级的粒子跃迁到第一激发态后,很快弛豫到三重态激发态能级,由于三重态激发态能级的寿命很长,在强光作用下,大量粒子积累在三重态激发态能级上,结果达到饱和。

1967 年,Gulllano 等人在研究瓷染料及其同类时最早发现反饱和吸收现象。此后,大量反饱和吸收化合物如酞菁[3-5]、五氮齿[6]和富勒烯[7,8]等被发现,使反饱和吸收光限幅在光子学器件中得到极大应用,推动了材料反饱和吸收特性的深入研究。

反饱和吸收主要是由激发态能级非线性引起的,一般用五能级模型解释。如图 7-1 所示,在入射光作用下,分子吸收光子跃迁到单重态第一激发态振动能级上,并迅速弛豫到单重态第一激发态最低能级,处于单重态第一激发态最低能级的分子可经系间跃迁到三重态 T_1,在 S_1 和 T_1 的分子可再吸收光子激发到 S_2 和 T_2。如果 S_1 吸收截面大于 S_0 吸收截面或者 T_1 吸收截面大于 S_0 吸收截面,就有可能实现反饱和吸收。这种模型适用于长脉冲或者连续激光与介质的作用。在皮秒时域,由 S_1 跃迁到 T_1 的时间远大于脉冲持续时间,在脉冲作用时间内,第一激发态单重态的电子来不及跃迁到第一激发态三重态,此时对反饱和吸收起作用的主要是第一激发态单重态。

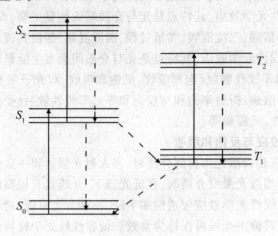

图 7-1 反饱和吸收五能级模型图

3. 非线性光限幅材料

随着非线性光学现象和理论研究的日益深入，非线性光学材料的研究也得到了蓬勃的发展。具有非线性光学性质的材料有许多种，不同材料的非线性机制各不相同，见表 7-2。早期研究的非线性材料主要以无机晶体为主，如铌酸锂、磷酸二氢钾、β-偏硼酸钡等。由于高质量的单晶价格昂贵、难以生长，因此无法满足迅速发展的光通信、光信号处理所需的高容量、高速度、高频宽和易加工性等一系列要求，在实际应用中受到极大限制[9,10]。

表 7-2 非线性光学材料及其非线性机制

非线性光学材料	非线性机制
半导体材料	电子机制、激子机制
有机高分子材料	电子、分子极化与分子取向
电光晶体	外电光效应
光折变材料	内电光效应
液晶材料	分子取向
团簇材料（C_{60}等）	分子极化
纳米复合材料	表面等离子体激元
手性分子材料	分子电、磁极距
等离子体	电子、离子

目前非线性光学材料研究的发展趋势是从晶体材料到非晶体材料；从无机材料到有机材料；从对称材料到非对称材料（手性材料）；从单一材料到复合材料；从高维材料到低维材料，如从三维的体块到二维的表面、薄膜材料；从宏观材料到纳米材料，如导体量子线和量子点；光子晶体以及纳米管、纳米球和团簇材料等。光限幅材料研究的发展可以分为两个主要阶段：自 1969 年 Geusie 发现硅的光限幅特性至 20 世纪 90 年代以前，是光限幅材料的理论发展阶段，这一阶段以半导体材料研究为主体。而 90 年代以后，以含大 π 共轭体系有机物的研究为主体。这类材料具有非线性系数大，响应速度快，光限幅阈值低，激光损伤阈值高等特点，而且光限幅波段从可见光扩展到近红外，这些优点是半导体材料所不及的，从而使光限幅材料的研究进入一个新的发展时期。目前研究的非线性光限幅材料主要有以下几种：

3.1 半导体材料

半导体材料有较大的非线性三阶极化率,其限幅机理为自散焦(或自聚焦)。即通过光致折射率变化使光束发散(或会聚)来降低输出光的强度,利用介质非线性折射的特性引起限幅作用。其缺点是防护波段较窄,不能防护可调谐激光器。

3.2 富勒烯及其衍生物

C_{60} 因其独特的结构而表现出较强的反饱和吸收性能和强的光限幅效应,目前已成为一种基准的光限幅材料。由于 C_{60} 本身的溶解度较低,但可以通过化学修饰引入取代基形成 C_{60} 衍生物,这使其溶解性大大增强[11-14]。同时也可通过改变电子结构,从而使 C_{60} 分子的光学特性得以改变。

3.3 无机过渡金属簇合物

无机过渡金属团簇可表现出强的非线性光学特性以及对可见光波段有较低的线性吸收,可实现较强的光限幅特性[15,16]。例如,金属与硫形成的簇合物的 $\chi^{(3)}$ 和 γ 值都较大,在 $10^{-12}\sim10^{-9}$ esu 和 $10^{-32}\sim10^{-28}$ esu 之间,且依赖于簇合物的组成、结构及配体的种类等因素。如 Cu—S—M 簇合物的光限幅性能随配体中卤素相对原子质量的增大而增强[17]。

3.4 有机/无机纳米材料

当有机小分子形成有机纳米材料后,其光学性质(吸收光谱、荧光发射光谱等)表现出量子尺寸效应和量子限域效应,其三阶非线性光学性质也产生类似的效应[18-20]。另外,金、银等贵金属纳米材料也具有很强的光限幅性能,不仅表现出强的非线性散射效应,也具有高的非线性吸收效应[21]。

3.5 聚炔类高分子

聚炔类共轭高分子材料由于具有高度离域的π电子,因此具有很大的非线性光学系数[22-24]。聚乙炔、聚丁二炔衍生物在多数溶剂中具有很好的溶解性,且具有优异的加工性能,在 1064 nm 和 1500 nm 处都具有高的 $\chi^{(3)}$ 值。

3.6 有机材料

有机化合物的非线性系数要比已经使用的无机晶体高 1~2 个数量级,激光损伤阈值较高。有机非线性光限幅材料最突出的优点是能在分子的水平上进行结构设计,以取得最佳的光学非线性响应和特定的光电性质[25-27]。针对光限幅的应用目标主要是可见—近红外区激光,特别是 532 nm 和 1064 nm 的激光。目前有机光限幅材料主要是对在可见—近红外区具有强非线性吸收、折射或散射效应的分子结构与非线性光学性质间关系研究,从而作为制备新型光限幅材料的理论基础。

第二节 卟啉非线性光学材料的分子设计

1. 卟啉非线性光学材料的分子设计

卟啉及金属卟啉已被广泛应用于生物化学、分析化学、催化化学、药物化学、石油化学等众多领域。近十几年来,化学工作者们也相继研究了卟啉的非线性光学效应[28,29],由于卟啉为大环π电子离域体系,外环上可接任意取代基,中心金属可以改变,甚至扩展环的大小,即分子具有可修饰性,特别适合于非线性光学材料的分子设计,而且其热稳定性好,易于成膜,容易做成波导结构,因此在光电子技术和集成光学领域有着潜在应用价值。

卟啉分子呈中心对称，但其 5,10,15,20 位上可以连接不同取代基，而使分子为非中心对称结构，极化后材料在宏观上呈非中心对称，从而表现出二阶非线性。依据简单紧束模型[30]，计算出 $\chi^{(3)}$ 与电子离域长度 N_d、能隙 E_g 的关系，表示如下：

$$\chi^{(3)} \propto (N_d/E_g)^a \tag{7-4}$$

$a=6$，即电子离域长度越长、能隙越小，$\chi^{(3)}$ 越大。但用微扰法计算时，链长有限制[31]，当链长达到一定值后，非线性光学效应达到饱和。Brédas 研究其饱和性时，发现了如下关系[32]：

$$\gamma \propto N^{a(N)} \tag{7-5}$$

$$\lg\gamma \propto a(N)\lg N \tag{7-6}$$

N 为链方向单元数，幂 a 为 N 的函数。$\chi^{(3)}$ 除与链长有关外，还与分子结构、取代基、分子取向等因素有关。卟啉分子为大环离域 π 电子体系，外环上可接各种取代基，可以形成各种金属卟啉，与酞菁形成二元化合物，环的大小可以改变，可以聚合等，分子的可修饰性非常强，因而成为三阶非线性光学材料的有力候选。

2. 卟啉的二阶非线性光学效应

从双能级结构模型出发，分子为 D(给体)—π—A(受体)型有利于分子在电场作用下产生分子内电荷转移(CT)，从而产生二阶非线性光学效应。Kenneth 等人[33]在四苯基卟啉(TPP)的苯环上连上不同取代基，如硝基、氨基，形成了一系列具有给受体结构的推拉型卟啉，在 1.91 μm 处运用电场诱导二次谐波产生(EFISH)方法测得它们在 $CHCl_3$ 中的 β 值，并由 β 值说明有效分子内电荷转移程度越大，分子的 β 值越大，上述结构尽管表现出了二阶非线性光学效应，但与一维 π 电子体系相比，β 值较小，这是由于苯环与卟啉环之间存在扭曲，产生了二面角，因此阻碍了分子内电荷转移。受启发，Avijit 等[34]研究了用 F 取代苯的邻间位、—NMe_2 取代苯的对位、—NO_2 取代卟啉的 2 位给受体型卟啉和其金属卟啉，发现中心金属 Cu、Zn 有利于电荷转移，因而 β 值增大。为了解决二面角问题，Karki 等合成了推拉型 5,15-二芳炔基卟啉[35]，给体为—NMe_2，受体为—NO_2；Ming 等[36]也合成了类似结构卟啉，前者用超瑞利散射技术(HRS)测得 β(1064 nm)=4933×10^{-30} esu。Li 等还对卟啉膜作了研究，5,10,15,20-四吡啶基卟啉与硅氧化物表面层键合，形成了自组装单层膜体系，该体系宏观上非中心对称，且有大环 π 电子离域，其 SHG(二次谐波产生)响应 X_{zzz} 约为 10^{-8} esu。由于受制于仅有的几种底物，Sunao 等将四吡啶基卟啉固定在聚氯乙烯上定向形成两种不同的膜，Jiang 等[37]将四吡啶基卟啉—C_{60} 磺酸根复合物在脱乙酰壳多糖聚合物上通过静电自组装形成薄膜。

3. 卟啉的三阶非线性光学效应

与卟啉的二阶非线性光学效应相比，对其三阶非线性光学效应的研究深入得多，其材料大致可以概括为以下几类：TPP、TBP 及其金属卟啉、卟啉酞菁二元化合物、八乙基卟啉、低聚卟啉、三明治形卟啉、卟啉聚合物及其他。

3.1 TPP、TBP 类及其金属卟啉

人们为了发现结构与非线性之间的关系，更好地发挥分子工程学原理，对卟啉中 TPP 类进行了深入的研究，包括苯环上取代基对三阶非线性光学效应的影响、非线性机制及各种测试手段等的研究。因取代基能改变卟啉环上的电子云密度，其电子离域程度也随之改变，故能影响其三阶非线性光学效应。

而且，研究也发现吸电子基团对 $\chi^{(3)}$ 的影响顺序为：对位>邻位；推电子基团对 $\chi^{(3)}$ 的影响顺序为：邻位>对位>间位。这为探索取代基种类和适当取代位置来提高 $\chi^{(3)}$ 提供了方向。

人们对卟啉及金属卟啉的非线性机制也进行了研究，发现了其反饱和特性[38,39]，即从非平衡的激发态向更高能级的激发态跃迁大于第一激发态的跃迁，表现为吸收系数随激光光强增大而增大。

Toru 等还研究了金属卟啉的介晶性质，在苯环对位上引入长脂肪链，在这种介晶态中，激子在相邻分子中的迁移和分子迁移率将影响非线性光学过程。与 TPP 相似，TBP 与四芳炔基卟啉也有非线性，人们用 Z-扫描技术[40]、简并四波混频(DFWM)、光学克尔门技术研究它们，其中 5，10，15，20-四对甲氨基四苯并卟啉的 $\chi^{(3)}=2.8\times10^{-8}$ esu。

3.2 类卟啉

线性共轭分子的 $\gamma_{ijkl}(\omega_4;\omega_1,\omega_2,\omega_3)$ 大小与共轭链长相关，已被理论和实践所证实，由线性变为二维环状，尽管多个张量起作用，但维数增加却有减小 γ 的趋势，不过对于实现波导结构全光源型装置的应用，各张量对 $\chi^{(3)}$ 都有贡献是必需的，因此，类卟啉环上原子的变化和环的大小成了人们研究的对象[41,42]。

3.3 二元化合物及其他卟啉小分子

卟啉、酞菁各自都有较大的三阶非线性极化系数 $\chi^{(3)}$，其二元化合物也引起了广泛的兴趣[43,44]。由于卟啉、酞菁二元分子在激光作用下发生了电荷转移过程，造成卟啉、酞菁电荷分离，而产生了三阶非线性。

3.4 卟啉聚合物

长的共轭长度及低的能隙有助于增大 $\chi^{(3)}$，卟啉提供了较长的可极化 π 电子体系，在近 IR 区通过 DFWM 测量其 $\chi^{(3)}=7.3\times10^{-8}$ esu，而且成为第一个有光学活性的可溶性卟啉聚合物，基本上能满足实用要求[45]。除均聚物外，卟啉也能形成共聚物，其 $\chi^{(3)}$ 值明显大于各单体的 $\chi^{(3)}$ 值，卟啉衍生物的一种氮杂卟啉也可制备共聚物，表现出了较大的三阶非线性效应。卟啉聚合物作为较活跃的分支，除了要得到较大的 $\chi^{(3)}$ 值外，还要解决它的溶解性和加工性问题，Arun 等[46]将卟啉联结在聚硅氧烷上作为侧链，使以上问题得到了较为满意的结果，DFWM 测得其膜的 $\chi^{(3)}=3.25\times10^{-10}$ esu。卟啉掺杂在各种聚合物中形成薄膜，可以使光学均匀性及化学稳定性得到改善，也因此推动了它的研究。1991 年，人们就把叶绿素掺杂在聚合物中形成薄膜，非线性响应由基态到激发态的跃迁产生。1999 年，Dou 等人研究了四苯基卟啉磺酸钠掺杂于硅铝酸盐玻璃纤维凝胶体系[47]。最近，Rao[48] 对四苯基卟啉掺杂硼酸玻璃进行了实验，这表明对卟啉掺杂聚合物为非线性光学材料的研究越来越广泛。

卟啉主要应用于光限幅材料，其光限幅机制为激发态反饱和机制。另外还可应用于激光调制、光双稳、相位共轭、光开关。虽然卟啉非线性光学效应的研究取得了可喜的进展，但真正实际应用还需时日。因为卟啉还不能达到 $\chi^{(3)}\geq10^{-8}$ esu 的器件化要求，我们还需寻找更大 $\chi^{(3)}$ 值的卟啉化合物；另外在满足实际应用时需要综合考虑各种因素，除了 $\chi^{(3)}$ 之外，还有高的激光损伤阈值、超快响应时间、溶解性、透明性及加工性能等方面。因此有必要从以下两方面着手：

(1) 完善理论，包括分子设计、材料加工、器件化等，使其对卟啉的研究具有指导作用；

(2) 从材料的综合优化着手，这样就需要化学工作者与物理、材料等方面的工作者们相互促进。

第三节 卟啉光限幅材料研究进展

研究表明,金属卟啉配合物有其独特的优点:

(1)卟啉环具有刚性结构,周边官能团的方向和位置可较好地得到控制;

(2)卟啉分子有较大的表面,对金属卟啉分子轴向配体周围的空间容积和相互作用方向的控制余地较大;

(3)金属卟啉配合物具有多样性。

因此,通过改变中心金属离子、外围取代基以及卟啉的分子骨架等方法可以提高卟啉化合物的三阶非线性极化率$\chi^{(3)}$值,从而实现对光限幅性能的优化。近十几年来,科研工作者通过瞬态吸收技术、光克尔技术、简并四波混频(DFWM)和Z-扫描技术等实验手段对卟啉及金属卟啉配合物的光限幅性能进行了广泛研究。

1993年,Bhyranpa系统研究了以Zn^{2+}、Cd^{2+}、Cu^{2+}、Co^{2+}、Pb^{2+}为中心离子的八溴四苯基卟啉配合物的光限幅性能。结果表明,配合物的光限幅性能顺序为:$Zn^{2+}>Cd^{2+}>Pb^{2+}>Cu^{2+}\approx Co^{2+}$,这个顺序可以用中心离子的电子自旋态效应解释。$Cu^{2+}$和$Co^{2+}$为开壳层的顺磁性离子,d轨道有空轨道,由此产生的自旋—轨道耦合可导致三重激发态寿命的下降,进而降低了分子在此激发态的分布而减弱了非线性效应。但是,Zn^{2+}、Cd^{2+}、Pb^{2+}为闭壳层反磁性离子,没有上述的自旋—轨道耦合效应,所以具有较强的限幅效果。不过,由于没有自旋—轨道耦合效应,所以就观察不到酞菁化合物中存在的重原子效应。

在四甲苯基卟啉中也观察到与八溴四苯基卟啉类似的现象,在Zn^{2+}、Ni^{2+}、Cu^{2+}、Co^{2+}、Fe^{2+}和VO^{2+}等金属离子配位的TPP中,只有以Zn^{2+}和VO^{5+}为中心离子的配合物表现出与C_{60}相当的光限幅性能[49]。中心金属能有效增加该类分子的电子离域能力,可以提高四氮杂卟啉配合物对纳秒激光的光限幅能力,同时重原子效应有利于分子的系间窜跃,增加第一激发三重态的布居数密度,从而增强对纳秒激光的限幅能力。

2004年,研究者[50]对大分子有机化合物钴卟啉的非线性光学特性在457.19 nm、488 nm和514.15 nm三个连续激光波长下进行了研究。用单光束Z-扫描技术测得三个曲线均是先出现峰值后出现谷值,样品的非线性折射率为负值。实验结果表明,钴卟啉在三个波长下均具有光限幅特性,限幅阈值随着波长的减小而减小,其光学限幅效果好且限幅阈值低,存在较大潜在应用价值。2005年,Martin等[51]合成了两种简单金属卟啉衍生物,在532 nm处的纳秒激光脉冲下显示出优良的光限幅性能,其基态和激发态参数表明该卟啉衍生物为反饱和吸收模式。2006年,Sendhil等[52]利用四苯基锌卟啉的非线性光学性质设计出一种新型的低阈值光限幅器,并在632.8 nm氦氖激光上进行Z-扫描测得非线性折射率为1.4×10^{-7} cm^2/W,实验表明,理想阈值的光限幅可以有效地选择孔径大小和实验几何数值。

最近,研究者[53]利用纳秒脉冲激光下的Z-扫描技术对两种新型卟啉衍生物在450~490 nm范围内的反饱和吸收进行了研究。通过与四苯基卟啉比较发现,在Soret和Q吸收带之间,两种新型卟啉衍生物都具有较大的反饱和吸收。利用五能级系统模型对实验结果进行分析,并比较了各个样品在不同波长范围内的优势。结果表明,两种新型卟啉衍生物具有比四苯基卟啉更好的光限幅应用前景。研究者[54]从理论和实验两方面对新型卟啉化合物5,10,15,20-(对十六烷氧基苯基)卟啉的非线性吸收特性进行了研究。实验采用开孔Z-扫描方法，使用可调谐纳秒光参量振荡器(OPO)作为光源，获得了 710~820 nm 波长范围内的 Z-扫描曲线。研究表明，该卟啉化合物在较宽的光谱范围内具有较大的非线性吸收系数，是一种具有潜在应用价值的新型光限幅材料。2008 年，研究者[55]也研究了卟啉—蒽化合物的双光子和三光子吸收特性。结果表明，卟啉—蒽化合物具有很好的非线性光学吸收效应，其光学吸收谱可以覆盖很宽的波长范围，将卟啉—蒽光限幅材料用于航天器的光学器件，可以抵御远红外激光武器的攻击，提高航天器的生存能力以及空间系统的防御能力。

近年来，国内外研究者从实际应用的角度出发，通过物理、化学的方法将卟啉衍生物引入固体基质中，实现对其性能的综合优化。常用的方法有溶胶—凝胶法、接枝聚合法、旋涂成膜法、悬浮成膜法及提拉成膜法等。在以上成膜方法中，经测试，提拉成膜的反饱和吸收效果最佳，但存在分散程度差、薄膜牢固性差等缺点。为克服以上缺点，也有研究者采用以玻璃为基体的自组装技术，提高分子成膜的性能[56]。

目前对非线性光学规律的研究趋势是：从稳态转向动态；所用光源从连续、宽脉冲转向纳秒、皮秒及飞秒甚至阿秒超短脉冲；从强光非线性的研究转向弱光非线性的研究；从基态—激发态跃迁非线性光学研究转向激发态—更高激发态跃迁非线性光学研究；从研究共振峰处的现象转向研究非共振区的现象；从二能级模型的研究转向多能级模型的研究；研究物质的尺度从宏观尺度(衍射光学)到介观(纳米)尺度(近场光学)，再到微观尺度(量子光学)。

参考文献

[1] 李淳飞. 非线性光学. 哈尔滨：哈尔滨工业大学出版社，2005.

[2] 张正华，徐伟箭，刘含茂. 反饱和吸收光限幅效应与激光防护. 精细化工中间体，2005，35(5): 51-54.

[3] Su Xinyan, Xu Hongyao, Yang Junyi, et al. Soluble functional polyacetylenes for optical limiting: relationship between optical limiting properties and molecular structure. Poly. Lett., 2008, 49(17): 3722-3730.

[4] Unnikrishnan K P, Thomas J, Nampoori V P N. Nonlinear absorption in certain metal phthalocyanines at resonant and near resonant wavelengths. Opt. Commun., 2003, 217: 269-274.

[5] Venkatram N, Rao D N, Giribabu L, et al. Femtosecond nonlinear optical properties of alkoxy phthalocyanines at 800nm studied using Z-Scan technique. Chem. Phys. Lett., 2008, 464(4-6): 211-215.

[6] 郭丰启，赵理曾. 五氮齿金属配合物溶液对皮秒激光的光限幅. 功能材料，2000，21(1): 87-88.

[7] 王婷婷，曾和平. 富勒烯金属有机配合物的研究进展. 有机化学，2008，28(8): 1303-1312.

[8] 周云山，韩瑞雪，张立娟，等. {Mo102}型纳米多孔无机富勒烯衍生物的合成、表征及三阶非线性光学性质. 高等学校化学学报，2008，29(7): 1299-1303.

[9] 周晓莉，魏振枢，宋静文. 配合物非线性光学材料研究进展. 中州大学学报，2007，(3): 116-119.

[10] 常新安，陈丹，臧和贵，等. 非线性光学晶体的研究进展. 人工晶体学报，2007，36

(2): 327-333.

[11] Cagle D W, Kennel S J, Mirzadeh S, et al. In vivo studies of fullerene-based materials using endohedral metallofullerene radiotracers. Proc. Natl. Acad. Sci., 1999, 96 (9): 5182-5183.

[12] Kato H, Suenaga K, Mikawa M, et al. Synthesis and EELS characterization of water-soluble muti-hydroxyl Gd@-82 fullerenols. Chem. Phys., 2000, 324(4): 255-256.

[13] Sun D, Huang H, Yang S. Synthesis of the first water-soluble endohedral metallofullerol. Chem. Mater., 1999, 11(4): 1003-1004.

[14] Lezzi E B, Cromer F, Stevenson P, et al. Synthesis of the first water-soluble trimetallic nitride endohedral metallofullerols. Synthesis. Metals., 2002, 128(3): 289-290.

[15] 宋龄瑛,谭建欣,蒋笃孝,等. 金属簇合物的研究与发展概况. 广州大学学报: 自然科学版, 2007, 6(5): 37-41.

[16] Zhang C, Song Y, Xin X, et al. A new approach to superior optical limiting materials-planar "open" heterothiometallic clusters. Chem. Commun., 2001, (9): 843-844.

[17] Zhang H, Zelmon D E, Deng L, et al. Optical limiting of nanosized polyieosahedral gold-silver clusters based on third-order nonlinear optical effects. J. Am. Chem. Soe., 2001, 123(45): 11300-11301.

[18] LittyIrim P, Bindu K, Nampoori V P N, et al. Luminescence tuning and enhanced nonlinear optical properties of nanocomposites of ZnO-TiO2. Journal of Colloid and Interface Seienee, 2008, 324(1-2): 99-104.

[19] Li Cun, Shi Guang, Song Yinglin, et al. Third-order nonlinear optical properties of Bi2S3 and Sb2S3 nanorods studied by the Z-scan technique. Journal of Physiesand Chemistry of Solids, 2008, 69(7): 1829-1834.

[20] Han Y P, Ye H A, Wu W Z, et al. Fabrication of Ag and Cu nanowires by a solid-state ionic method and investigation of their third-order nonlinear optical properties. Mater. Lett., 2008, 62(17-18): 2806-2809.

[21] Ispasoiu R G, Balogh L, Varnavski O P, et al. Large optical limiting from novel metal-dendrimer nanocomposite Materials. J. Am. Chem. Soe., 2000, 122(44): 11005-11006.

[22] 汪新,徐洪耀,胡大乔. 含恶二唑侧链的新型三阶非线性聚炔材料的制备及性能. 功能材料, 2007, 38(11): 1919-1920.

[23] 尹守春,李刚,徐洪耀,等. 新型含非线性生色团的聚炔共聚物的制备和光限幅. 功能材料, 2006, (3): 496-499.

[24] 光善仪,徐洪耀,李村,等. 偶氮苯取代功能聚炔的制备及光限幅性能. 功能材料, 2006, 1(37): 102-105.

[25] 蔡志彬,浦忠威,高建荣. 有机低分子三阶非线性光学材料. 材料导报, 2008, 22(3): 1-6.

[26] 于世瑞,赵有源,李潞瑛. 有机材料ZnTBP-CA-PhR的非线性吸收和光学限幅性能. 物理学报, 2003, 52(4): 860-863.

[27] Anandhababu G, Bhagavannarayana G, Ramasamy P. Synthesis, crystal growth, structural, optical, thermal and mechanical properties of novel organic NLO material: Amnlonium malate. Journal of Crystal Growth, 2008, 310(6): 1228-1238.

[28] Rarikanth M, Kumar G R. Cheminform abstract: nonlinear optical properties of porphyrins.

Curr. Sci., 1995, 68(10): 1010-1017.

[29]Kamdasamy K, Puntamberkar P N, Singh B P, et al. Resonant nonlinear optical studies on porphyrin derivatives. J. Nonlinear Opt. Phys. Mater., 1997, 6(3): 361-375.

[30]Nalwa H S. Organic materials for third-order nonlinear optics. Adv. Mat, 1993, 5(5): 341-358.

[31]Marder S R, Beratan D, Cheng L T. Approaches for optimizing the first electronic hyperpolarizability of conjugated organic molecules. Science, 1991, 252: 103-106.

[32]Brédas J L, Adant C. Third-order nonlinear optical response in organic materials: theoretical and experimental aspects. Chem. Rev., 1994, 94(1): 243-278.

[33]Kenneth S S, Chen Chin-Ti, et al. Push-pull porphyrins as nonlinear optical materials. J. Am. Chem. Soc., 1992, 114(17): 6928-6930.

[34]Avijit S, Pa resh C R, Puspendu K P, et al. Metalloporphyrins for quadratic nonlinear optics. J. Phys. Chem.,1996, 100(50): 19611-19613.

[35]Karki L, Vance F W, Hupp J T, et al. Electronic stark effect studies of a porphyrin-based push-pull chromophore displaying a large first hyperpolarizability: state-specific contributions to β. J. Am. Chem. Soc.,1998, 120(11): 2606-2611.

[36]Ming Y, Anthony C H Ng, Michael G B D, et al. Facile synthesis and nonlinear optical properties of push-pull 5,15-diphenylporphyrins. Org. Chem., 1998, 63 (21): 7143-7150.

[37]Jiang H, Su W, Hazel J, et al. Electrostatic self-assembly of sulfonated C60-porphyrin complexes on chitosan thin films. Thin Solid Films, 2000, 372(1,2): 85-93.

[38]Mishra S R, Rawat H S, Laghate M. Nonlinear absorption and optical limiting IN metalloporphyrins. Opt. Commun., 1998, 147(4,5,6): 328-332.

[39]Ono N, Ito S, Wu C H, et al. Nonlinear light absorption in meso-substituted tetrabenzoporphyrin. Chem. Phys., 2000, 262(2-3): 467-473.

[40]Wang J, Blau W J. Nonlinear optical and optical limiting properties of individual single-walled carbon nanotubes. Applied Physics B: Lasers and Optics, 2008, 91 (3-4): 521-524.

[41]Masahiro H, Tatsuo W, Anothony F G, et al. Third-order optical nonlinearities in porphyrins with extended π-electron systems. Jpn. J. Appl. Phys., 1992, 31(3A): L249-L251.

[42]Sun Wengfang, Byeon C C, Lawson C M, et al. Third-order nonlinear optical properties of an expanded porphyrin cadmium complex. Appl. Phys. Lett., 2000, 77(12): 1759-1765.

[43] 刘志斌, 张新夷, 田宏健, 等. 不同链长二元分子卟啉酞菁 (TTP-O-$(CH_2)_n$-O-Pc)的三阶光学非线. 发光学报, 1994, 15(3): 233-238.

[44]Bruno D, Ccile M, Isebelle J, et al. A peculiar excited electronic state of allene (1,2-propadiene). Chem. Phys. Lett., 1999, 300(1,2): 169-176.

[45]Barlow T W. Self-organizing maps and molecular similarity. Journal of Molecular Graphics, 1995, 13(1): 24-27.

[46]Arunk K S, Bipin B, Mandal B J, et al. Nonlinear optical properties of a new porphyrin-containing polymer. Marcomolecules, 1995, 28(16): 5681-5683.

[47]Dou Kai, Sun X D, Wang Xiaojun,et al. Optical limiting and nonlinear absorption of excited states in metalloporphyrin-doped Sol-gels. IEEE. J. Quantum. Electron., 1999, 35(7):

1004-1014.

[48] Bezerra Jr A G, Borissevitch I E, De Araujo R E, et al. Investigation of picosecond optical nonlinearity in porphyrin metal complexes derivatives. Chemical Physics Letters, 2000, 318(6): 511-516.

[49] 陈志敏, 吴谊群, 左霞, 等. 四叔丁基四氮杂卟啉配合物的合成、热稳定性及光限幅特性研究. 无机化学学报, 2006, 22(1): 47-52.

[50] 尹延锋, 王秀如, 慧, 等. 钴卟啉溶液的光限幅特性研究. 光谱学与光谱分析, 2004, 24(1): 33-35.

[51] Martin R B, Li Huaping, Gu Lingrong, et al. Superior optical limiting performance of simple metalloporphyrin derivatives. Opt. Mater., 2005, 27: 1340-1345.

[52] Sendhil K, Vijayan C, Kothiyal M P. Nonlinear optical properties of a porphyrin derivative incorporated in Nafion polymer. Opt. Mater., 2005, 27: 1606-1609.

[53] 张冰, 刘智波, 陈树琪, 等. 新型卟啉衍生物反饱和吸收研究. 物理学报, 2007, 56(9): 5252-5256.

[54] 姚文杰, 徐海军, 朱著, 等. meso-四(对十六烷氧基苯基)卟啉的非线性吸收研究. 光电子·激光, 2007, 18(9): 1089-1092.

[55] 何远航, 惠仁杰, 等. 卟啉材料在激光防护上的应用. 北京理工大学学报, 2008, 28(3): 271-273.

[56] 陈红祥, 严煤, 孙文博. 卟啉非线性光学材料研究进展. 化工新型材料, 2002, 30(11): 35-38.

第八章 卟啉传感器及金属卟啉模拟生物酶研究

卟啉为大环π电子离域体系,外环上可接多种取代基,中心金属可以改变,甚至扩展环的大小,即分子具有可修饰性。通过改变卟啉化合物的周边取代基可以改变卟啉分子的荧光,发射波发射出很强的磷光,其热稳定性好,其分子具有刚柔性、电子缓冲性、光电磁性和高度的化学稳定性。而且,由于卟啉含有一大共轭体系,拥有生色基团,所以利用紫外、红外、荧光、磷光、拉曼等光谱技术都可以检测到它的微小变化,成为传感器研究的理想模型化合物。

另外,从生物无机化学的角度来看,一个细胞有上千种化学反应,人体有1000多种酶,每一种酶控制着一种化学反应。人体新陈代谢过程就是许许多多种酶的催化反应,每一种酶都有特定的活性部位和生物功能。但活性部位相似的金属卟啉配合物,却具有不同的生物功能[1,2]。例如:血红蛋白、肌红蛋白、细胞色素单加氧酶、过氧化氢酶、过氧化物酶和细胞色素c以及一氧化氮合成酶等,这些酶的活性部位相似,功能不同,是因为这些酶的活性中心——铁卟啉周围的蛋白环境不同,特别是轴向配体不同,引起作用机制不同,它的性质也随之发生变化。因此,研究金属卟啉模型化合物的结构—功能—作用机制之间的规律性就成为当今生物无机化学的一个新热点,它是揭开生命奥秘的重要途径。本章第二节根据生物大分子的特点,从研究作用机制入手,简要介绍某些金属酶模拟的研究方法、特点、规律及其进展。

第一节 卟啉传感器

1. 卟啉氧传感器

从环境保护、工业生产到许多生化过程等诸多方面都涉及氧的分析测定。卟啉类物质有很稳定的荧光和磷光,一些金属卟啉的荧光能被氧有效地猝灭。磷光(RTP)传感器利用氧对磷光体发光信号的猝灭可以定量分析氧。基于磷光强度变化的传感器避免了激发光背景的干扰,对氧的检测有较高的灵敏度。现有的金属卟啉的荧光猝灭的氧传感器可用于气态氧、溶解氧的检测。Knor等人[3]对醋酸基四苯基卟啉铪Hf(TPP)(oac)$_2$和乙酰丙酮基四苯基卟啉钍Th(TPP)(acac)$_2$进行了研究,配合物表现出很强的荧光和由于重金属原子表现出很强的磷光,在最大波长处有很强的发光强度,在氧气存在的条件下,磷光被有效地猝灭,所以,可检测溶液中痕量的氧。

Victor等人[4-6]制备了基于分子氧磷光猝灭的氧传感器,阳离子为Pt(Ⅱ)、Pd(Ⅱ)和Rh(Ⅲ)的水溶性卟啉[H$_2$TMPyP]$_4^+$和[H$_2$TTMAPP]$_4^+$,并固定在Nafion离子交换膜上。在膜内金属卟啉的磷光很容易被分子氧猝灭,磷光强度随O$_2$体积分数增加而减小,并且,它们的i_0/i_{100}>3(i_0为纯氩气流过时的磷光强度,i_{100}为纯氧气流过时的磷光强度),值越大灵敏度越高。当[PdTTMAPP]$_4^+$的i_0/i_{100}=84时,响应时间为8 s;当[PdT$_2$MPyP]$_4^+$的i_0/i_{100}=70时,响应时间为7 s,猝灭呈线性关系,且可以检测空气或溶液中的氧。这类传感器灵敏度高,结构简单,成本低廉,从而备受关注。

除此之外,Basu 等人以甲基—三乙氧基硅烷基(MTEOS)为前驱体、盐酸为催化剂,用 Sol-gel 法制备了卟啉氧传感器。分别采用四乙氧基卟啉铂(PtOEP)和五氟四苯基卟啉铂(PtFPP)为敏感物质,研究了加水量、H_2O 与 MTEOS 的物质的量比、pH 值和陈化时间对薄膜灵敏度的影响,测得氧气的灵敏度 i_0/i_{100} 分别为 50 和 28。

2. 卟啉二氧化碳传感器

光学 CO_2 传感器分为两种类型:一类传感器是检测指示剂颜色的变化;另外一类是检测 CO_2 导致发光物质的荧光变化。例如:研究者报道了全-(2,6-二-O-异丁基)-β-环糊精对固定于 PVC 膜中的 meso-四(4-甲氧基苯基)卟啉(TMOPP)有明显的荧光增强效应,且该荧光可被溶液中的 CO_2 可逆猝灭。TMOPP 受紫外光激发后能发出紫色荧光,但与水溶液中 CO_2 作用后形成不发荧光的配合物。在 pH 为 9.0 条件下与空白溶液和 CO_2 水溶液接触后的发射荧光光谱(423 nm 激发)显示,随着 CO_2 溶液体积分数的增加,敏感膜的荧光强度逐渐减小,可用于 CO_2 含量的测定。

在此基础上,Yutaka 等人[7,8]制备了有指示剂颜色改变的 α-萘酚酞和内反射荧光卟啉的 CO_2 传感器,对传感器灵敏度和响应时间等进行了研究。将 α-萘酚酞溶于乙基纤维素里涂到玻璃的一面,在室温下晾干;将四苯基卟啉溶于聚苯乙烯里涂到玻璃的另一面,得到检测 CO_2 的敏感膜。当 CO_2 的体积分数不同时,其发光强度随 CO_2 的体积分数增加而增强,灵敏度 $i_0/i_{100}>53.99$(i_0 为纯氮气流过时的荧光强度,i_{100} 为纯 CO_2 流过时的荧光强度),其响应时间和恢复时间都小于 5 s。当 α-萘酚酞溶于聚三甲基色氨酸丙炔(TMSP)中时,涂到玻璃的一面,在室温下晾干;将四苯基卟啉溶于聚苯乙烯里涂到玻璃的另一面,得到检测 CO_2 的敏感膜。$i_{10}/i_{100}>10.3$(i_{10} 为纯氩气流过时的荧光强度,i_{100} 为纯 CO_2 流过时的荧光强度),响应时间和恢复时间都小于 3 s。此外,Vilar 等人采用钴卟啉对检测 CO_2 进行了研究,结果表明:检测 CO_2 的体积分数达 10^{-6} 数量级。

3. 卟啉氯化氢传感器

目前,检测空气中 HCl 的传感器已经取得较大进展,主要采用烷氧基四苯基卟啉和羟基四苯基卟啉作为检测 HCl 的传感器。例如:Nakagawa 等人[9,10]对四苯基卟啉的衍生物检测 HCl 进行了相关的研究。采用烷氧基四苯基卟啉($TP(OR)PH_2$)(R=—CH_3,—OC_4H_5,—2Et,—C_8H_{17})衍生物和甲基丙烯酸甲酯(BuMA)为聚合膜,对检测 HCl 的光谱性质进行研究。HCl 的体积分数在 $(0\sim11)\times10^{-6}$ 范围内,检测 HCl 灵敏性高低的顺序依次为 $TP(OC_4H_9)PH_2>TP(OCH_3)PH_2TP(O-2ET)PH_2>TP(OC_8H_{17})>TPPH_2$,表明链的长度影响膜的敏感性。

Itagaki 等人[11]研究了将四羟基四苯基氢化卟啉($TP(OH)_4PH_2$)分散在不同物质的量比的甲基三乙基硅烷(MTEOS)和四乙基原硅酸盐(TEOS)溶胶—凝胶中,利用旋涂法制得薄膜,利用光谱技术检测 HCl 气体。当甲基三乙基硅烷(MTEOS)和四乙基原硅酸盐(TEOS)比(M/T)分别为 4、8、16 和 32 时,检测 HCl 气体的体积分数范围为 $(0.5\sim5.0)\times10^{-6}$。其灵敏度的高低取决于 M/T 的值,当 HCl 气体的体积分数为 0.5×10^{-6} 时,灵敏度顺序为 $M/T=8>M/T=4$,$M/T=4$ 与 $M/T=16$ 相当,$M/T=16>M/T=32$;在高体积分数范围内($HCl\geqslant2.8\times10^{-6}$),$M/T=8$ 和 $M/T=4$ 时,灵敏度相当。但是,若存在 SO_2,并且当 SO_2 的体积分数达到 0.5×10^{-6} 时,检测 HCl 时几乎没有响应,所以,检测 HCl,应在无 SO_2 的条件下进行。

4. 卟啉氮氧化物传感器

用卟啉传感器检测氮化物具有代表性的有 Nizam 等人[12]将 Pt 电极表面用锰卟啉(Mn(Ⅱ)triOMeTCPPyP)修饰，分别采用安培法和循环伏安法检测生物体系统内的 NO。结果表明：采用安培法比循环伏安法检测 NO 的灵敏度高得多。

Richardson 等人[13,14]报道了分别采用卟啉(EHO)L-B 膜和卟啉(EHO)/杯芳烃 L-B 膜，对卟啉 L-B 膜 NO_2 传感器进行了研究。结果表明：膜的质量影响传感器的灵敏度和响应时间，加入杯芳烃可提高膜的质量。当 L-B 膜与 NO_2 接触时，吸收光谱发生了显著变化。采用卟啉(EHO)/杯芳烃 L-B 膜，且体积分数比(EHO:杯芳烃)为 1:2，L-B 膜为 20 层，检测温度在 20~90 ℃时，检测 NO_2 体积分数的范围为$(0.13~4.6)\times 10^{-6}$；在无甲苯存在的条件下检测 NO_2，响应时间 t_{50} 为 15 s；通甲苯后检测 NO_2，响应时间 t_{50} 为 48 s，恢复到原来状态条件，再检测 NO_2，响应时间 t_{50} 为 16 s，并不再发生变化，说明其性能稳定。

根据 NO 的体积分数，医学上可诊断哮喘病。正常人周围空气中 NO 的体积分数为 10×10^{-9} 左右，哮喘病患者周围空气中 NO 的体积分数为$(70~80)\times 10^{-9}$，且病情的严重程度和 NO 的体积分数成正比。Vilar M 等人[15]发明了一种预防哮喘的 NO 传感器。该传感器具有灵敏度高、专一性强、可重复使用等优点。

5. 卟啉生物传感器

目前，卟啉生物传感器有酶生物传感器、葡萄糖生物传感器、SPR 生物传感器等。研究者以双羟基乙基二茂铁、单羟基乙基二茂铁和羧基二茂铁三种二茂铁衍生物为介质，在过氧化物模拟酶-meso-四-(4-磺苯)卟啉锰存在下测过氧化氢，得到了很好的效果[16]。

Jaime 等人[17]用谷氨酸和氮氧化合物的选择性电极，在电极表面用 Ni 卟啉修饰，可以检测从细胞释放的谷氨酸和氮氧化合物。当电压为 750 mV 时，可作为检测 NO 的传感器；当电压为 -55 mV 时，可作为检测谷氨酸的传感器。即对化学物质进行检测时，可根据细胞化验来取代动物试验，且灵敏度和选择性都较高。另外，卟啉生物传感器在食品包装测氧和细胞生存耗氧方面，Papkovsky 等人[18]进行了大量研究。在食品包装测氧上，以铂卟啉为传感探头，同肉类食品一同包装，然后，通过相调制技术进行扫描，不损坏包装袋即可获知其中氧的含量。

6. 其他卟啉传感器

除了上面常见物质的卟啉传感器外，还有检测氰化物、磷化物、氟化物和硫化物等一些物质的卟啉传感器。其中具有代表性的有 Huantian[19]等人将四苯基卟啉溶到二甲基甲酰胺里，利用光谱性质检测二嗪农。将 TPP 溶液的光谱与二嗪农加入 TPP 溶液的光谱进行比较，结果表明：光谱发生红移，说明卟啉溶液对二嗪农有响应；又将不同体积分数的二嗪农加入卟啉溶液中的光谱进行了比较，结果显示：体积分数对最大吸收波长和最小吸收波长无影响，但在最大吸收波长和最小吸收波长处的吸光度差别很大。

Legako 等人[20]报道了将 CuTPPT 固定在纤维素膜上，检测溶液中 NaCN 的含量。当溶液中无 NaCN 时，最大吸收波长为 416 nm；当溶液中含 NaCN 时，最大吸收波长为 417 nm。若在 411 nm 处有一峰出现，表明只是 CuTPPT 的吸收；在 421 nm 处有强吸收峰，表明溶液中存在 NaCN，且吸光度与 NaCN 浓度的增加呈曲线性关系，检测下限为 0.7×10^{-9} mol/L，上限为 75×10^{-9} mol/L。当检测 HCN 气体时，检测体积分数的最大值为 10×10^{-6}。这种传感器的显著优点是检测的速度极快，且膜放置数年仍能保持活性。将卟啉分散在乙基纤维素(EC)里，利用光谱技术可检测HCl、NO_2和SO_2等一些有毒气体。在TPP里引入给电子基团烷氧基或羟基可提高灵敏度;相反,引入受电子基团灵敏度降低。在卟啉乙基纤维素(EC)里添加增塑剂邻苯二甲酸二辛酯,灵敏度和响应时间都显著提高。检测HCl时检测限达10^{-6}级,检测NO_2和SO_2时可使检测限达10^{-6}以下。检测NO_2比SO_2略灵敏,但检测NO_2恢复性略差[21]。

Lukasz等人[22]制备出用Zr(Ⅳ)OEP或Zr(Ⅳ)$_2$TPP修饰聚亚安酯膜检测氟化物的传感器,并和传统的以Zr(Ⅳ)$_2$OEP或Zr(Ⅳ)$_2$TPP修饰聚氯乙烯敏感膜检测氟化物的传感器进行了比较,研究表明:以聚亚安酯为敏感膜不影响卟啉的性质。当检测电压为-64.6 mV时,对氟离子的响应时间($t_{95}<120$ s)比以聚氯乙烯为敏感膜的响应时间更短;当检测电压为-58.3 mV时,对氟离子的响应时间$t_{95}<12$ s,检测限为10^{-5} mol/L。

Awawdeh等人[23]根据光谱性质用可溶性的卟啉检测溶液中的五氯苯酚。当卟啉与氯酚或其他有机氯杀虫剂相互作用时,光谱性质发生了改变。TPPS, Zn-TPPS, TPPS1, C1TPP, C4TPP和Cu-C4TPP在400-450 nm范围内有很强的吸收,吸收波长分别为413, 418, 403, 405, 407, 404 nm, 当与五氯苯酚相互作用后B带红移,吸收波长分别变为421, 427, 431, 416, 417, 416 nm, 对五氯苯酚的检测限分别为1×10^{-9}, 0.5×10^{-9}, 1.16×10^{-9}, 1×10^{-9}, 0.5×10^{-9}和0.5×10^{-9} mol/L。

近年来,卟啉类分子合成发展较快,卟啉传感器开发还落后于材料的制备和合成。要充分发挥卟啉分子材料独特的性能,开发新的传感器和新的用途。卟啉传感器的专一性很强,单只传感器在检测混合气体或有干扰气体存在的情况下,难以得到较高的检测和识别精度。要解决上述问题,需构建卟啉敏感阵列,利用同一敏感阵列对多种目标物质进行同时检测。同时,由于卟啉具有专一性和特殊性,利用光谱技术构成微型化的传感器系统,是卟啉传感器的发展方向[24]。卟啉传感器处于刚刚发展阶段,具有很好的发展前景。

第二节 金属卟啉模拟生物酶研究

1. 金属卟啉模拟血红素酶研究

1.1 血红素酶作用机制的研究

血红蛋白和肌红蛋白载氧的活性部位是血红素辅基——铁卟啉。目前对血红蛋白载氧的活性部位的作用机制从结构化学方面已经作了相当详细的研究。在Fe(Ⅱ)离子和氧分子键合过程中,与Fe(Ⅱ)离子轴向配位的咪唑基发挥了十分重要的作用,它促成了Fe(Ⅱ)卟啉辅基与球蛋白之间的直接作用,同时又影响Fe(Ⅱ)离子卟啉辅基—氧分子复合物中的电子分布。咪唑是一个良好的电子给体,可将其电子给予Fe(Ⅱ),从而提高了轨道的电子给予能力,有利于Fe(Ⅱ)离子和氧分子之间的反馈键的形成,促进了分子氧的键合。研究者在20世纪50年代已经开始了人工合成铁卟啉氧载体,但一直未能成功。后来,从研究作用机制入手发现这是由于两种人工合成的铁卟啉在与氧分子作用时生成$Fe^{Ⅱ}—O_2—Fe^{Ⅱ}$,该分子中间的O_2容易断裂,分别与另一个未氧铁卟啉配合物再配位生成不可逆载氧的μ氧二聚体。而血红蛋白和肌红蛋白的铁卟啉辅基处于多肽链的盘绕之中,正是这种位阻效应阻止了两个血红素的Fe(Ⅱ)离子互相靠近到能同时和一个O_2结合的距离,抑制了生成不可逆载氧的μ氧二聚体$Fe^{Ⅲ}—O—Fe^{Ⅲ}$途径。

1.2 血红素酶模型化合物的设计和合成

在血红蛋白的作用机制的研究基础上,研究者先后合成了在室温下能载氧的围栅型Fe

(Ⅱ)卟啉配合物、帽式铁卟啉配合物[25]、以横跨卟啉环的长链桥基作障碍基团的吊带卟啉、带有半封闭状的障碍基团的袋式卟啉和用冠醚类结构作横跨卟啉环的桥基的冠状卟啉等多种铁卟啉以模拟血红素酶的载氧功能。

计亮年等[26]通过长链 4-(间-吡啶氧基)丁氧基的轴向配体与卟啉环相连形式尾式钴卟啉,用紫外—可见光谱测定了尾式钴卟啉配合物的载氧能力,该模型化合物在室温、苯溶液中呈现了可逆载氧能力,通过红外光谱进一步证明在氧合金属卟啉分子中的氧分子是以端基方式与中心金属 Co 配位。

通过以上的研究,研究者发现血红蛋白的主要生理功能是把氧从肺部运送到组织,再把二氧化碳通过和血红蛋白中的自由氨基结合生成氨基甲酰血红蛋白,将组织产生的二氧化碳运送到肺部,血红蛋白的生理功能问题在生物界争论近半个世纪之后终于在化学家合作之下通过血红蛋白活性中心的模拟得到了圆满解决。

2. 金属卟啉配合物模拟细胞色素 P450 酶研究

2.1 细胞色素 P450 酶作用机制

在生物体内广泛存在着细胞色素 P450 单加氧酶,它可以在常温常压即温和的条件下实现对氧分子的活化,并使烃类发生选择性氧化[27],总反应式如下:

$$R-H+O_2+H^+ \rightarrow R-OH+H_2O$$

大量研究表明细胞色素 P450 的活性中心为铁卟啉,它能使氧分子活化并能将其中一个氧原子加合到烃类底物中,在卟啉环的轴向上第五配位的配体是一个半胱氨酸残基硫,$Fe^{Ⅲ}-S$ 距离约为 2.2 Å,卟啉环的轴向上第六配体至今仍未确定,研究推测第六轴向配体可能是水、羟基或氨基酸中的氧,并且有可能在不同条件下有不同的基团配位在此位置上[28]。

众多的科学家经过 20 多年的努力,总结出细胞色素催化底物氧化反应的循环图[29],如图 8-1 所示。静态的细胞色素 P450 处于两种铁卟啉配合物的平衡中,这两种铁卟啉配合物一种是六配位低自旋态的 P450,另一种是五配位高自旋态的 P450,当靠近蛋白质疏水部位的底物与活性中心铁结合时,平衡则移向五配位高自旋态的 P450,高自旋态 Fe(Ⅲ)配合物接受 NADPH 传递的电子从而被还原至五配位的 Fe(Ⅱ)配合物,接着再结合分子氧形成六配位氧合物,该氧合物继续得到一个电子使分子氧活化而形成高价活性中间体,该活性中间体将自身氧原子转移至底物,从而完成催化循环。由图 8-1 可以看出,细胞色素 P450-Fe(Ⅲ)卟啉也可以使用单氧给体 PhIO(亚碘酰苯)等直接经过一个催化短循环过程转变为活性中间体,避免了电子供体的参与。

以上循环中,除了高价中间体外,其余几种中间体都已通过谱学手段得到证实[29],这些都有助于人们更好地了解 P450 催化循环,关于高价中间体的存在,至今未得到直接证据的证实,对此科学家正在积极进行研究。

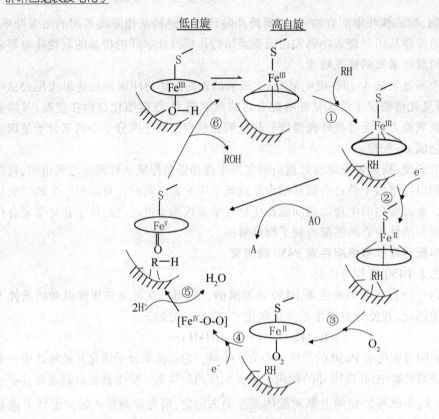

图 8-1 P450 催化底物单加氧作用[29]

2.2 细胞色素 P450 酶模拟体系的组成

细胞色素 P450 在温和条件下高效、专一地催化活化分子氧与选择性氧化烃类的能力引起了研究者们的兴趣。目前主要工作分为两个方面:(1)进一步了解细胞色素酶 P450 的作用机制,从而对其催化循环有更为清晰的认识;(2)在目前已经了解其催化循环作用机制的基础上合成金属卟啉配合物,作为模拟细胞色素 P450 活性中心,从而组装该模拟体系,通过模拟研究去部分或全部再现细胞色素 P450 的各种反应,以期开发在温和条件下烃类选择性氧化催化剂。

1979 年,Groves 等[30]报道了一个简单的细胞色素 P450 单加氧酶模拟体系,首先利用 PhIO 作为单分子氧源,FeTPPCl 作为催化剂,实现了烯烃环氧化和烷烃羟基化反应。由图 8-1 也可以看出,一个细胞色素 P450 模拟体系由以下几部分组成。

2.2.1 催化剂

催化剂即具有催化氧化还原能力的金属卟啉配合物。大量研究表明,具有多种价态且能发生两电子氧化过程的金属离子是良好的模拟酶金属活性中心,因此中心金属离子 d 电子数、自旋态及能级高低都很重要。目前,常用于卟啉化合物的过渡金属有 Fe、Mn、Co、Mo、Ru 和 Rh[31]。金属卟啉配合物大致可分为以下几类。

(1)对称金属卟啉

研究者合成了在金属卟啉苯环的邻位、间位和对位具有各种推电子和拉电子取代基团的四苯基卟啉衍生物和对称的 Fe、Mn、Co 金属卟啉配合物[32-34],以分子氧作为氧源,抗坏血酸作为还原剂,硫醇化合物作为轴向配体,构成了细胞色素 P450 的模拟体系,实现了在常温常压下将苯一步氧化为苯酚,转化率达到 40%[32],或将环己烷氧化为环己醇和环己酮等[35],这在细胞色素 P450 的模拟中是一个重要的进展。

(2)不对称金属卟啉

研究者进一步合成了在苯环上具有各种推电子和拉电子取代基团的不对称四苯基卟啉衍生物及其各种金属配合物[36],研究了其作为细胞色素 P450 模拟酶对环己烷的羟基化作用,进一步证明不对称金属卟啉的催化活性比结构类似的对称金属卟啉稍大。

(3)尾式金属卟啉

考虑到金属卟啉还需要轴向配体作为第五配位体才能具有细胞色素 P450 的单加氧酶功能,研究者根据卟啉分子结构的特点,合成了末端基团为吡啶基、咪唑基[36-38]、苯并咪唑基[39]、苯并噻唑基[40]、对羟基苯基、对羧基苯基、对亚硝基苯基、苯并恶唑基[41]和噻吩基[42]经烷基链—$(CH_2)_n$—接于卟啉环苯基的对位、间位、邻位的尾式金属卟啉,由 MCD 谱证明了咪唑基和苯并咪唑基与铁卟啉中心配位,用顺磁共振技术系统研究了不同尾端基团的尾式铁卟啉配合物及其与含氮配体加合物的磁学性质,发现由于尾端基团的配位,一些加合物给出异常的 ESR 信号,如末端基团为咪唑基的尾式卟啉具有新的配位模式和独特的磁性质[43,44],并研究了在温和条件下,对环己烷转变为环己酮和环己醇的催化作用。由于尾端基团对中间体具有稳定作用,结果表明上述尾式卟啉配合物均比其他简单的相应的金属配合物对烷烃的羟基化作用更为有效。在低自旋六配位的血红蛋白配合物中,只有当其中的一个配体是含硫配体时,它的光谱才和生物体内低自旋细胞色素 P450 的 EPR 谱最吻合,具有一个可极化特性的 S 配体可以将电荷经中心铁推到对位的配体上,在研究细胞色素 P450 催化硝基甲烷到醛的重排反应中,发现 S 可以稳定活性中间体,并且可以防止卟啉 π 正离子自由基的形成。研究者也发现含硫的末端基团配体在类似催化剂中具有最高的催化活性[35,41],这主要是由于尾端基团含硫的尾式铁卟啉的结构类似于天然细胞色素 P450 的活性部位,含硫的尾端基团是通过一些柔性或者刚性链连接于卟啉环的苯基上,这些尾端基团在一定的条件下可以配位于金属中心从而起到外加轴向配位作用,推电子作用又可削弱 O—O 键,加速了 O—O 键的异裂或均裂,有利于活性中间体 $Fe^V=O$ 的形成。

(4)高分子固载的金属卟啉

以键合于高分子载体的金属卟啉配合物作为细胞色素 P450 模拟体系的研究也引起了研究者广泛的关注,这是由于金属卟啉配合物可以模拟细胞色素 P450 的活性中心,而高分子固载可以模拟活性中心周围的多肽链和疏水环境,且在高分子化合物中存在高分子效应[45],对金属卟啉的催化活性的提高也有影响,以通过胺、羰基或酯键键合于高分子载体的铜(Ⅱ)、钴(Ⅱ)等卟啉配合物催化硫醇、对二苯酚、氢醌等底物的氧化也早已有报道[46-48]。研究者以新的合成方法使铁(Ⅲ)卟啉通过醚键键合于聚苯乙烯上,中心铁离子处于高自旋态(S=5/2)。由于铁卟啉处于聚苯乙烯载体表面各种基团所造成的类似周围蛋白的疏水环境中,在同样条件下与未固载的相同卟啉作对比,其催化活性提高了 2~4 倍[49],而在锰卟啉体系时提高了 5.9 倍。另外,研究者还系统研究了模拟酶立体构型与催化活性的关系。

目前用作固载金属卟啉的载体主要有离子交换树脂、高分子聚合物、硅藻土、氧化铝、分子筛或无机黏土材料等,此外,可以用作金属卟啉载体的高分子材料还有炭黑[50]、石墨[51]、无机矿土[52]、有机生物膜材料和某些无机—有机高分子复合材料[53,54]等。

(5) 金属双卟啉

事实上,细胞色素 P450 活性部位——金属卟啉是处于蛋白链的包围之中,而底物的键合最终也发生在一段靠近血红素的疏水蛋白链上,缺少这种疏水环境则限制了反应中催化活性的提高。同时,由细胞色素 P450 单加氧酶催化循环可知,铁卟啉中铁(Ⅲ)离子的还原是最初的也是最重要的步骤,因此立体疏水环境以及电子和能量转移都是极为重要的因素,但对于双卟啉这方面的模拟研究的文献报道并不多。金属双卟啉二聚体及多聚体的研究起源于发现天然卟啉二聚体及聚合体在光合作用中的生物效应。Deisenhofer[55]首次测定了细菌光合作用反应活性中心的射线三维结构,并因此获得了 1988 年诺贝尔化学奖。

自然界除了光合作用系统外,还存在许多以卟啉二聚体或多聚体作为活性中心的金属蛋白及蛋白酶,它们在氧的输送、活化、光催化及光电导方面有重要的生物效应,对于生物活性双卟啉结构及功能模拟一直是生物无机化学领域重要的研究方向。研究者设计和合成了一系列具有各种推电子和拉电子基团的以柔性烷氧链相连的 p-Fe(Ⅲ)TPPCl/p-H$_2$TPPCl,p-Fe(Ⅲ)TPPCl/p-Zn(Ⅱ)TPP 和 p-Fe(Ⅲ)TPPCl/p-Fe(Ⅲ)TPPCl 双卟啉各种金属配合物作为细胞色素 P450 模拟体,研究了苯环上取代基的性质对双卟啉分子内能量转移的影响,并进行了催化烃类反应的研究。这些模拟体能使烷烃和 O_2 在烷氧基为桥的"面—面"之间疏水环境的空穴中反应,在同样条件下与相应单核未修饰卟啉相比,发现其具有较高的催化活性,有的提高了近 6 倍[56]。利用一维核磁共振氢谱、二维核磁共振氢谱等手段进一步研究发现双卟啉溶液中存在开放式和闭合式构象平衡,随着烷氢链长度的增加,构象平衡由开放式向闭合式移动,当碳原子为 4 时,最有利于自由双卟啉形成闭合构象[57]。实验证明,双卟啉闭合式构象的形成,使分子内 π—π 相互作用增大,双卟啉催化活性依次增加[58],加入一种二齿轴向配体 DABCO,能进一步提高双卟啉的催化活性,这是由于形成双齿 DABCO 配体夹层于双铁卟啉平面之间的特殊结构(图 8-2)[59]。

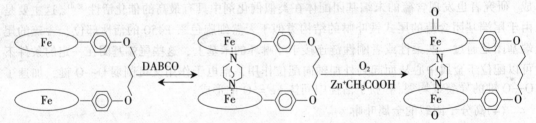

图 8-2　DABCO、O_2 和还原剂的之比等于 1 时 DABCO 与铁(Ⅲ)双卟啉之间的双重配位[59]

2.2.2　氧源

细胞色素 P450 模拟酶氧源体系分为单氧给体和分子氧给体两类。从细胞色素 P450 催化循环图 8-1 可以看出,使用单氧给体时,不用分子氧和还原剂就可以直接将金属卟啉氧化到高价中间体从而完成催化短循环,由于克服了分子氧反应惰性,该类反应一般产率较高,单氧给体包括 PhIO、次氯酸钠、过氧化氢、有机过氧化物(ROOH)、氨基 N—氧化物、过硫酸氢钾(KHSO$_5$)等。根据上述单氧给体进行的烃类选择性氧化反应得到了令人满意的结果,但单氧结体来源少,价格昂贵,有机过氧化物容易发生爆炸,过硫酸氢钾的合成十分复杂,且条件苛刻;NaOCl 价廉易得,但需要相转移催化剂参与。

2.2.3　还原剂(针对分子氧源)

在以分子氧为氧源的模拟体系中,为了催化剂的再生和闭合循环还需加入还原剂,还原剂的性质对底物氧化的产物组成发挥了决定性的作用,还原剂包括化学还原剂和其他形式的还原剂,前者包括硼氢化物、抗坏血酸钠、H_2/胶态 Pt、锌/醋酸等,后者有电化学还原及光化学还原,值得注意的是,在顺式 1,2-二苯乙烯为底物的环氧化反应中,使用 $MnTPPCl/O_2$/抗坏血酸钠体系,产物中顺式、反式环氧化物的比例为 33:67,与 PhIO 及 NaOCl 为氧源的产物比例基本相同,这可能是由于金属卟啉在上述三种不同来源的氧源中所形成的金属中间体结构十分相近之故。

2.2.4 轴向配体

一般轴向配体对金属卟啉的影响表现在电子因素及立体位阻效应两个方面,研究者[60]对三类轴向配体进行了研究。第一类轴向配体含 SH 基团,如邻疏基苯甲酸、苯硫酚、硫醇、巯基乙酸、巯基丙酸和四氢噻吩等,这类配体具有给电子效应,具有较高的催化活性,它们不仅作为轴向配体,还作为一种助催化剂,以 σ 配位键转移部分电荷到中心金属离子上对氧分子进行活化,其中疏基苯甲酸效率最高。第二类为含氮有机配体,如吡啶、咪唑及其衍生物,它们以氮作为配位原子通过 σ 键或 π 键与中心金属离子结合。这类配体由于与分子氧对金属离子产生竞争,催化效率较差。第三类在催化体系中既加入疏基化合物又加入含氮配体,催化效率则显著提高,并且该体系具有与细胞色素 P450 相似的电子光谱,说明体系较接近细胞色素 P450 单加氧酶的电子状态。

2.2.5 其他因素

其他影响因素包括催化体系介质的 pH 值、体系各组成的浓度(催化剂、还原剂、轴向配体等用量)、反应时间、反应温度和溶剂效应等方面,当模拟体系介质为弱酸性时催化效率相对较好,pH 值过高或过低都会影响催化剂的活性。一般而言,催化剂用量越大,催化产率越高,当还原剂用量过大时,还原剂会与底物竞争氧化反应。同时,反应时间并非越长越好,体系所用溶剂性质也会对催化效果有影响,到目前为止,这方面的研究报道并不很多。

参考文献

[1] 计亮年. 生物无机化学导论. 广州: 中山大学出版社, 1992.

[2] 林定夷. 科学·社会·人才. 广州: 中山大学出版社, 2001.

[3] Gunther K, Andreas S. Coexisting intraligand fluorescence and phosphorescence hafnium (IV) and thorium (IV) porphyrin complexes in solution. Inorganic Chemistry Communications, 2002, 5: 993-995.

[4] Victor V, Vasil'ev, Borisov S M. Optical oxygen sensors based on phosphorescent water-soluble platinum metals porphyrins immobilized in perfluorinated ion-exchange membrane. Sensors and Actuators B, 2002, 82: 272-276.

[5] Basu B J. Optical oxygen sensing based on luminescence quenching of platinum porphyrin dyes doped in ormosil coatings. Sensors and Actuators B, 2007, 123: 568-577.

[6] 杨荣华, 王柯敏, 肖丹, 等. 基于环糊精/卟啉包络物荧光猝灭的二氧化碳光学敏感膜的研究. 高等学校化学学报, 2001, 1: 38-42.

[7] Yutaka A, Naoki N. Optical CO_2 sensor with the combination of colorimetric change of naphtholphthalein and internal reference fluorescent porphyrin dye. Sensors and Actuators B, 2004, 100: 347-351.

[8] Yutaka A, Tasuku K. Optical CO_2 sensor of the combination of colorimetric change of

naphtholphthalein in poly (isobutyl methacrylate) and fluorescent porphyrin in polystyrene. Talanta, 2005, 66: 976-981.

[9] Katsuhiko N, Yoshihiko S, Heru S, et al. Optochemical HCl detection using alkoxy substituted tetraphenyporphyrin-polymer composite films effects of alkoxy chian length on sensing characteristics. Sensors and Actuators B, 2001, 76: 42-46.

[10] Katsuhiko N, Tomochika A, Gen U, et al. Development of an eco-friendly optical sensor element based on tetraphenyporphyrin derivatives dispersed in biodegradable polymer Effects of tetraphenyporphyrins on HCl detection and biodegradation. Sensors and Actuators B, 2005, 108: 542-546.

[11] Yoshiteru I, Katsuyuki D, Shun-Ichi N, et al. Development of porphyrin dispersed sol-gel films as HCl sensitive optochemical gas sensor. Sensors and Actuators B, 2006, 117: 302-307.

[12] Nizam, Joshua O, Wolfgang S. Electrochemical nitricoxide sensor preparation: A comparison of two electrochemical methods of electrode surface modification. Bioelectrochemistry, 2005, 66: 105-110.

[13] Richardson T H, Brook R A, Davisb F, et al. The NO_2 gas sensing properties of calixarene/porphyrin mixed L-B films. Colloids and Surfaces A., 2006, (284-285): 320-325.

[14] Dunbar A, Richardson T H, McNaughon A J, et al. Understanding the interacttions of porphryin L-B films with NO_2. Hunter Colloids and Surfaces A., 2006, (284-285): 339-344.

[15] Rei Vilar M, El-Beghdadi J, Debontridder F, et al. Development of nitric oxide sensor for asthma attack prevention. Materials Science and Engineering C, 2006, 26: 253-259.

[16] 章咏华, 杨赛鹏, 汪尔康. 过氧化物模拟酶生物传感器——二茂铁衍生物为介质. 分析化学, 1995, 23 1: 9-13.

[17] Jaime C, Sonnur I, Andrea B, et al. Simultaneous detection of the release of glutamate andnitric oxide from adherently growing cells using an array of glutamate and nitric oxide selective electrodes. Biosensors and Bioelectronics, 2005, 20: 1559-1565.

[18] Papkovsky D B, Riordan T O, Soini A. Phosphorescent porphyrin probes in biosensors and sensitive bioassays biochem. Soc. Trans., 2000, 28: 74-77.

[19] Huantian C, Jinhee N, James H H, et al. Spectrophotometric detection of organophosphate diazinon by porphyrin solution and porphyrindyed cotton fabric Spectrophotometric detection of organophosphate diazinonby porphyrin solution and porphyrin-dyed cotton fabric. Dyes and Pigments, 2007, 74 : 176-180.

[20] Legako J A, White B J, Harmon H J. Detection of cyanide using immobilized porphyrin and myoglobin surfaces. Sensors and Actuatoes B, 2003, 91: 128-132.

[21] Yoshiteru I, Katsuyuki D, Shun-Ichi N, et al. Toxic gas detection using porphyrin dispersed polymer composites. Sensors and Actuators B., 2005, 108: 393-397.

[22] Ukasz G, Elzbieta M. Fluoride-selective sensors based on polyurethane membranes doped with Zr (IV) porphyrins. Analytical Chemical Acta, 2005, 540: 159-165.

[23] Awawdeh A M, Harmon H J. Spectrophotometric detection of pentachlorophenol (PCP) in water using immobilized and water-soluble porphyrins. Biosensors and Bioelectronics, 2005, 20:1595-1601.

[24] 侯长军, 张红英, 霍丹群, 等. 卟啉传感器研究进展. 传感器与微系统, 2008, 27

(3): 1-4

[25] Froidevaux J, Ochsenbein P, Bonin M, et al. Side selection of the fifth coordinate with a single strapped zinc(II) porphyrin host: full characterization of two imidazole complexes. J. Am. Chem. Soc., 1997, 119: 12362-12363.

[26] 计亮年, 覃夏, 黄锦汪. 尾式金属卟啉配合物的催化和载氧功能. 中山大学学报: 自然科学版, 1993, 32(4): 1-5.

[27] White R E, Coon M J. Oxygen activation by cytochrome P450. Ann. Rev. Biochem., 1980, 4: 315-399.

[28] Dawson J H, Andersson L A, Sono M. Spectroscopic investgations of ferric cytochrome P450-CAM ligand complexes. J. Biol. Chem., 1982, 257: 3606-3617.

[29] Mansuy D. Cytochrome P450 and synthetic models. Pure. Appl. Chem., 1987, 59: 759-770.

[30] Groves J T, Nemo T E. Epoxidation reactions catalyzed by iron porphyrins, Oxygen transfer from iodosylbenzene. J. Am. Chem. Soc., 1983, 105: 5786-5791.

[31] 曹锡章, 宋天佑, 王杏乔. 无机化学. 3版. 北京: 高等教育出版社, 1994.

[32] 郑颖, 曾添贤, 计亮年. 细胞色素P450模拟酶的合成及其用于常温常压下催化氧化苯为苯酚的研究. 无机化学学报, 1988, 4(4): 54-57.

[33] 曾添贤, 黄锦汪, 陈特强, 等. 碘化(4-三甲胺苯基)卟啉铁(III)配合物的聚合及其轴向配位的研究. 化学学报, 1989, (47): 128-131.

[34] Ji L N, Liu M, Kang H A. Synthesis and characterization of some porphyrins and metalloporphyrins. Inorg. Chim. Acta, 1990, 178: 59-65.

[35] Ji L N, Liu M, Hsieh A K, et al. Metalloporphyrin-catalyzed hydroxylation of cyclohexane with molecular oxygen. J. Mol. Catal., 1991, 70: 247-257.

[36] 计亮年, Hsieh A K. 不对称卟啉的合成及其模拟细胞色素P450对环己烷的羟化作用. 中山大学学报: 自然科学版, 1992, 31(1): 52-55.

[37] 黄锦汪, 刘军峰, 焦向东, 等. 尾式金属卟啉配合物的研究III咪唑基尾式卟啉及其铁(III)配合物的合成和性质. 高等学校化学学报, 1995, 16(2): 136-139.

[38] 焦向东, 黄锦汪, 计亮年. 尾式金属卟啉配合物的研究IV含氮配体与尾式铁(III)卟啉轴向配位反应热力学及其配位模式. 化学学报, 1995, 53: 861-863.

[39] He H S, Huang J W, Ji L N. Studies of tailed metalloporphyrins IV synthesis, characterization and catalysis of benzimidazole-linked iron (III) porphyrins. Tran. Met. Chem., 1997, 22: 113-120.

[40] 何宏山, 黄锦汪, 计亮年, 等. 尾式金属卟啉配合物的研究V苯并噻唑基尾式铁卟啉的合成、表征及其轴向配位和催化性质. 高等学校化学学报, 1996, 17(12): 1817-1961.

[41] He H S, Huang J W, Ji L N. Synthesis, characterization and catalytic studies of new benzthiazolethiol and benzazole-linked iron porphyrin. Tran. Met. Chem., 2000, 14: 2337-2343.

[42] 焦向东, 黄锦汪, 计亮年. 尾式噻唑基卟啉及其铁配合物的合成与表征. 化学通报, 1997, 1: 25-30.

[43] Jiao X D, Huang J W, Liu J F, et al. The novel magnetic properties of ferric ion on the addition of pyridines to tailed porphyrin iron complexes as studies by EPR spectroscopy. Polyhedron, 1995, 14: 2595-2597.

[44] 焦向东, 段新华, 黄锦汪, 等. 含氮配体与尾式咪唑基铁卟啉轴向加合物的低温EPR

研究. 高等学校化学学报, 1997, 18(3): 333-336.

[45] Tsuchida E J, Macromol. Aproaches to artificial macromolecular oxygen carriers. Sci-Chem. A, 1979, 13: 545-571.

[46] Rollman L D. Porous, polymer-bonded metalloporphyrins. J. Am. Chem. Soc., 1988, 97: 2132-2136.

[47] 林厚斌. 聚合物键联金属卟啉化合物的合成及其催化性能研究. 有机化学, 1988, 8: 45-53.

[48] Yamakita H, Hayakawa K. Preparation of polymer-bonded metalloporphyrins. J. Poly. Lett. Ed., 1980, 18: 529-536.

[49] Liu Z L. The preparation of some polystyrene-supported porphyrinatoron and their catalysts in hydroxylation of cyclohexane with molecular oxygen. J. Mol. Cata. A. Chem., 1966, 104: 193-200.

[50] Battioni P, Bartoli D, Mansuy Y S, et al. An easy access to polyhalogenated metalloporphyrins covalently bound to polymeric supports as efficient catalysis for hydrocarbon oxidation. J. Chem. Soc. Chem. Comm., 1992, 15: 1050-1051.

[51] Parton R F, Neys P E, Jacobs P A. Iron-phatholocyanine immobilized on activated carbon black: Aselective catalyst for alkane oxidation. J. Catal., 1996, 164(2): 341-345.

[52] Shi C, Anson F C. Electrocatalysis of O_2 to H_2O by tetraruthenated cobalt meso-tetrakis (4-pyridyl) porphyrin absorbed on graphite electrodes. Inorg. Chem., 1992, 31(24): 5078-5083.

[53] Anikin M V, Borovkov V V, Ishida A, et al. Observation of conformational relaxation hindrance in the singlet excited state for porphyrin incorporated in a lipid membrane. Chem. Phy. Lett., 1994, 226: 337-353.

[54] Liu H G, Feng X S, Yang K Z, et al. The incorporation of C_{60} in cavities of 5, 10, 15, 20-tetra-4-oxy (2-stearic) acidphenyl porphyrin in L-B film. Chin. J. Chem., 1995, 13(5): 415-422.

[55] Deisenhofer J, Epp O, Mikiet K, al. Structure of the protein subunits in the photo-synthetic reaction center of rhodopseudomonas virids at 3 resolution. Nature, 1985, 318: 618-624.

[56] Huang J W, Liu Z L, Gao X R, et al. Hydroxylation of cyclohexane catalyzed by iron metal-porphyrin dimmer with molecular oxygen, the effect of the steric hindrance and the intramolecular interaction between the two porphyrin rings. J. Mol. Catal. A. Chem., 1996, 111: 261-263.

[57] 任奇志, 黄锦汪, 林翠悟, 等. 双卟啉化合物的构象平衡及π—π作用研究. 高等学校化学学报, 1999, 20(3): 333-335.

[58] Ren Q Z, Huang J W, Liu Z L, et al. The effect of conformation on the catalytic activity of iron metal-free porphyrin dimmers. S. Afr. J. Chem., 1997, 50(4): 181-183.

[59] Ren Q Z, Liu Z L, Ji L N, et al. Hydroxylation of cyclohexane catalyzed by iron-iron porphyrin dimmers and DABCO with molecular oxygen: evidence for the conformation effect of porphyrin dimmers on the catalytic activity. J. Mol. Catal. A. Chem., 1999, 1-2(148): 9-14.

[60] 计亮年, 彭小彬, 黄锦汪. 金属卟啉配合物模拟某些金属酶的研究进展. 自然科学进展, 2002, 12(2): 120-129.

第九章 卟啉超分子化合物

自 1987 年诺贝尔化学奖获得者 Lehn 教授[1]首次提出超分子化学的概念以来,超分子化学作为包含物理和生物现象的化学科学前沿领域,已得到迅速的发展。超分子是由分子间的弱相互作用(氢键、配位键、静电作用、范德华力、疏水作用等)形成的分子聚集体。超分子化学的一个重要发展方向是研究分子组装过程和组装体,并且通过分子组装形成超分子功能体系。利用超分子化学,人工开发和创造的超分子体系,如:功能材料与智能器件、DNA 芯片、分子器件与机器、导向及程控药物释放与催化抗体、高选择性催化剂等,在诸多科学和技术领域中都展示了良好的应用前景[2,3]。

以卟啉为砌块,加入小分子(有机、无机分子)、大环分子(冠醚、富勒烯、环糊精等)及聚合物等,通过分子间的弱相互作用可以形成各种各样的卟啉超分子。按其空间结构,卟啉超分子可以分为二维构型(如直链形、网格形、树枝形等)和三维构型(如侧臂形、矩形、面对面形、多面体形等)。根据卟啉基本砌块中含有的卟啉环个数,卟啉类超分子还可分为单卟啉、二聚卟啉、多聚卟啉砌块构筑的超分子等几种类型。

卟啉及其衍生物如血红素(铁卟啉)、血蓝素(铜卟啉)、维生素 B_{12}(钴卟啉)、叶绿素(镁卟啉)等广泛存在于生物体内与催化、氧的输运和能量转移等相关的重要细胞器中[4]。运用卟啉单体、二聚体或多聚体砌块单元,通过分子间非共价键作用自组装而成的卟啉超分子具有优异的光、电、仿生等性能[5,6],在光学材料、化学催化、电致发光材料、分子靶向药物等不同领域均有潜在的应用前景。

第一节 卟啉超分子化合物的组装合成

卟啉及金属卟啉配合物的超分子组装研究已成为仿生化学的热门课题。运用卟啉构建的超分子化学体系,可展示出有意义的光、电、电化学等多种特性,在生命、信息、材料科学等许多相关学科均有潜在的应用价值。因此,卟啉及金属卟啉配合物在各方面所显示出的多样性越来越多地吸引人们对卟啉类化合物进行功能分子的设计,用它来构建功能多聚物体系,详细研究它的功能与性质[7-9]。在构筑卟啉功能多聚物体系时,最常用的有两种方法:共价键连接和自组装。

1. 共价键构筑卟啉组装体

1.1 利用炔键构筑组装体

为了模仿生物体内的能量诱导系统和电子转移系统,多采用炔键构筑卟啉组装体。炔键桥连的卟啉能导致强的电子耦合,通过光激发会产生大的激发能变化。在利用炔键连接卟啉获取光电特性时,有四个因素至关重要:

(1)共轭的构造单元有电子激发态特征;

(2)富电子单元和缺电子单元的交替结构;

(3)σ 单电子削弱大环组成部分,有效地降低 HOMO 轨道能;

(4) 在共轭体系间有强的电子耦合[10]。

Youngblood 等[11]利用不同金属卟啉组装了 Zn—pbp—Mg（pbp：卟啉—桥—卟啉）和 Zn—pbp—Fb（Fb：自由碱卟啉）两种聚合体(图 9-1)，目的是探索能量传递的速率、效率和机理。研究结果显示，二聚体中能量传递速率分别为 11 ps^{-1} 和 37 ps^{-1}，能量传递效率分别是 0.995 和 0.983。能量通过键传递的速率与通过空间传递的速率之比大于 97，由此可见，在这两种聚合体中能量传递的机理主要是通过键而不是通过空间位置。

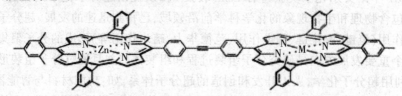

图 9-1　不同金属卟啉组装成的二聚卟啉(M=Mg,H)[11]

分子光线圈作为光合反应中光接收系统的仿生模式，在制造分子器件的材料化学中有潜在的应用价值。Kirmaier 等[12,13]合成了四个分子光线圈，每个光线圈都由输入单元、发射单元、输出单元三部分构成。研究结果表明，每个线圈都展示了可见吸收光谱的特征，能量通过炔键进行快速传递，最终使光能从输入单元传到输出单元。

若在单一能量接受体周围增加发色团的数目，就增加了捕获质子和对中心发色团传递能量的可能性。Solladie 等[14]利用手性核苷将自由碱卟啉和 Zn 卟啉进行连接，通过紫外光谱与荧光发射光谱发现该组装体具有物理、化学的双重性质，手性核苷阻止了电子从 Zn 卟啉到自由碱卟啉的转移，但在诱导能量传递方面起到了重要的作用。同时，中心自由碱卟啉又显示出了氧化还原的性质。

1.2　利用富勒烯构筑组装体

富勒烯与卟啉在基态和激发态都有氧化还原性质，富勒烯在可见区有弱的吸收，卟啉在可见区一般有强的吸收。为了设计新颖的人造光合反应系统，大多数以 C_{60} 为基础的供给体—接受体系统中，都利用卟啉作为有序地捕获光能的天线[15]。Fukuzumi 等[16]设计合成了 Zn 卟啉—C_{60} 组装体。对光激发来说，与以往的卟啉—C_{60} 组装体不同，该组装体产生了一个不寻常的长时间停留的激发离子对，这对于获取长时间停留的电荷独立态是十分重要的。如此的光电效应在打开或关闭光脉冲方面具有转换光诱导电子转移的能力，有潜在的应用价值[17]。

Ikeda 等[18]利用静电力和有效光电流的产生合成出一个稳定存在的卟啉—C_{60} 组装体。该组装体具有双层薄膜结构，对可见光能产生光电流效应。此方法对发展光电流发生器是新颖而有益的。该组装体的主要特征是：

(1) 没有加任何修饰的 C_{60} 能组装成一个超薄分子层，使足够的 π 电子堆积系统保留在分子层内；

(2) 所有的 C_{60} 分子相互独立存在。

人造光合反应系统的发展目的是把可见光转变成电能、化学能，许多种卟啉—C_{60} 组装体作为一个理想的构成元件可以获取长时间停留的电荷独立状态[19]。为了加快电子转移的速率，一个有效的方法是利用纳米结构的 SnO_2 薄膜。Imahori 等[20]运用电化学方法，采用金属电极，将合成的卟啉—C_{60} 组装体聚集在 SnO_2 薄膜上。此分子组装体相对于单分子层系统来说能有效地改善光吸收,在可见光激发条件下产生阳极光电流,揭示了光诱导多步电子转移和能量转移的过程。该研究为发展光电化学电池(如:太阳能电池)提供了一个可供借鉴的方法。

另外,最近关于光诱导电子转移的许多事例都揭示出当有第三个组成部分作为选择性催化剂存在时会加速反应,更重要的是此时光化学氧化还原反应还会正常发生。但到目前为止,选择性的催化剂还很有限。Fukuzumi 等[21]合成了一个三聚卟啉组装体(图 9-2),开发使用分子氧作为催化剂而不是氧化剂。研究结果表明,O_2 作为催化剂加速了电子转移,在 C_{60} 的激发阴离子和 Zn 卟啉的激发阳离子之间,产生了两个激发离子对,并且电子转移与 O_2 浓度的增加呈线性关系。该研究给发展人造光合反应系统、获取长时间停留的电荷独立态提供了一个十分重要的方法,给理解天然光合反应电子转移机理提供了一个十分有力的佐证。

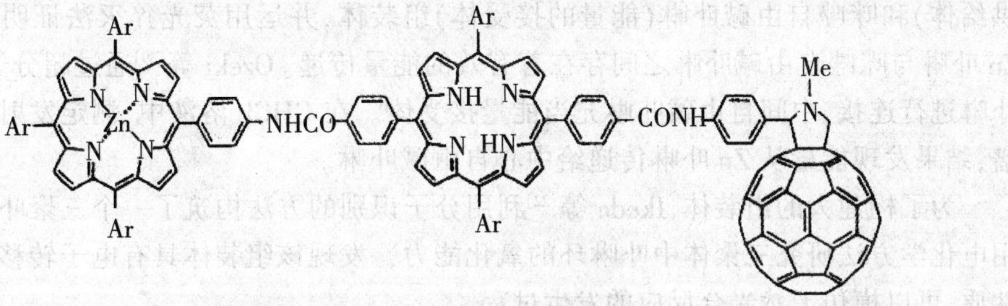

图 9-2 卟啉和 C_{60} 组装成的三聚卟啉[21]

2. 非共价键构筑卟啉聚合物

自组装方法以多种非共价键相互作用为基础,能自发地结合小的结构单元,产生大尺度的组装体,这是共价键方法难以实现的。就自组装体的形状来说,有一维线性的、二维平面结构的、三维环状的或圆形结构[22]。卟啉的基本组织结构和性能使其超分子组装可以有氢键、配位键、堆积效应、静电相互作用、疏水相互作用及模驱动自组装等多种形式。通过非共价键相互作用建造的卟啉组装体,在功能性物质的设计中是最有发展前途的[23]。

2.1 氢键自组装

氢键是超分子自组装的基本作用力之一。在生物系统中,DNA 的双螺旋和三螺旋结构是在氢键作用的基础上构筑和稳定存在的。由于氢键的形成具有方向性和选择性,近年来化学工作者普遍采用氢键来构筑超分子体系。

在生物系统中,为了更好地理解非共价键在调整电子和能量传递中所起的作用,几个非平面、非刚性的超分子卟啉电子受体已经被研究。Drain 等[24]报道了通过自身氢键形成的共平面的多卟啉排列组装体,其主要目的是研究自组装过程和卟啉自组装体的排列方式。

2.2 配位键自组装

自组装中最有效的合成方法是使用配位键。通过配位化学完成自组装的类型有:不可逆自组装、辅助自组装、直接自组装、修饰母体的自组装、修饰位置的自组装和反应间歇性的自组装[25]。因为金属与配体的相互几何位置可以自由改变,在构建二维或三维超分子结构时,如大环分子、分子器件、管件分子、分子内开关结构,改变金属与配体相互作用是实现自组装的一种有效手段[26]。

金属卟啉的轴向配位化学在构建超分子实体方面提供了一个广阔的研究领域。通常所见的金属卟啉有:Co、Mg、Mn、Ru、Zn、Fe、Sn、Ce、Eu 等多种金属卟啉[27-32],但使用最多的是 Zn

卟啉。这些金属卟啉都具有良好的光电性质。Ru 在构建大的卟啉组装体时是一种合适的金属,在与含 N 化合物进行轴向配位时,其轴向配位能力大于 Zn、Mg,易形成大尺度的卟啉组装体[33,34]。

Hunter 等[35]利用刚性分子吡啶和多个配位点相互作用,合成了几种具有稳定结构的环状卟啉。通过荧光光谱法,确定这些组装体具有光物理性质,在大环与轴向配位物之间存在光诱导电子转移和能量传递的基本过程,这有助于理解在自然界中 LH_2 的细菌叶绿素功能。

在 LH_1 和 LH_2 天线化合物中[36,37],接收光的发色团通过非共价键作用依附于蛋白质分子上,呈近乎平面结构排列。为模仿光合反应,人们设计并合成了多个包含金属和自由碱卟啉的供给体—接受体系统,进一步研究能量传递的过程。Paul 等[38]合成了 Zn 卟啉(能量的供给体)和咪唑自由碱卟啉(能量的接受体)组装体,并运用荧光猝灭法证明在该组装体中 Zn 卟啉与咪唑自由碱卟啉之间存在着有效的能量传递。Ozeki 等[39]通过超分子方法,将几个卟啉进行连接,中间自由碱卟啉充当能量接受体。在 $CHCl_3$ 溶液中,测定发射光谱和激发光谱,结果发现能量从 Zn 卟啉传递给中心自由碱卟啉。

为了构建大的组装体,Ikeda 等[22]利用分子识别的方法构筑了一个三聚卟啉组装体。运用电化学方法研究三聚体中卟啉环的氧化能力,发现该组装体具有电子转移和能量传递的性质,可以模仿天然光合反应的发生过程。

3. 卟啉功能材料的最新研究进展

以卟啉配体作为平台,借助分子组装技术,构建具有特殊光、电性能的低维功能配位聚合物,对于开发新型功能材料具有重要意义。

构建非线性光学材料,卟啉是一种理想的分子。通过丁二炔连接成的线性卟啉显示了大的电极化率[40]。Oyawa 等[41]将两个 N-甲基咪唑 Zn 卟啉通过中位进行连接,在 $CHCl_3$ 的溶液中形成了约 800 个结构单元连接在一起的轴向配位组装体。π—π 堆积系统使得该组装体有更高的电极化率,形成了一个三维非线性光学材料。

为发展有机分子感光剂系统,运用 L-B 薄膜技术,Jia 等[42]将卟啉化合物进行单分子层组装,沉积在 SnO_2 电极上,测定卟啉的光电行为。实验结果显示,卟啉连接的化合物明显具有高的开路电压和短路光电流性质,这正是有机感光剂的光电行为。

开发具有纳米级分子的传导材料在分子电器件方面的应用也是一个十分重要的研究课题。Scolaro 等[43]利用水溶性卟啉,通过一个简单的方法,将卟啉聚合体堆积在硅表面上,使用卤化试剂改变金属离子和周边的组成部分,结果光电性质得到很大调整。这个方法在发展光电器件领域有很大的应用潜力。Segawa 等[44]利用电化学方法合成了不同构型的噻吩卟啉衍生物,然后将不同构型的噻吩卟啉衍生物进行电化学氧化反应,得到具有纳米级的一维和二维噻吩卟啉衍生物聚合体。进一步研究聚合体的性质,一维衍生物聚合体具有电化学的传导性能,可以用来制作光转换设备;二维衍生物聚合体具有有机半导体材料的性质,可用作有机导电材料。

树枝状卟啉合成是近几年新兴起的热点,该领域的主要研究目的是利用大分子物质固有的结构特征,开发大分子化合物在能量转移、催化剂、传感器方面的应用。Hest 等[45]报道了用聚苯乙烯树枝状共聚物自组装形成胶束状和泡状超分子结构。卟啉和富勒烯分子进入空穴,构成具有物理和化学双重功能的分子。Sun 等[46]使用库仑相互作用和疏水相互作用制造聚电解质表面活性剂化合物,通过层与层组装成薄膜结构。Bo 等[47]利用非共价键库仑力和疏水相互作用合成的树枝状卟啉具有新颖的特征,树枝状卟啉充当了球状蛋白质的合成。

树枝状卟啉中一般有两个或多个电子接受体和供给体,这些电子接受体和供给体依靠非共价键连接形成刚性或柔性结构。Cramer 等[48,49]利用自由卟啉和苯醌以柔性链连接成树枝状卟啉。卟啉的光激发导致电子转移速率常数相当高,暗示着电子通过空间转移的机理。通过静电吸收、静电荧光及半经验理论方法,Capitosti 等[50]对自由卟啉和苯醌合成的一系列刚性树枝状卟啉进行了研究。结果发现,在卟啉环的单一激发态和苯醌末端基团之间进行了有效的电子转移。

卟啉及金属卟啉配合物的超分子组装及功能性质的研究已成为当代超分子化学的重要分支。在卟啉环不同部位引入不同生色团、配位基等基团,可以赋予它们更多的性质。随着合成卟啉组装体形状越来越多样化,其功能也越来越广泛,在材料科学和化学生物学领域内有很好的发展前景。目前,科研工作者正在不断地探索合成新的卟啉组装体,寻求开发卟啉组装体新的功能与性质。虽然卟啉组装体具有诱人的性能和应用前景,但它们的研究也存在相当大的难度。在合成过程中,一些副反应往往难以控制,从而造成分离上的困难,但这类组装体的分子设计、合成及应用十分重要,具有很大的创新性和挑战性。

第二节 金属卟啉超分子催化剂与分子识别

1. 金属卟啉超分子催化剂

超分子化学中的催化研究是当前生命科学和分析科学的边缘学科前沿中最重要的领域之一,分子识别和酶催化过程与生命现象及自然环境密切相关。酶是超分子化学家的重要灵感来源,其过渡态的识别性质与位于酶活性位置的催化功能相结合,降低了反应的活化能。诺贝尔奖获得者 Cram、Pederson 与 Lehn 相互发展了对方的经验,提出了主客体化学[51]和超分子化学[52],奠定了超分子催化和模拟酶的重要理论基础。

金属卟啉是一类非常重要的生命物质,它广泛存在于生物体内,如细胞色素、血红素、叶绿素、维生素 B 等中[53],在生命体系中的催化氧化、太阳能贮存[54]、氧的运输和贮存、免疫反应和蛋白质合成等许多生物化学过程中,都起着重要作用。同时,其化合物在光电、生物模拟催化氧化、气体传感器、模拟天然产物、微量分析等领域的应用前景也非常诱人。由于金属卟啉配合物在溶液中不稳定,容易发生氧化降解或不可逆的二聚反应,使催化活性降低,限制了它们的应用。解决这些问题的有效办法之一就是构筑超分子体系,以金属卟啉为基础构建微反应器,使反应活性中心处在一个特定的微环境中,并在一定条件下通过超分子作用对反应底物进行选择性识别,促进反应快速或定向进行,从而实现更高的催化效率、更高的反应专一性以及温和的反应条件。

金属卟啉超分子根据催化剂母体结构及作用机制分为:
(1)借助环糊精构筑的金属卟啉类超分子催化剂;
(2)形成环状空腔的金属卟啉类超分子催化剂;
(3)仿酶金属卟啉类超分子催化剂;
(4)基于光、电催化的金属卟啉类超分子催化剂;
(5)基于模板的金属卟啉类超分子催化剂以及动态的金属卟啉类超分子催化剂等[55]。

关于金属卟啉超分子已有大量相关文献报道,限于篇幅,这里不再赘述。

2. 金属卟啉与分子识别

分子识别可理解为底物与给定受体的选择性结合。识别过程可能引起体系的电学、光学性质及构象的变化,也可能引起化学性质的变化。这些变化意味着化学信息的存储、传递及处理。因此,分子识别在信息处理及传递、分子及超分子器件制备过程中发挥着重要作用[56]。

酶的高效率、高选择性和高速度等生理功能与酶对底物分子有特殊的识别能力有密切的关系。卟啉配合物是许多酶的活性中心,它同肽链靠共价键和非共价键作用产生的三维结构是卟啉类酶实现其特殊分子识别的结构基础,也是理解酶的特殊生理功能的关键[57]。

卟啉具有特殊的结构:
(1) π-电子共轭的平面结构,很适合作为设计立体结构分子的框架;
(2) 卟啉化合物具有较大的平面,对轴向配体周围的空间容积和相互作用方向的控制余地较大;
(3) 卟啉化合物的种类具有多样性,母体卟啉具有 4 个中位和 8 个 β-吡咯位可用于立体分子设计;
(4) 可以在光谱上感应卟啉与环绕分子间相互作用的微小变化。

基于以上优点,以卟啉类配合物为主体的分子识别研究,正引起人们日益浓厚的兴趣。

Sun 等[58]合成的"上下颚"超分子金属卟啉化合物能识别富勒烯(如图 9-3:R=Ag、Ni、Cu、Zn、Co、Rh)。研究发现在 C_{60} 中加入少量更大的富勒烯(C_{70}、C_{76}、C_{78}、C_{84})与 Cu_2JawsP 作用时,MALDI 质谱显示有更多的 C_{84} 配合物生成,说明 Cu_2JawsP 能很好地选择识别 C_{84}。这种识别与传统的 π—π 堆积作用不同,它是通过紧密靠近的弯曲分子表面与平面分子之间的亲和力来完成的。而且这种识别体系可以通过精巧的配位作用、电荷转移、静电作用及溶剂效应很好地调节与卟啉中心金属的相互作用。同时这种识别方法为设计、合成离散型的超分子配合物和固态超分子配合物提供了新的途径。可以预期这一分子设计原理在调控分子磁体和分子导体的光物理性质、电荷转移以及研制新型多孔金属—有机骨架等方面具有非常重要的应用前景。目前,由于卟啉超分子具有强配位能力及分子识别功能,其衍生的固定相已成功地应用于 HPLC[59]、CE[60]等分析设备中。详细情况可参考相关文献[61]。

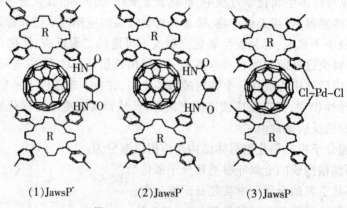

(1) JawsP″　　(2) JawsP′　　(3) JawsP

图 9-3　卟啉超分子识别富勒烯[58]

第三节 卟啉超分子化合物在分子器件中的应用

1981年,Carter[62]提出了分子器件的概念。1988年Lehn[63]再次阐述了分子器件的概念、原理和应用前景。此后分子器件的研究引起了各国科学家的重视[64]。分子器件是指分子水平上由光子、电子或离子(总称为介体mediator)操纵的器件[65]。

分子器件应有以下几个条件:
(1)元件应含有光、电或离子活性功能基;
(2)元件分子必须能按特定需要组装成组件,大量的组件有序排列能形成信息处理的超分子体系,即微型分子器件;
(3)分子器件的输出信号必须易于检测。

卟啉化合物因具有独特的性质,已成为化学家们首选的分子器件材料之一。

1. 卟啉分子导线

分子导线是分子器件和外界连接的桥梁,对于它一般有以下几个要求:
(1)能够导电;
(2)有确定的长度,足以跨越诸如类脂单层膜或双层膜;
(3)含有能够连接到系统功能单元的连接端点;
(4)允许在连接端点进行化学反应;
(5)导线必须与周围绝缘以阻止电子的任意传输。

由双卟啉四酮合成的准一维、全共轭的四卟啉衍生物[65],长度约为6.5 nm,并且在主链的周围有叔丁基作为保护套,以保证共轭核心与周围绝缘,并使分子在大多数溶剂中有较好的溶解度。四卟啉的结构如图9-4所示。它在一定程度上组装成了绝缘的导线。

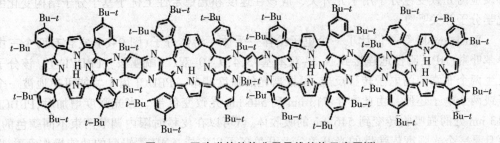

图9-4 四卟啉的结构分子导线结构示意图[65]

Lindsey等[66]合成的分子光子导线,其结构如图9-5,这种导线不同于分子电子导线,它支持激发态能量运输的不是电子或空穴运输过程。它用光来产生激发态,如光合作用中发生的一样,完成输入一个信号到分子导线的过程。荧光染料能用来输出光信号,由硼—二吡咯亚甲基染料在一端提供光输入,两个锌卟啉的线性排列作为信号的输运单位,而原卟啉在另一端提供光输出。用波长为485 nm的光照射化合物所产生的发射光谱,荧光量子产率(500~800 nm)为0.107。在这些发射光中,3%来自输入染料,5%出自锌卟啉,另外92%由原卟啉发射。与相同取代的硼—二吡咯亚甲基模型化合物相比,输入染料的荧光猝灭提高了26倍,输运到相邻的锌卟啉的能量达到96%。用波长为649 nm的光照射化合物,这里只有原卟啉吸收，结果表明，原卟啉的 Φ_f 值为 0.121，此值比单分子原卟啉的 Φ_f(Φ_f=0.114)稍大。原卟啉的发射强度比预期的大 6.8 倍，这是由于它在 485 nm 有 12%的吸收以及较高的量子产率。这些结果只能解释为能量从输入染料端和输运单元高效地输运到分子另一端的原卟啉上。

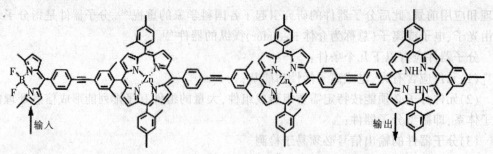

图 9-5　Lindsey 等合成的分子光子导线结构示意图[66]

为了从这些结果中求出信号输运系数，用各组分的定量光谱数据和各个能量传递步骤的各种数值对化合物的发射光谱作了模拟，传输系数分别为 95%、93%、93%和 93%。这些数据和实验结果完全符合。由这些数据计算，信号从输入到输出中的运输系数为 76%。

2. 卟啉分子开关

分子开关是分子计算机的重要部件，它的主要优点就是组合密度高、响应速度快和能量效率高。应用分子开关有可能使储存密度达到 10^{18} bit/cm^2。然而，尺寸的骤然变小将受到量子统计学因素方面的限制。不过，如果用光来控制分子器件，其优点是补偿尺寸缩小带来的统计学方面的限制。因为分子内的能量转移和电子转移过程能够在皮秒时限内进行，有可能制得响应非常快的高效器件。光驱动的分子器件的基本要求就是光稳定性。基于可逆电子转移反应的光致变色分子用于光开关，应该在速度和光稳定性上优于基于分子结构变化的光化学分子开关[67-69]。

Wasielewski 等[70]设计了一个电子给体—受体—给体分子(图 9-6)：它含有两个三戊基单苯卟啉(HP)，刚性地连接到 N，N-双苯基-3,4,9,10-双(二酰亚胺)(PBDCI)上。该分子可发生两个非常快的电子转移反应，可根据光的强度调节。PBDCI 衍生物能可逆地被一个电子或两个电子还原，由此产生 713 nm 到 546 nm 光致变色带。当光强度增加时，PBDCI 从 713 nm 处的强吸收转变到 546 nm 的吸收体，就可以在皮秒时限内调节两束不同颜色的光，而且是一个光强度依赖性的光开关。此开关还可以完成 AND 逻辑门的功能操作[71,72]，当用一束一定波长的飞秒激光脉冲激发分子时，一个电子发生转移生成 PH$^+$—PBDCI$^-$—PH，再用另一波长来激发，发生二次光致电子转移，生成 PH$^+$—PBDCI^{2-}—PH$^+$，也就是说，其中任意一种波长都会有输出，可完成 AND 逻辑门的功能操作。开关速度是其他开关的 10~100 倍，一天内进行了 8600 次开关试验，没有发生失灵。Wasielewski 等[73]还合成了下面结构的化合物(图 9-7)：它由两个光电开关共价连接而成，当对其中一端开关进行激发时，发现另一端的开关也会有响应。

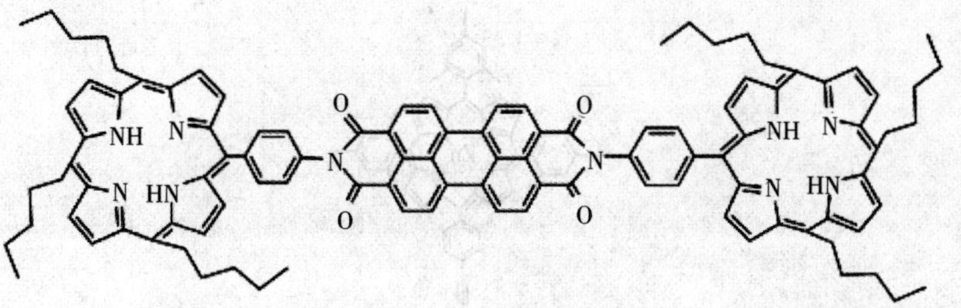

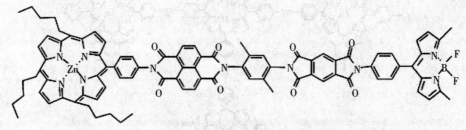

图 9-6 Wasielewski 等设计合成的电子给体—受体—给体分子结构示意图[70]

图 9-7 Wasielewski 等设计的两个光电开关分子结构示意图[73]

3. 光合作用中的卟啉分子器件

太阳能是一个巨大的能源，植物的光合作用通过将水和二氧化碳转化为复杂的能量丰富的化合物的方式将能量储存起来。因此人们一直想实现人工光合作用，来储存太阳能代替煤和石油等不可再生能源[74,75]。

光合作用包括以下几个步骤：

(1) 通过分子天线对太阳辐射进行宽带吸收；
(2) 吸收的能量传递到反应中心；
(3) 光能转变为电能；
(4) 电能进一步转化成化学能的电化学过程；
(5) 系统恢复到原始状态。

这个过程中有几种天然分子器件，其中，分子天线和分子能量转换器是最基本的两个。

3.1 卟啉超分子天线

分子天线是光合作用中最基础的部分，它的主要作用就是光吸收和能量转移。当它受到光照后，吸收一个光子，并将能量转移到一个专门的位点。

Lindsey 等[76-78]经过大量的研究，总结了光能转化器的以下几个特点：

(1) 树枝状和环形的比线形的效率高；
(2) 激发态能量低更有利于能量传递；
(3) 当独立的电子给体都由不同的能量传递途径到达能量收集部分时，这种分子天线是最有效的。

下面的卟啉超分子体系分子天线就是由 Lindsey[79-81]等合成的(图 9-8)：它是由四个锌卟啉和一个原卟啉构成的，锌卟啉作为电子给体，原卟啉作为能量收集部分，形成星状超分子结构。其量子产率为 76%。

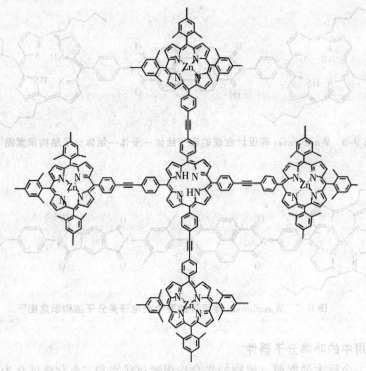

图 9-8 Lindsey 等合成的卟啉超分子体系分子天线结构示意图[79]

Gust 等[82]也将 C_{60} 引入超分子体系,设计合成了一种多卟啉光—能转换器,其量子产率达到 70%。最近,Choi 等[83]合成的卟啉树状大分子分子天线可以更有效地捕捉光子,这种树状大分子以锌卟啉为能量给体,以原卟啉为能量受体,含有 28 个光吸收单元,不仅能更有效地捕捉可见光光子,而且锌卟啉相互连接,使得长波范围的能量转移和传递更有效。

3.2 卟啉光—能转化器

光诱导电荷分离是植物光合作用的关键步骤。第一个成功模拟光合作用中光诱导电荷分离的例子[84]是由卟啉为光吸收单元(PS)、二甲苯氨基为电子给体(D)、萘醌基为电子受体(A)的 D—PS—A 体系。

用调制激光使卟啉激发,在不到 100 ps 内便发生电子转移,产生初步电荷分离,形成 D—PS$^+$—A$^-$自由基离子对。随后发生第二次电子转移,由二甲苯氨基将电子传递给卟啉自由基阳离子,实现可控电荷的有效分离,形成 D$^+$—PS—A$^-$电荷分离状态,寿命是 2.5 μs。金属配合物也可用于组装成 D—PS—A 体系。

4. 其他卟啉电子器件

越来越多的卟啉化合物被设计合成,其中包括直线的、四角的、轮状的、星形的、树枝形的、蔷薇形的等,它们通过共价键或自组装的方式连接而成。与此同时,越来越多的卟啉分子器件被启动,除了上述几种卟啉器件外,卟啉还用于光控制输入输出器[85]、光电子闸门[86]、癌治疗荧光探针[79]、生物传感器[87]等。

目前,有关分子器件的研究,已成为交叉于化学、材料学科的国际前沿课题和热点领域。以卟啉作为结构单元的分子器件越来越显示出其重要性。毫无疑问,超分子器件将在分子信息存储器件、模拟光合作用、纳米电子学中有实际的应用,但合成或组装这样的超分子体系在今后一段时期仍是十分重要和艰巨的任务。目前的开发研究,大部分处于起步阶段,离实用化、商业化还有一段距离。要想将它们组成真正的分子器件,并进一步组成功能电路,还存在很多问题。

第四节 卟啉超分子在光学领域中的应用

自 1985 年首次报道卟啉具有非线性光学(Nonlinear optical,NLO)性质以来[88],卟啉的非线性光学性质研究被广泛关注。然而由于卟啉单体的双光子吸收(two-photon absorption,TPA)截面小于 100 GM[89-90],其双光子吸收的实际应用受到限制。1990 年,Denk 等[91]提出将双光子激发现象应用到共聚焦激光扫描显微镜中,开辟了双光子荧光显微和成像这个崭新的领域。近年来,随着卟啉超分子化学的发展,科学家们发现卟啉分子经过修饰组装成超分子后会大大提高 TPA 截面[92]。如 Osuka 小组[93]总结了卟啉超分子外围基团及分子间弱相互作用对其 TPA 性质的影响因素,提出了许多行之有效的设计及合成策略。按这些策略设计合成双光子吸收截面大、转换荧光强的卟啉超分子能大大促进双光子荧光显微成像在生物系统中的应用,包括单分子检测、免疫测定、DNA 片断测定、化学和生物传感器、生物微芯片、毛细管分离检测等[94]。

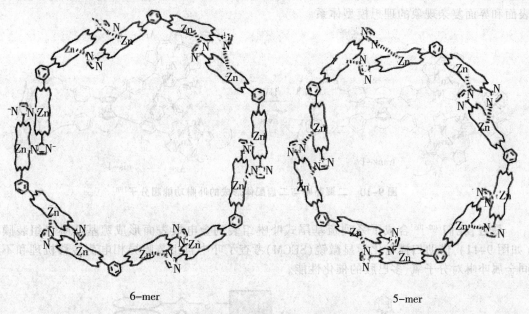

图 9-9 两种轮形卟啉超分子"光捕获天线"[97]

生物体的光合作用是非常有趣的光化学活动之一,它能够把太阳能转化为化学能。研究发现紫菌光合作用中的光捕获卟啉配合物是一种具有"轮"状结构的超分子体系,即光收集天线[95]。这种"轮"状结构卟啉超分子对太阳能的有效捕获以及随后在反应中心进行的反应都扮演着重要的角色[96]。Hwang 等[97]分析了图 9-9 所示的两种轮形卟啉超分子的能量转移速率,6-mer 和 5-mer 的能量转移速率分别为 5.3 ps 和 8.0 ps。而且 6-mer 的 Soret 带分裂更大,这说明 6-mer 的激子—激子耦合作用更强。这种自组装 6-mer 卟啉超分子是一个很好的 B850 的生物光收集天线模拟体系。卟啉衍生物叶绿素是天然光合作用中主要的反应中心,能够将太阳能转化成化学能。因此,化学家们利用卟啉超分子组装体来设计人工光收集天线系统,模拟光合成天线独特的纳米尺度三维结构,并成功地应用在光伏电池、场效应晶体管等分子器件中[98,99]。卟啉超分子在光伏电池研制中也有潜在的应用前景,沈珍等[100]最近综述了卟啉超分子在染料敏化太阳能电池中的实际应用。

第五节 卟啉类超分子的其他应用

卟啉类超分子具有光敏性好、性能稳定、易于修饰等优点,成为分子器件研究的理想模型。2006 年 Muraoka 等[101]设计合成出了剪刀形的二聚卟啉化合物主体分子与二齿配体客体分子形成的卟啉超分子(图 9-10)。这种卟啉超分子在不同光照下会像剪刀似的收缩或张开,在分子水平上实现了复杂机械功能的突破性进展。

卟啉类化合物因其自身的特殊结构,还被广泛应用于设计、合成自组装单分子膜方面的研究。自组装单分子膜是活性分子通过化学键相互作用自发吸附在固—液或气—液界面上而形成的有序分子组装体系,是近 20 年来发展迅猛的一种新型有机超薄膜。与其他单层膜相比,自组装膜有序度高、缺陷少、能量较低;易于用近代物理和化学表征技术进行研究;在分子水平上,可通过人为设计表面结构来获得预期的界面物理和化学性质。因此是研究各种表面和界面复杂现象的理想模型体系。

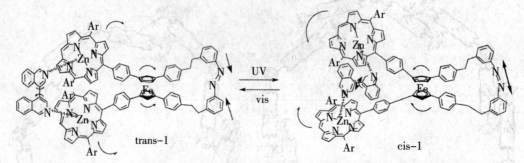

图 9-10 二聚卟啉与二齿配体形成的卟啉功能超分子[101]

最近,我们[102-105]合成了系列巯基尾式卟啉组装到金电极表面形成巯基卟啉自组装膜(如图 9-11),借助扫描电化学显微镜(SECM)考查了卟啉自组装膜异相电荷转移机理和不同金属卟啉对分子氧、多巴胺的催化性能。

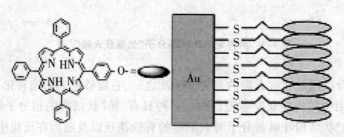

图 9-11 巯基卟啉在金电极上的 SAM 模型[103]

卟啉类超分子已广泛地用于光学、催化、仿生等方面的研究,部分研究成果已获得实际应用。随着卟啉超分子结构的变化,其性能也相应地变化。一些功能分子如富勒烯、冠醚、二茂铁等通过非共价键与卟啉自组装形成的超分子由于其表现出的特殊性越来越受到科学家的重视。然而从设计合成一个卟啉分子单元(包括单卟啉、二聚卟啉及多聚卟啉砌块),到分子自组装成超分子,其中有很多工作要做,而且结构越复杂的卟啉类化合物一般也越难合成。如何得到选择性高、方向性好、稳定的卟啉超分子是亟待解决的问题。在功能卟啉超分子研究方面,对细胞色素P450、光合反应中心和氧分子的输运等生命反应系统的仿生模拟仍然是重要的研究方向。利用卟啉配合物的特性,研究金属卟啉与氨基酸、多肽、核酸等生物功能分子的自组装正在引起人们的日益重视,借此可望人工合成出仿生活性高的自组装卟啉超分子体系。具有光、电、磁特性的卟啉超分子功能材料与器件的研究是卟啉超分子研究的另一主要方向。利用自组装技术制作亚微观尺度的卟啉超分子器件是具有广泛的应用前景的课题,还需科学家们深入研究。功能卟啉超分子器件的发展,不仅为卟啉超分子仿生化学的发展注入了新的活力,而且为具有实用价值的功能超分子的开发展示了诱人的前景[106]。

参考文献

[1] Lehn Jean-Marie. 超分子化学. 北京: 北京大学出版社, 2002.

[2] Rucareanu S, Mongin O, Schuwey A, et al. Supramolecular assemblies between macrocyclic porphyrin hexamers and star-shaped porphyrin arrays. J. Org. Chem., 2001, 66: 4973-4988.

[3] Veggel F C, Willem V, Reinhoudt D N, et al. Metallo-macrocycles: supramolecular chemistry with hard and soft metal cations in action. Chemical Reviews, 1994, 94: 279-299.

[4] 詹海鹰, 刘海洋, 胡军, 等. 卟啉超分子的组装合成及其应用新进展. 中国科学, B. 2009, 39(3): 253-268.

[5] 刘海洋, 胡希明, 应晓, 等. 金属卟啉配合物超分子自组装. 无机化学学报, 1998, 14(4): 371-387.

[6] 王树军, 阮文娟, 朱志昂. 卟啉组装体的结构、功能和性质. 化学通报, 2005, 3: 161-166.

[7] 刘育, 尤长城, 张衡益. 超分子化学. 天津: 南开大学出版社, 2001.

[8] 吴迪, 吴健. 卟啉的分子自组装. 化学通报, 2004, 10: 551-555.

[9] Piet J J, Taylo D N, Anderson H L, et al. Excitonic interactions in the singlet and triplet excited states of covalently linked zinc porphyrin dimers. J. Am. Chem. Soc., 2000, 122: 1749-1757.

[10] Susumu K, Therien M J. Decoupling optical and potentiometric band gaps in conjugated materials. J. Am. Chem. Soc., 2002, 122: 8550-8552.

[11] Youngblood W J, Gryko D T, Klammi R, et al. Glaser-mediated synthesis and photophysical characterization of diphenylbutadiyne- linked porphyrin dyads. J. Org. Chem., 2002, 67: 2111-2117.

[12] Ambroise A, Kirmaier C, Wagner R W, et al. weakly coupled molecular photonic wires: synthesis and excited-state energy-transfer dynamics. J. Org. Chem., 2002, 67: 3811-3826.

[13] Tomizaki K Y, Loewe R S, Kirmaier C, et al. Synthesis and photophysical properties of light-harvesting arrays comprised of a porphyrin bearing multiple perylene-monoimide accessory pigments. J. Org. Chem., 2002, 67: 6519–6534.

[14] Solladie N, Gross M, Gisselbrecht J D, et al. Pentaporphyrin with flexible, chiral nucleosidic linkers: unexpected duality of the physico-chemical properties of its core. Chem. Commun., 2001: 2206–2207.

[15] Imahori H, Tamaki K, Yaraki. et al. Stepwise charge separation and charge recombination in ferrocene-meso,meso-linked porphyrin dimer-fullerene tetrad. J. Am. Chem. Soc., 2002, 124: 5165–5174.

[16] Fukuzumi S, Ohkubo K, Imahori H, et al. Photochemical and electrochemical properties of zinc chlorin-C_{60} dyad as compared to corresponding free-base chlorin-C_{60}, free-base porphyrin-C_{60}, and zinc porphyrin-C_{60} dyads. J. Am. Chem. Soc., 2001, 123: 10676–10683.

[17] Liddel D A, Kodis G, Moore A L, et al. Photonic switching of photoinduced electron transfer in a dithienylethene-porphyrin-fullerene triad molecule. J. Am. Chem. Soc., 2002, 124: 7668–7669.

[18] Ikeda A, Hatano T, Shinkai S, et al. Efficient photocurrent generation in novel self-assembled multolayers comprised of [60]fullerene-cationic homoocacalix arene inclusion complex and anionic porphyrin polymer. J. Am. Chem. Soc., 2001, 123: 4855–4856.

[19] Guldi D M, Luo C, Swartz A. Self-organisation in photoactive fullerene porphyrin based donor-acceptor ensembles. Chem. Commun., 2001, 1066–1067.

[20] Imahori H, Hasobe T, Yamada H, et al. Spectroscopy and photocurrent generation in nanostructured thin films of porphyrin-fullerene dyad clusters. Chem. Lett., 2001, 784–785.

[21] Fukuzumi S, Imahori H, Yamada H, et al. Catalytic effects of dioxygen on intramolecular electron transfer in radical ion pairs of zinc porphyrin-linked fullerenes. J. Am. Chem. Soc., 2001, 123: 2571–2575.

[22] Ikeda C, Tannka Y, Fujihara T, et al. Self-assembly of a porphyrin array via the molecular recognition approach: synthesis and properties of a cyclic zinc (ii) porphyrin trimer based on coordination and hydrogen bonding. Inorg. Chem., 2001, 40: 3395–3405.

[23] Stulz E, Mak C C, Sanders J K M, et al. Matrix assisted laser desorption/ionisation (MALDI)-TOF mass spectrometry of supramolecular metalloporphyrin assemblies: a survey. J. Chem. Soc. Dalton Trans., 2001, 604–613.

[24] Drain C M, Shi X, Milic T. Self-assembled multiporphyrin arrays mediated by self-complementary quadruple hydrogen bond motifs. Chem. Commun., 2001, 287–288.

[25] Swiegers G F, Malefetse T J. ew Self-assembled structural motifs in coordination chemistry. Chem. Rev., 2000, 100: 3483–3537.

[26] Ayabe M, Yamashita K, Sada K, et al. Construction of monomeric and polymeric porphyrin compartments by a pd (ii)-pyridine interaction and their chiral twisting by a BINAP ligand. J. Org. Chem., 2003, 68: 1059–1066.

[27] Michelsen U, Hunter C A. Self-assembled porphyrin polymers. Angew. Chem. Int. Ed., 2000, 39: 764.–767.

[28] Stulz E, Mak C C, Sanders J K M, et al. Matrix assisted laser desorption/ionisation (MALDI)-TOF mass spectrometry of supramolecular metalloporphyrin assemblies: a survey. J.

Chem. Soc. Dalton Trans., 2001, 604-613.

[29] Collman J P, Boulatov R, Sunderland C J, et al. Electrochemical metalloporphyrin-catalyzed reduction of chlorite. J. Am. Chem. Soc., 2002, 124: 10670-10671.

[30] Gross T, Chevalier F, Lindsey J S. investigation of rational syntheses of heteroleptic porphyrinic lanthanide (europium, cerium) triple-decker sandwich complexes. Inorg. Chem., 2001, 40: 4762-4774.

[31] Agondanou J H, Georgios S A, Juris P. XAFS study of gadolinium and samarium bisporphyrinate complexes. Inorg. Chem., 2001, 40: 6088-6096.

[32] Chang C J, Chng L L, Nocera D G, et al. Proton-coupled O-O activation on a redox platform. J. Am. Chem. Soc., 2003, 125: 1866-1876.

[33] Campbell K, Mcdonald R, Tykwiaski R R, et al. Functionalized macrocyclic ligands for use in supramolecular chemistry. J. Org. Chem., 2002, 67: 1133-1140.

[34] Kedman J E, Feeder N, Teat S J, et al. Rh (III) porphyrins as building blocks for porphyrin coordination arrays: From dimers to heterometallic undecamers. Inorg. Chem., 2001, 40: 2486-2499.

[35] Hunter C A, Hyde R K. Photoinduced energy and electron transfer in supramolecular porphyrin assemblies. Angew. Chem. Int. Ed., 1996, 35: 1936-1939.

[36] Westerhuis W H J, Sturgis J N, Ratcliffe E C, et al. Isolation, Size Estimates, and Spectral Heterogeneity of an Oligomeric Series of Light-Harvesting 1 Complexes from Rhodobacter sphaeroides. Biochemistry., 2002, 41: 8698-8707.

[37] Timpmann K, Ellervee A, Pullerits T, et al. Short-range exciton couplings in LH2 photosynthetic antenna proteins studied by high hydrostatic pressure absorption spectroscopy. J. Phys. Chem. B, 2001, 105: 8436-8444.

[38] Paul D, Wytko J A, Koept M, et al. Design and synthesis of a self-assembled photochemical dyad based on selective imidazole recognition. Inorg. Chem., 2002, 41: 3699-3704.

[39] Ozeki H, Kobuke Y. Syntheses and reactions of the fluorinated cyclic thionylphosphazene $NSO(Ar[NPF_2]_2$ (Ar = $ButC_6H_4$) with difunctional reagents. Tetrahed. Lett., 2003, 44: 2287-2291.

[40] Kuebler S M, Denning R G, Anderson H J, et al. Large third-order electronic polarizability of a conjugated porphyrin polymer. J. Am. Chem. Soc., 2000, 122: 339-342.

[41] Wang Z, Kohen A. Thymidylate synthase catalyzed h-transfers: two chapters in one tale. J. Am. Chem. Soc. 2010, 132: 9820-9825.

[42] Jia J G, Xiao X Y, Xu J M, et al. Photoelectric behaviours of covalently linked porphyrin derivatives. Solar Energy Matericals and Solar Cells, 1995, 37: 25-31.

[43] Scolaro L M, Romeo A, Castriciano M A. Escherichia coli O86 O-antigen biosynthetic gene cluster and stepwise enzymatic synthesis of human blood group B antigen tetrasaccharide. J. Am. Chem. Soc., 2003, 125: 2040-2041.

[44] Segawa H, Wu F P, Nakayama N, et al. Approaches to conducting polymer devices with nano-structure: Electrochemical construction of one-dimensional and two-dimensional porphyrin-oligothiophene co-polymers. Synthetic Metals, 1995, 71: 2151-2154.

[45] Van Hest J C M, Delnye D A P, Baars M W P L, et al. Polystyrene-dendrimer

amphiphilic block copolymers with a generation-dependent aggregation. Science, 1995, 268: 1592-1595.

[46]Sun X Y P, Gao M L, Kong X X, et al. Effects of pH on the supramolecular structure of polymeric molecular deposition films. Macromolecular Chem. Phys., 1996, 197: 509–515.

[47]Bo Z S, Zhang L, Wang Z Q, et al. Investigation of self-assembled dendrimer complexes. Mater. Sci. & Eng., 1999, 10: 165–170.

[48]Rajesh C S, Capitosti G J, Cramer S C, et al. Photoinduced through-space electron-transfer within a series of free-base and zinc porphyrin- containing poly(amide) dendrimers. J. Phys. Chem. B, 2001, 105: 10175–10188.

[49]Capitost G J, Cramer S C, Rajesh C S. Intramolecular photoinduced electron-transfer within porphyrin-containing polyamide dendrimers. Org. Lett., 2001, 3(11): 1645–1648.

[50]Capitosti G J, Guerrero C D, Binkley D E, et al. The efficient synthesis of porphyrin-containing, benzoquinone-terminated, rigid polyphenyl dendrimers. J. Org. Chem., 2003, 68: 247–261.

[51]Cram D J. The design of molecular hosts, guests, and their complexes. Science, 1988, 240: 760–767.

[52]Lehn J M. Supramolecular chemistry — scope and perspectives molecules, supermolecules, and molecular devices (Nobel Lecture). Angew. Chem. Int. Ed. Engl., 1988, 27: 89–112.

[53]Biesaga M, Pyrzyńska K, Trojanowicz M. Porphyrin in analytical chemistry. Talanta, 2000, 51: 209–224.

[54]Milgrom L R. Synthesis of some new tetra-arylporphyrins for studies in solar energy conversion. J. Chem. Soc. Perkin Trans. 1, 1983, 10: 2535–2539.

[55] 杨再文, 杨进, 黄晓卷, 等. 金属卟啉类超分子催化剂. 化学进展, 2009, 21(4): 588–599.

[56]徐筱杰. 超分子建筑——从分子到材料. 北京: 科学技术文献出版社, 2000.

[57]游效曾, 孟庆金, 韩万书. 配位化学进展. 北京: 高等教育出版社, 2000.

[58]Sun D, Fook S T, Christopher A R, et al. Supramolecular fullerene-porphyrin chemistry fullerene complexation by metalated "Jaws Porphyrin" hosts. J. Am. Chem. Soc., 2002, 124(23): 6604–6607.

[59]Xiao J, Kibbey C E, Coutant D E. Immobilized porphyrins as versatile stationary phases in liquid chromatography. J. Liq. Chromatogr. Rel. Technol., 1996, 19 (17&18): 2901–2932.

[60]Charvátová J, Matejka P, Král V. Open-tubular electrochromatography of organic phosphates on a sapphyrin-modified capillary. J Chromatogr. A, 2001, 921(1): 99–107.

[61]饶明益, 罗国添. 超分子卟啉化合物的分子识别及其应用研究进展. 赣南师范学院学报, 2006, 3: 72–77.

[62]Latter C F. Moleccular electronic devices I. New York: Marcel Dkker Press, 1982.

[63]Lehn J M. Supramolecular chemistry — scope and perspectives molecules, supermolecules, and molecular devices (Nobel Lecture). Angew. Chem. Int. Ed. Engl., 1988, 27: 89–112.

[64]徐家业. 超分子化学发展简介. 有机化学, 1995, 15: 133–144.

[65]Crossley M J, Burn P L. An approach to porphyrin-based molecular wires: synthesis of a bis(porphyrin)tetraone and its conversion to a linearly conjugated tetrakisporphyrin system. J.

Chem. Soc., Chem. Commun., 1991,21: 1569-1571.

[66]Wagner R W, Lindsey J S. A molecular photonic wire. J. Am. Chem. Soc., 1994, 116: 9759-9760.

[67]Parthenopoulos D A, Rentzepis P M. Three-dimensional optical storage memory. Science, 1989, 245: 843-845.

[68]Lieberman K, Harush S, Lewis A, et al. A Light source smaller than the optical wavelength. Science, 1990, 247: 59-61.

[69]Feldstein M J, Vorhringer P, Wang W. Femtosecond optical spectroscopy and scanning probe microscopy. J. Phys. Chem., 1996, 100: 4739-4748.

[70]O'Nelil M P, Niemczky M P, Wasielewski M R, et al. Complexation of a tris-bipyridine cryptand with americium(III). Science, 1993, 25: 1115-1117.

[71]Wasielewski M R, O'Neil M P, Niemczyk M P, et al. Ultrafast photoinduced electron transfer reactions in supramolecular arrays: From charge separation and storage to molecular switches. Pure Appl. Chem., 1992, 64: 1319-1325.

[72]O'Neil M P, Niemczyk M P, Svec W A, et al. Picosecond optical switching based on biphotonic excitation of an electron donor-acceptor-donor molecule. Science, 1992, 257: 63-65.

[73]Debreczeny M P, Wasielewski M R, Svec W A, et al. Therapeutic antibodies: Magic bullets hit the target. Science, 1996, 274: 584-586.

[74]Ciamician G. The photochemistry of the future. Science, 1912, 36: 38-39.

[75]Kohen A, Jensen J. Boundary conditions for the swain-schaad relationship as a criterion for hydrogen tunneling. J. Am. Chem. Soc., 2002,124(15): 3858-3864.

[76]Van pattenP G, Shreve A P, Lindsey J S, et al. energy-transfer modeling for the rational design of multiporphyrin light-harvesting arrays. J. Phys. Chem. B, 1998, 102: 4209-4216.

[77]Prathapan S, Johnson T E, Lindsey J S. building-block synthesis of porphyrin light-harvesting arrays. J. Am. Chem. Soc., 1993, 115: 7519-7520.

[78]Wanger R W, ohnson J T E, Lindsey J S. A molecular photonic wire. J. Am. Chem. Soc.,1994, 116: 9759-9740.

[79]Yang S I, Seth J, Balasubramanian T, et al. Interplay of orbital tuning and linker location in controlling electronic communication in porphyrin-based nanostructures. J. Am. Chem. Soc., 1999, 121: 4008-4018.

[80]Li F, Yang S I, Seth J, et al. Design, synthesis, and photodynamics of light-harvesting arrays comprised of a porphyrin and one, two or eight boron-dipyrromethene accessory pigments. J. Am. Chem. Soc., 1998, 120: 10001-10017.

[81]Yang S I, Lammi R K, Riggs J A, et al. excited-state energy transfer and ground-state hole/electron hopping in p-phenylene-linked porphyrin dimers. J. Phys. Chem. B, 1998, 102: 9426-9436.

[82]Gust D, Moore T A, Moore A L. Mimicking photosynthetic solar energy transduction. Acc. Chem. Res., 2001, 34: 40-48.

[83]Chio M S, Aida T, Yamazaki T, et al. A large dendritic multiporphyrin array as a mimic of the bacterial light-harvesting antenna complex: molecular design of an efficient energy funnel for visible photons. Angew. Chem. Int. Ed., 2001, 40(17): 3194-3198.

[84] Gust D, Moore T A, Moore A L. Molecular mimicry of photosynthetic energy and electron transfer. Acc. Chem. Res., 1993, 26: 198–203.

[85] Birge R. Nanotchnology Research and Perspectives. Cambridge, MA: MIT Press, 1992.

[86] Wagner R W, Lindsey J S, Seth J, et al. Molecular optoelectronic gates. J. Am. Chem. Soc., 1996, 118: 3996–3997.

[87] De Silva A P, Gunaratne H Q N, Gunnlaugsson T, et al. Signaling recognition events with fluorescent sensors and switches. Chem. Rev., 1997, 97: 1515–1566.

[88] Blau W, Byrne H, Dennis W M. Reverse saturable absorption in tetraphenylporphyrins. Opt. Commun., 1985, 56: 25–28.

[89] Drobizhev M, Karotki A, Kruk M. Resonance enhancement of two-photon absorption in porphyrins. Chem. Phys. Lett., 2002, 355: 175–182.

[90] Karotki A, Drobizhev M, Kruk M. Enhancement of two-photon absorption in tetrapyrrolic compounds. J. Opt. Soc. Am. B, 2003, 20: 331–333.

[91] Denk W, Strickler J H, Webb W W. Two-photon laser scanning fluorescence microscopy. Science, 1990, 248 (4951): 73–76.

[92] Kadishi K M, Smith K M, Guilard R. The Porphyrin Handbook. Oxford: Academic Press, 2003.

[93] Kim K S, Lim J M, Osuka A. Various strategies for highly-efficient two-photon absorption in porphyrin arrays. J. Photochem. Photobiol. C: Photochem. Rev., 2008, 9: 13–28.

[94] 黄池宝, 樊江莉, 彭孝军. 双光子荧光探针研究及其应用. 化学进展, 2007, 19(11): 1806–1812.

[95] Dermott G M. Planetary science: How mercury got its spin. Nature, 1995, 374: 517–520.

[96] Oijen A M V, Ketelaars M, Kohler J. Unraveling the electronic structure of individual photosynthetic pigment-protein complexes. Science, 1999, 285: 400–402.

[97] Hwang I W, Park M, Ahn T K, et al. Excitation-energy migration in self-assembled cyclic zinc (II)-porphyrin arrays: A close mimicry of a natural light-harvesting system. Chem. Eur. J., 2005, 11: 3753–3536.

[98] Wasielewski M R. Photoinduced electron transfer in supramolecular systems for artificial photosynthesis. Chem. Rev., 1992, 92: 435–438.

[99] Gust D, Moore T A, Moore A L. Mimicking photosynthetic solar energy transduction. Acc. Chem. Res., 2001, 34: 40–43.

[100] 吴迪, 沈珍, 薛兆力. 卟啉类光敏剂在染料敏化太阳能电池中的应用. 无机化学学报, 2007, 23(1): 1–10.

[101] Takahiro M, Kazushi K, Takuzo A. Mechanical twisting of a guest by a photoresponsive host. Nature, 2006, 440: 512–515.

[102] Lu X Q, Nan M, Zhang H R, et al. Investigation of the antioxidant property of ascorbic acid. J. Phys. Chem. C, 2007, 111: 14998–15002.

[103] Lu X Q, Li M R, Yang C H, et al. Electron transport through self-assembled monolayer of thiol-end functionalized tetraphenylporphines and metal tetraphenylporphines. Langmuir, 2006, 22: 3035–3039.

[104] Lu X Q, Zhang L M, Li M R, et al. Electrochemical characterization of self-assembled thiol-porphyrin monolayers on gold electrode by scanning electrochemical microscopy (SECM). Chem. Phys. Chem., 2006, 7: 854–862.

[105] Lu X Q, Zhang H R, Hu L N, et al. Investigation of the effects of metalloporphyrin species containing different substitutes on electron transfer at the liquid/liquid interface. Electrochem. Commun., 2006, 8: 1027–1032.

[106] 刘育, 张衡益, 李莉. 纳米超分子化学——从合成受体到功能组装体. 北京: 化学工业出版社, 2004.

第十章　卟啉自组装膜概述

卟啉及金属卟啉是血红素、细胞色素和叶绿素等生命活性物质的核心部分，其分子具有较大的表面积和特殊的刚性结构。卟啉自组装单分子膜是最接近天然生物膜的理想模型，已被广泛应用于自组装膜领域的研究。卟啉自组装膜是将功能化的卟啉或金属卟啉直接组装到基底表面或间接将卟啉化合物作用于自组装膜上。卟啉类自组装膜在电化学领域显示出巨大的优势，主要表现在：利用自组装膜的卟啉分子设计和结构可控的特点，在分子水平上预先设计膜结构，获得特殊功能，达到模拟生物膜的目的，是研究界面电子转移的理想模型；此外，卟啉作为一种光电活性物质，将其自组装于基底表面可以有效地研究其光电性质，用于模拟植物体内光合反应过程和能量传递，在光电分子器件的设计和合成上有潜在的应用；卟啉及金属卟啉分子在均相和异相化学反应中是良好的催化剂，可用于构建各种不同的电化学生物传感器，在生命系统以及在仿生学中发挥着良好的功能，备受研究者的关注；卟啉化合物由于结构的特殊性，通过在卟啉环周边进行化学修饰，引入特定官能团，使之与底物产生多重相互作用，可进行分子大小、形状、官能团和手性异构体的识别；另外，利用卟啉作为分子机能材料对未来情报信息的表示、传递、识别，超导材料的制备，有机电致发光以及太阳能电池的开发研究已成为国内外十分活跃的研究领域。

第一节　卟啉自组装膜的制备方法

卟啉自组装膜是以自组装技术为基础，将功能化的卟啉或金属卟啉直接自组装于基底表面或间接将卟啉化合物作用于自组装膜上。其制备方法主要包括以下几种。

1. 直接法

直接法是以巯基卟啉作为组装分子，在基底上直接形成高度有序的单分子膜[1-3]。显然这种方法需要预先合成巯基卟啉试剂，其制备、分离、纯化的步骤烦琐，而且巯基卟啉的巯基易发生氧化、配位等化学反应，且储存的困难也较大。直接以巯基卟啉合成金属卟啉时，由于巯基与金属离子的配位，对其分离和纯化也造成了极大困难。

Porter研究小组合成了单巯基尾式金属卟啉和对位的尾式二巯基金属卟啉(图10-1)，通过将此卟啉直接在金电极表面自组装成膜，研究了该修饰电极的表面红外光谱性质及其电化学响应。通过控制巯基基团的数目可以定位卟啉平面与电极表面的夹角。利用这一特性，Murray等合成了对称的四巯基尾式钴卟啉(图10-2)，并将此化合物自组装于金电极表面，制得了卟啉分子平行定位于电极表面的自组装膜，用X-光电子能谱(XPS)、吸收光谱(UV-vis)以及电化学手段对修饰电极进行了表征[4]。四巯基卟啉在电极表面的空间取向决定了卟啉与电极表面形成配位键的数目，直接用巯基金属卟啉在自组装过程中易通过巯基与金属的轴向配位而形成自组装多层膜。Uosaki研究小组合成了自由巯基尾式卟啉(图10-3)，并自组装修饰于金电极表面，通过将卟啉修饰电极在含有金属离子的溶液中回流成功制得了金属卟啉自组装膜修饰电极，使用这种方法避免了不规则单层膜和多层膜的形成[3]。另外，在考察卟啉自组装膜分子信息储存过程中，Lindsey等合成了大量巯基保护的卟啉试剂，研究表明，巯基保护基团在溶液中与金接触时很容易发生断裂，从而以硫—金化学键的形式在金表面形成了金属卟啉自组装膜。Abrantes等人[5,6]则合成了双硫键连接的金属卟啉，这种卟啉结构在自组装过程中可以有效地避免不规则膜的生成(图10-4)。我们也合成了系列巯基尾式卟啉，并将其自组装于金电极表面，详细考察了自由巯基卟啉和金属巯基卟啉自组装膜的制备、表征及其光谱和电化学性质[7-10]。

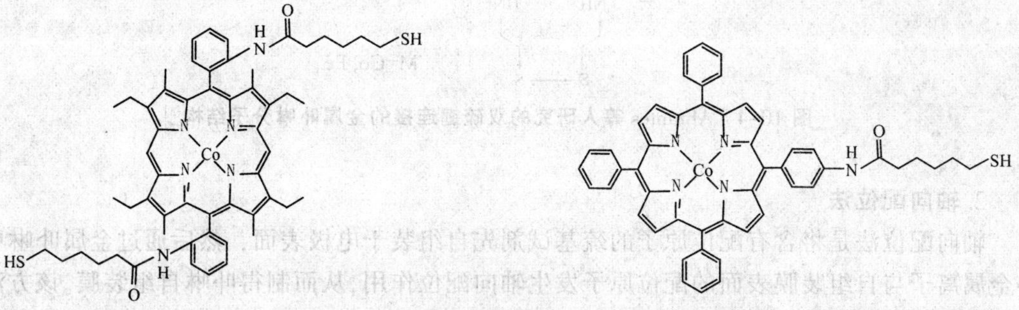

图 10-1 Porter 研究小组合成的对位尾式二巯基金属卟啉和单巯基尾式金属卟啉分子结构[2]

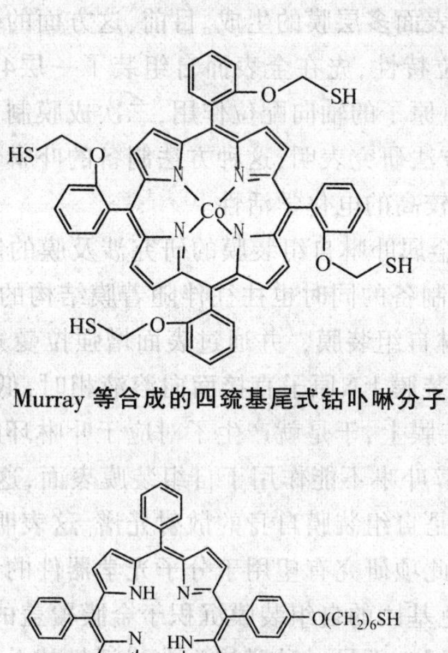

图 10-2 Murray 等合成的四巯基尾式钴卟啉分子结构[4]

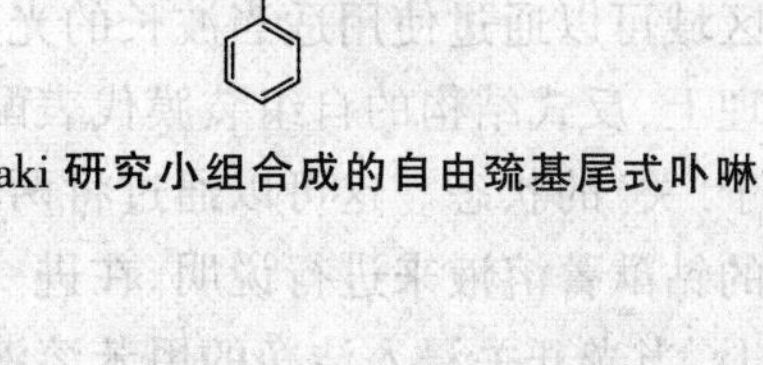

图 10-3 Uosaki 研究小组合成的自由巯基尾式卟啉分子结构[3]

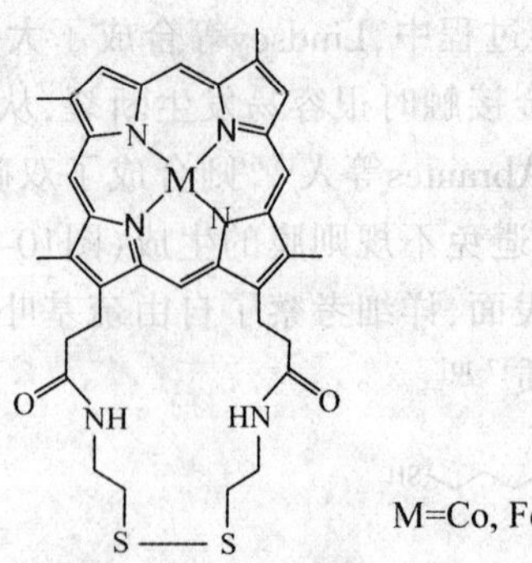

图 10-4 Abrantes 等人研究的双硫键连接的金属卟啉分子结构[5]

2. 轴向配位法

轴向配位法是将含有配位原子的巯基试剂先自组装于电极表面，然后通过金属卟啉中心金属离子与自组装膜表面的配位原子发生轴向配位作用，从而制得卟啉自组装膜。该方法不需要合成巯基卟啉试剂，只需制得相应的金属卟啉或含有配位原子的巯基试剂，大大简化了操作步骤，也避免了电极表面多层膜的生成。目前，这方面的研究报道已有很多。如刘忠范等利用金属卟啉的轴向配位特性，先在金表面自组装了一层4-巯基吡啶，然后通过金属卟啉中心钴离子与吡啶环上氮原子的轴向配位作用，二次成膜制备了金属卟啉的自组装膜。表面增强拉曼光谱和电化学方法研究表明，这种方法制备的卟啉修饰电极，卟啉平面与电极表面基本呈平行定位，且具有较高的电化学活性[11]。

关于通过轴向配位的金属卟啉自组装膜的研究涉及膜的制备、结构表征及金属卟啉自组装膜的性质和应用等，而制备的同时也往往伴随着膜结构的表征。例如，Kitano 等人通过配位键制备了有序金属卟啉自组装膜，并通过表面增强拉曼光谱对其进行了表征（图10-5）。研究发现，当吡啶自组装膜上N原子直接面向溶液相时，低浓度水溶性锌卟啉可以通过轴向作用配位于吡啶自组装膜上，于是就产生了对应于卟啉环的拉曼散射光谱；但当吡啶环上N原子面向基底表面时，锌卟啉不能作用于自组装膜表面，这时就不能观察到相应卟啉环的散射光谱，观察到的仅仅是自组装膜自身的散射光谱。这表明卟啉基团作用于自组装膜表面是通过轴向配位联系的。此项研究有望用于分子光学器件的设计[12]。类似的研究是将两种包含有4-吡啶基偶氮苯发色基团的自组装膜沉积于金膜覆盖的玻璃基底表面。其中一种膜包含的发色基团作为单一成分，而另一种膜掺入非光活性成分形成1:1的混合膜。两种自组装膜光开关可逆行为消失的区域可以通过使用适当波长的光照射，以膜产生的吸收光谱和电化学行为进行监控。在原理上，反式结构的自组装膜代表配位原子面向溶液，即处于"开"的状态，而其顺式结构，则处于"关"的状态。这可以通过将两种形式的自组装膜分别浸入钴和锌卟啉以及一个辛基取代的钴酞菁溶液来进行说明。在进一步的研究中，将金属大环化合物结合于反式结构自组装膜上，当光开关浸入洁净的甲苯溶液中时，大环化合物将从自组装膜表面释放进入溶液中，这种释放可以通过吸收光谱对保留在膜结构中的材料进行监控来测量。另外，这项研究也被扩展去发展一个在线的释放/配位循环。例如，使用365 nm的光照射结合在反式结构混合自组装膜上的金属卟啉时，ZnTPP将脱落进入甲苯溶液中。进一步用439 nm的光照射自组装膜，其顺式结构将转变为反式结构，即反式结构获得再生，又可以重新结合溶液中的ZnTPP。表明使用一定波长的光去控制自组装膜的分子结构具有潜在的应用。

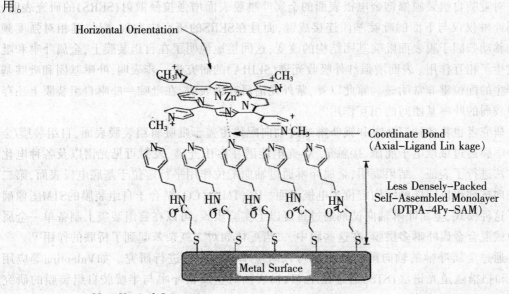

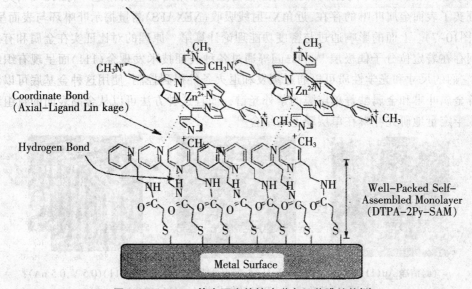

图 10-5 Kitano 等人研究的锌卟啉自组装膜结构[12]

研究者也将锌卟啉通过轴向配位于硅玻璃基底表面的3-胺基丙基-三硅氧基自组装膜表面,并通过X-光电能谱、紫外—可见光谱对膜的结构及膜对光、酸、碱以及与吡啶的反应活性进行了研究[13]。通过轴向配位,并结合L-B膜和自组装膜组织,研究者将钴卟啉作用于长脂肪链C18NHPy的自组装膜上,通过红外光谱、紫外光谱、线性二向色性、以及X-衍射对膜结构进行了研究,发现1/1/5混合的CoPTMS、C18NHPy和methyl eicosanoate分散于硝酸银相中,可以形成银—乙炔—聚螯合的卟啉结构。将此卟啉转至固体基底上,大环在基底表面上呈现平面定位[14]。另外,研究也发现,金属离子钴Co(Ⅱ)卟啉中心并不能通过轴向配位作用于自组装膜表面;但是钴的氧化态为Co(Ⅲ)时,即使Co(Ⅲ)被还原为Co(Ⅱ),也可以在轴向试剂的自组装膜表面形成非常稳定的连接,这可以通过吸收光谱证明[15]。

对吡啶自组装膜修饰金电极表面的金属卟啉膜表面增强拉曼散射(SERS)的研究表明，金属卟啉仅仅与下面的吡啶基团连接成键，而且在SERS的研究中发现，配位键相对强度频率的移动归因于膜表面吡啶基团结构的改变，这间接地说明了在自组装膜上，金属卟啉和吡啶发生了相互作用。表面增强红外吸收光谱(SEIRA)的研究进一步表明，卟啉基团和吡啶基团产生的配位键非常明显。除此以外，紫外光谱研究也表明，在吡啶—卟啉自组装膜上还存在着较弱的卟啉基团间的相互作用[16]。

研究者也将各种钌卟啉和锇卟啉通过轴向配位覆盖于吡啶自组装膜表面，自组装膜/金属卟啉膜通过俄歇电子能谱、接触角、X-光电能谱、FT-IR光谱、发射可见光谱以及各种电化学方法进行了表征。结果表明，金属卟啉通过轴向配位作用平行定位于基底电极表面，第二个轴向试剂也以同样的结构定位于电极表面。Ru(TMP)(CO)结合于自组装膜的STM图像确证了这种模式，金属卟啉轴向试剂通过重复以上实验步骤，可以在自组装膜上制备单一金属卟啉或混合金属卟啉多层膜，在这些膜中，二齿配体如对二氮杂苯起到了桥联的作用[17]。

通过金属卟啉的轴向配位作用也可以对金单晶的结构进行研究。如Valentina等应用STM和扫描隧道光谱法(STS)通过在Au(111)表面对金属钴卟啉与半胱胺自组装膜的研究表明，Au(111)的人字形的表面形貌结构经轴向配位作用后被重新构建[18](图10-6)，光电子能谱证实了表面金属卟啉的存在，近角X-射线吸收(NEXAFS)测量揭示卟啉环与表面呈70°夹角(图10-7)。上面的影响通过与密度功能理论计算第一原理的对比证实在金属和有机材料之间存在着定位分子偶极层[19]。Israel等通过各种表征技术发现金(111)面呈现有织纹的排列，它们的尺寸和光学性质可以通过蒸发和退火条件得到控制。使用这种金基底可以清晰地解释金属卟啉和金属酞菁单层膜的紫外光谱，而且这种方法可以从金晶表面的自组装单分子膜半定量地研究卟啉在单层膜上的结合[20]。

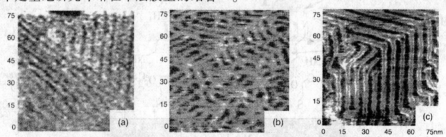

(a)洁净 Au(111)(0.5 V,0.5 nA)；(b)4-巯基吡啶功能化的 Au(111)(0.5 V,0.5 nA)；
(c)4-巯基吡啶-CoTBPP 功能化的 Au(111)(0.2 V,1 nA)
图10-6 典型的未经滤过的 STM 图像[18]

Umezawa等人另辟蹊径，将4-巯基吡啶自组装修饰于STM金探针表面，从溶液/石墨界面自发形成的锌卟啉、自由卟啉和镍卟啉混合膜中成功地分辨出了锌卟啉。STM图像中的卟啉中心用修饰探针观察显示出了桥点，而未修饰的探针对卟啉中心观察显示出暗点。锌卟啉的中心比自由卟啉和镍卟啉更亮。这可以从卟啉中心和探针上的吡啶基团的配位相互作用来进行解释[21]。

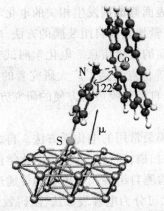

图10-7 金属钴卟啉与Au(111)表面吡啶自组装膜相互作用的结构模型[19]

3. 其他方法

方法之一是表面合成法。这种方法首先将尾端含有功能基团($-NH_2$,$-COOH$,$-OH$等)的组装分子自组装于电极表面,然后将尾端连有功能基团($-NH_2$,$-COOH$,$-OH$等)的卟啉试剂通过表面合成的方法以共价键的形式作用于自组装膜表面。

除此之外,Dietrich小组研究了一种简单有效的通过电化学阴极扫描得到金属卟啉自组装膜的制备方法[22],Rubinstein利用静电相互作用将卟啉化合物作用于自组装膜表面(图10-8)。值得一提的是,随着纳米技术的发展,利用纳米粒子良好的表面结构和较大的比表面积,研究者也将其自组装膜修饰电极用于卟啉的固定和组装。另外,光电材料的发展也促使制备含有多种功能基团,多π电子的单分子薄膜。许多混合卟啉自组装膜的制备及其性质也得到了研究者的关注。研究表明,混合卟啉自组装膜在电催化、光电转换等方面均表现出一定的特性[23,24]。

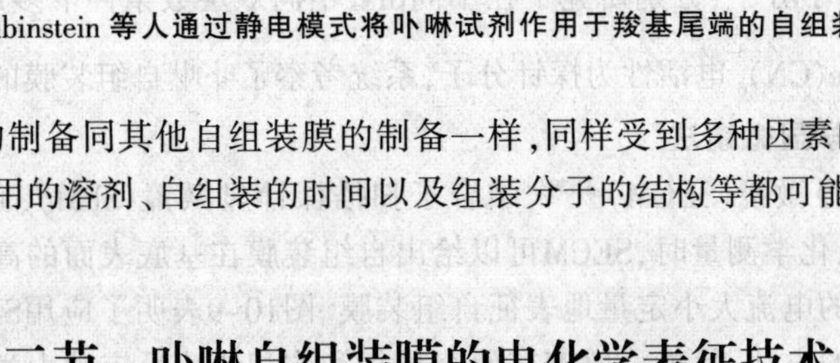

图10-8 Rubinstein等人通过静电模式将卟啉试剂作用于羧基尾端的自组装膜上[23]

卟啉自组装膜的制备同其他自组装膜的制备一样,同样受到多种因素的影响[25,26]:卟啉自组装液的浓度、选用的溶剂、自组装的时间以及组装分子的结构等都可能使膜的结构和性质发生改变。

第二节 卟啉自组装膜的电化学表征技术

随着自组装技术的迅速发展,相应的表征方法也日臻完善。目前,有很多技术和方法被广泛应用于自组装膜的表征,大致可分为以下两种,即电化学方法和物理方法。电化学方法

卟啉自组装膜电化学

表征自组装膜是通过研究电极表面修饰剂发生相关的电化学反应的电流、电量、电位和电解时间等参数间的关系来定性、定量地表征自组装膜的方法。电化学方法可以方便快捷地给出有关自组装膜的界面结构和性质的直接信息。电化学测试方法由于其灵敏度高、快捷方便、仪器便宜、操作简单、选择性好等优点受到了广大研究者的关注，几乎所有的成膜体系都可以用电化学的方法来表征，它是自组装膜最为方便的研究方法。

1. 循环伏安法

循环伏安法是研究自组装膜最常用的电化学方法。自组装膜形成以后，让其在含有"探针"分子的底液中进行循环伏安扫描，通过探针分子电化学信号的变化来判断电极自组装膜的组装程度和缺陷程度，可以检测自组装膜中电子的传递过程即通过探针分子来表征膜的电化学性质。循环伏安法表征又可分为电容表征法、阻碍效应表征以及还原解析表征等。对于电活性多层膜，一般采用循环伏安法考察膜的增长行为[27,28]。利用循环伏安法结合膜的阻碍效应对金电极表面的自组装膜进行了电化学表征[27-29]。由于膜电极的电容随膜的厚度线性减小，所以可利用这一关系对电极表面形成的自组装膜进行表征[30]。Popense等[31,32]利用烷基硫醇的解析还原对其在金电极表面形成的自组装膜进行了表征，通过连续扫描可以得出烷基硫醇在KOH溶液中的逐步解离，该法不仅可以用来表征膜的形成，还可用于处理电极等进一步的应用。由于$Fe(CN)_6^{3-/4-}$灵敏的氧化还原性，所以经常作为氧化还原探针。循环伏安法在自组装膜中更重要的应用是研究膜的分子识别功能，并利用具有分子识别功能的膜构建电化学传感器，通过特定的待测物质的电化学响应来定量测定某些物质。

2. 电化学交流阻抗法

电化学交流阻抗法是常用的一种暂态电化学测试技术，属于交流信号测量的范畴，是目前研究自组装膜中针孔和缺陷的最为有效和准确的方法。它具有频率范围广、对体系扰动小等特点，是研究电极过程动力学、电极表面现象以及测定固体电解质电导率的重要工具，因此在实际科研工作中，电化学阻抗技术的应用范围非常广泛。通过分析探针分子$Fe(CN)_6^{3-/4-}$在不同膜层引起的阻抗谱行为的变化，可以得出电荷传递电阻(R_{ct})和双电层电容(C_{dl})以及膜层数之间的关系。在膜层较少时，拟合参数呈现不规律性变化，而随层数增加，拟合参数显示出很好的线性规律[33]。电化学阻抗谱也可用于监测多层膜组装的生长情况[34]，以及膜的渗透性和结构[35]的变化。Bruening课题组[36]利用循环伏安法及阻抗谱，以电活性的$Fe(CN)_6^{3-}$和$Ru(NH_3)_6^{3-}$为探针小分子，分别研究了在不同pH、不同双层数条件下多层膜的渗透性。我们也结合CV法，以$Fe(CN)_6^{3-}$电活性为探针分子，系统考察了卟啉自组装膜的缺陷及针孔效应[37]。

3. 扫描电化学显微镜法

扫描电化学显微镜(SECM)[38,39]，结合了扫描探针显微镜(SPM)、微电极和薄层池的特点，所以在进行电化学测量时，SECM可以给出自组装膜在基底表面的高分辨率的形貌图，可以靠探针感应到的电流大小定量地表征自组装膜。图10-9表明了应用SECM测定电活性SAM的电子传递速率的方法。另外，SECM在研究异相和均相快速反应动力学、探测微区电化学活性和几何形貌方面也具有很大的应用潜力。SECM不但可以研究探头与基底上的异相反应动力学及溶液中的均相反应动力学，分辨电极表面微区的电化学不均匀性，给出导体和绝缘体表面的形貌，而且还可以对材料进行微加工，研究许多重要的生物过程等。SECM的应用，使许多重要的生物、化学体系,如化学传感器,药理学中的药物释放,相转移催化,模拟生物膜等研究中的快速、动态过程的检测成为现实。

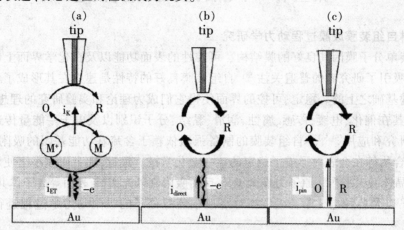

(a)通过中介;(b)通过 SAM 的直接电子隧穿;(c)通过针孔缺陷的电子传递
图 10-9　穿越电活性 SAM 电子传递速率的 SECM 过程示意图[39]

除以上电化学方法表征自组装膜以外,几乎所有的光谱法、能谱法和表面分析方法均可用于自组装膜的表征,其中被广泛应用于自组装膜的研究主要有以下方法:

(1)傅立叶变换红外光谱(FTIR)

红外光谱是研究自组装膜中分子堆积和取向的常用手段,用红外光谱技术研究自组装膜主要有衰减全反射红外光谱,通过 IR 吸收峰的位置和强度,在分子水平上研究烷基硫醇自组装膜的结构与链长的关系[40]。

(2)表面增强拉曼光谱(SERS)[41]

拉曼光谱是研究金属表面吸附的有力工具,巨大的增强因子(一般在 $10^4 \sim 10^6$ 左右)使其具有亚单分子层的检测灵敏度,大量的振动光谱信息可用来确定吸附分子的种类、吸附状态取向等,在单分子膜的研究中已有广泛应用。

(3)X 射线光电子能谱(XPS)

XPS 在表征有机单分子膜时不仅可给出有机膜中存在的元素,而且还可以提供有关元素的原子组成分析,此外,通过考察光电子的衰减波长,还可用于检测单分子膜的厚度[42]。

(4)原子力显微镜(AFM)

AFM 借助针尖与所观察材料中逐个原子发生作用,成为从原子水平上研究自组装体系最有力的工具[43]。

(5)扫描隧道显微镜(STM)[44-47]

STM 不仅能够观察导体和半导体材料表面的形貌,而且能够研究材料局域的物理和化学特性,并且进一步实现纳米量级与原子水平的改性以及加工,从而在纳米科学与技术中发挥着越来越重要的作用。

此外,还有椭圆光度法[48,49]、润湿接触角[50,51]、表面等离子体共振[52]等研究方法。

第三节 卟啉自组装膜研究概述

1. 卟啉自组装膜成膜过程动力学研究

自组装单分子膜因其良好的膜结构、可控性的表面功能以及在化学界面上制备方法的不断发展，吸引了研究者的普遍关注[53]。自组装膜良好的特性是建立在其形成了稳定的硫—金属化学键基础之上的。稳定、可控的界面使得它们成为理论和实验研究的理想模型平台，迄今为止，其在催化、电镀、传感、腐蚀、润滑、黏结、分子识别以及电子—能量传递领域得到了广泛的研究和应用[54-56]。自组装膜的制备强烈依赖于含巯基功能基团的吸附分子，如硫醇、双硫化合物等在金表面形成了稳定的硫—金化学键。最近，几个研究小组使用了多种技术，如石英晶体微天平[57,58]、原子力显微镜[59]、反射吸附红外光谱[60]、表面等离子体共振波谱[61]、二次谐波[62]以及各种电化学方法[63-68]对溶液中硫醇分子的吸附动力学过程进行了相关研究。

在早期作为代表性的巯基化合物自组装成膜过程研究中，Bain 等人[69]通过椭圆偏振光测量吸附层的厚度以及通过接触角测量湿润角研究了乙醇溶液中的 1-巯基癸烷在金表面的成膜动力学过程。他们发现，整个成膜过程分为两个阶段：第一个阶段是快速吸附过程，膜的厚度快速增长，在最初的 2 min 即达到饱和膜的 80%~90%；第二阶段是膜的重组过程，自组装膜缓慢重组，约 10~20 h 达到其稳定值。许多研究者也考察了组装溶液浓度对自组装成膜过程的影响，巯基化合物的浓度从 1 到 1000 μmol/L 之间均可观察到吸附的两步动力学过程。例如，Shimazu 等人[70]报道了 500 μmol/L 巯基—十一烷基二茂铁在正己烷溶液中的两步吸附过程，在最初的 10 s 内，膜的厚度即达到其最大值的一半，然后是一个慢速吸附过程，达到稳定值约需 10~100 min。Bensebaa 等人报道了在 5 μmol/L 溶液中，廿二硫醇在 45 s 内即形成了单层膜。De Bono 等人[61]研究了烷基硫醇的浓度从 10 μmol/L 到 100 μmol/L 范围变化时对自组装成膜的影响，他们也证实了两步吸附过程的存在，在最初的 1000 s，第一步快速吸附即达到单分子膜的 50%~80%，随后的第二步吸附过程超过 8 h。Dannenberger 等人[62]使用二次谐波技术从详细的成膜过程吸附动力学角度测量了硫醇分子的浓度在 1 μmol/L~10 μmol/L 范围内硫醇分子的吸附速率常数并且报道了吸附过程可以被 Langmuir 吸附动力学过程描述。Hong 等人[71]调查了浓度对自组装成膜过程的影响，并且也证实了吸附分子的典型成膜过程包括两个步骤：快速吸附过程和随后的慢速重组过程，对于所有的吸附分子，其浓度的范围从 8×10^{-6} 到 1×10^{-5} mol/L 时，吸附过程可以被简单的 Langmuir 吸附动力学过程所描述。

然而，迄今为止，在吸附分子成膜过程的动力学模式研究中，在诸如吸附速率常数的测定、吸附分子浓度对吸附过程定量的研究等方面仍然存在着大量矛盾的结果。最近，我们应用电化学方法对不同烷基链长巯基卟啉自组装膜成膜过程动力学进行了系统性的研究。吸附分子 $H_2TPPO(CH_2)_nSH$ 包含较大的电活性尾式基团——卟啉环，并经由不同数目的亚甲基链与金表面进行化学作用。这种结构的卟啉分子，有利于对不同链长以及较大尾端基团对自组装分子成膜的动力学和热力学过程进行考察，进而为吸附分子的成膜过程动力学研究领域提供实验和理论依据。研究结果表明：从自组装膜成膜过程的表面覆盖度、界面电容分析，巯基苯基卟啉的自组装存在两步成膜过程，即初始阶段的快速成膜和随后的表面重组过程。在整个吸附过程中,实验体系满足简单的 Frumkin 吸附平衡模式,并且在吸附过程的最初阶段,可以用简单的 Langmuir 吸附平衡模式描述成膜过程,而不必考虑吸附分子烷基链长的影响。根据简单的 Langmuir 吸附平衡模式,分析得到了初始阶段的吸附分子 $H_2TPPO(CH_2)_nSH$ 的成膜过程的速率常数。这些常数值随着尾式烷基链长的增加呈减小趋势,并且在吸附过程的初始阶段,与不同链长烷基硫醇相比,较大的尾式基团——卟啉环对成膜过程也产生了较大的影响[72]。

2. 卟啉自组装膜长程电子转移研究

在自然界中,由蛋白质分子和液膜组成的分子组装系统是控制能量(EN)和电子传递(ET)效率的关键因素。自然界中分子聚集的重要意义促使我们在没有蛋白质分子的帮助下,采取人工的方式去排列功能性的分子。迄今为止,各种实验技术,如 L-B 膜和液膜等已经被应用于分子的自组装。然而,出于稳定性和表面缺陷等方面的原因限制了这些技术在实际分子装置中的应用。

基于自组装单分子膜高度有序、更少缺陷、密集排列的表面分子结构系统,被广泛地应用于界面性质的研究。在这些研究中,大量的报道涉及自组装膜链长的影响。自组装膜的烷基链长和膜结构之间具有一定关系。而这其中最引起研究者关注的是电子转移与距离之间的关系,其重要性主要是体现在生物电子转移方面,尤其是对于生物能学研究,从本质上说,均为长程电子转移[72]。

近期,出现了大量将氧化还原电对引入体系的定量电化学研究报道。在这些系统中,烷基硫醇(或它们的衍生物)通过金属—硫共价键的形式结合于电极表面。这其中有两种类型的反应物被应用于此领域:一种是将氧化还原活性基团通过共价或非共价的形式结合于功能性的有机单分子自组装膜上[73-84];另一种是将多电荷离子电对溶解在溶液中[85-100]。对于前者,迄今有将二茂铁基团[82,101,102]、偶氮苯基团[103-105]、卟啉基团[106-108]、富勒烯[109-111]以及其他功能基团[77,112,113]应用于自组装膜体系。而对于后者,由于它们的高电荷性和强亲水性,从而既不能吸附于疏水性的有机膜上,也很难渗透进入膜内[95,98]。

然而,相比于较小的氧化还原活性基团,如二茂铁基团,相对较大的功能基团,如卟啉和富勒烯等,链长对功能性自组装膜的影响依然很模糊。这是因为包含这些功能基团的自组装膜体系中,除了烷基链之间存在着相互作用外,尾端的电活性基团之间可能也同样存在着相互作用。

为了进一步在分子水平上设计电子装置,我们设计了卟啉自组装膜体系,并应用循环伏安法(CV),对氧化还原电活性基团——卟啉环被共价固定于自组装膜表面体系以及应用扫描电化学显微镜(SECM)对氧化还原电对为处于膜外溶液中的 $Fe(CN)_6^{3-}$ 体系的长程电子转移进行了研究[114-117]。组装分子结构如图 10-10 所示,$H_2TPPO(CH_2)_nSH$ 包含电活性基团——卟啉环以及头基上连有巯基与基底表面结合。自组装膜均相和异相电子传递与距离之间的关系可通过巯基卟啉尾端烷基链的长度来控制。卟啉分子的这种结构特点,无论是对于研究共价作用于自组装膜表面的电活性基团,还是对于研究溶解在溶液中的多电荷离子电对与基底间的电子传递都是非常有利的,借此还可研究不同空间链长以及较大尾端基团的存在对其电子传递的影响。

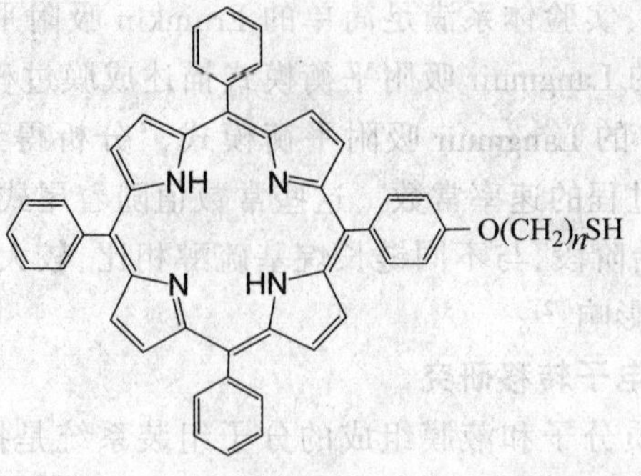

$(H_2TPPO(CH_2)_nSH(n=3,4,6,9,10 和 12))$

图 10-10 巯基苯基卟啉结构示意图[117]

研究结果显示：两种测定方法均表明卟啉自组装膜电子转移速率常数的对数随空间烷基链长的增加而成比例减小。两种体系测定的电子隧穿系数 β 分别为 0.21 $Å^{-1}$ 和 0.095 $Å^{-1}$。电子隧穿系数 β 值与相似的研究体系相比明显偏小。氧化还原电对的位置、膜的结构、斜置角以及溶剂等均对其电化学动力学性质产生重要影响。

参考文献

[1] Zak J, Yuan H, Ho M, et al. Thiol-derivatized metalloporphyrins: monomolecular films for the electrocatalytic reduction of dioxygen at gold electrodes. Langmuir, 1993, 9: 2772–2774.

[2] Postlethwaite T A, Hutchison J E, Hathcock K W, et al. Optical, electrochemical, and electrocatalytic properties of self-assembled thiol-derivatized porphyrins on transparent gold films. Langmuir, 1995, 11: 4109–4116.

[3] Nishimura N, Ooi M, Shimazu K, et al. Post-assembly insertion of metal ions into thiol-derivatized porphyrin monolayers on gold. J. Electroanal. Chem., 1999, 473: 75–84.

[4] Hutchison J E, Postlethwaite T A, Murray R W. Molecular films of thiol-derivatized tetraphenylporphyrins on gold: film formation and electrocatalytic dioxygen reduction. Langmuir, 1993, 9: 3277–3283.

[5] Hu B C, Zhou W Y. Simple and efficient method for synthesis of metallodeuteroporphyrin derivatives bearing symmetrical disulphide bond. Chinese Chemical Letters, 2011, 22(5): 527–530.

[6] Viana A S, Leupold S, Montforts F-P, et al. Self-assembled monolayers of a disulphide-derivatised cobalt-porphyrin on gold. Electrochimica Acta, 2005, 50: 2807–2813.

[7] Lu X Q, Jin J, Kang J W, et al. Electrochemical behavior of unitary or binary self-assembled monolayers with thiol-derivatized cobaltous porphyrin. Mater. Chem. Phys., 2002, 77: 952–957.

[8] Lu X Q, Lv B Q, Xue Z H, et al. Self-assembled monolayers of a thiol-derivatized porphyrin on gold electrode: Film formation and electrocatalytic dioxygen reaction. Thin Solid Films, 2005, 488: 230–235.

[9] Lu X Q, Li M R, Yang C H, et al. Electron transport through self-assembled monolayer

of thiol-end functionalized tetraphenylporphines and metal tetraphenylporphines. Langmuir, 2006, 22: 3035-3039.

[10] Lu X Q, Zhang L M, Li M R, et al. Electrochemical characterization of self-assembled thiol-porphyrin monolayers on gold electrode by scanning electrochemical microscopy (SECM). Chem. Phys. Chem., 2006, 7: 854-862.

[11] Zhang Z J, Hu R S, Zhu Z, et al. Preparation and characterization of a porphyrin self-assembled monolayer with a controlled orientation on gold. Langmuir, 2000, 16: 537-540.

[12] Kanayama N, Kanbara T, Kitano H. Complexation of porphyrin with pyridine moiety in self-assembled monolayer on metal surfaces. J. Phys. Chem. B., 2000, 104: 271-278.

[13] Zhang Z J, Hu R S, Liu Z F. Formation of a porphyrin monolayer film by axial ligation of protoporphyrin IX zinc to an amino-terminated silanized glass surface. Langmuir, 2000, 16: 1158-1162.

[14] Armand F, Albouy P-A, Cruz F D, et al. Interconnection of porphyrins in Langmuir-Blodgett and self-assembled monolayers by means of silver acetylide bridges. Langmuir, 2001, 17: 3431-3437.

[15] Zou S Z, Clegg R S, Anson F C. Attachment of cobalt "picket fence" porphyrin to the surface of gold electrodes coated with 1-(10-mercaptodecyl)imidazole. Langmuir, 2002, 18: 3241-3246.

[16] Zhang Z J, Imae T. Surface-enhanced raman scattering and surface-enhanced infrared absorption spectroscopic studies of a metalloporphyrin monolayer film formed on pyridine self-assembled monolayer-modified gold. Langmuir, 2001, 17: 4564-4568.

[17] Offord D A, Sachs S B, Ennis M S, et al. Synthesis and properties of metalloporphyrin monolayers and stacked multilayers bound to an electrode via site specific axial ligation to a self-assembled monolayer. J. Am. Chem. Soc., 1998, 120: 4478-4487.

[18] Arima V, Blyth R I R, Della Sala F, et al. Long-range order induced by cobalt porphyrin adsorption on aminothiophenol-functionalized Au (111): the influence of the induced dipole. Materials Science and Engineering C, 2004, 24: 569-573.

[19] Arima V, Fabiano E, Blyth R I R, et al. Self-assembled monolayers of cobalt(ii) (4-tert-butylphenyl)-porphyrins: the influence of the electronic dipole on scanning tunneling microscopy images. J. Am. Chem. Soc., 2004, 126: 16951-16958.

[20] Kalyuzhny G, Vaskevich A, Ashkenasy G, et al. UV/vis spectroscopy of metalloporphyrin and metallophthalocyanine monolayers self-assembled on ultrathin gold films. J. Phys. Chem. B., 2000, 104: 8238-8244.

[21] Ohshiro T, Ito T, Bu1hlmann P, et al. Scanning tunneling microscopy with chemically modified tips: discrimination of porphyrin centers based on metal-coordination and hydrogen bond interactions. Anal. Chem., 2001, 73: 878-883.

[22] Nan T, Kielman U, Dietrich C. Electrochemical metallization of self-assembled porphyrin monolayers. Anal. Bioanal. Chem., 2002, 373: 749-753.

[23] Imahori H, Hasobe T, Yamada H, et al. Concentration effects of porphyrin monolayers on the structure and photoelectrochemical properties of mixed self-assembled monolayers of porphyrin and alkanethiol on gold electrodes. Langmuir, 2001, 17: 4925-4931.

[24] Imahori H, Arimura M, Hanada T, et al. Photoactive three-dimensinal monolayers:

porphyrin alkanethiolate-stabilized gold clusters. J. Am. Chem. Soc., 2001, 123: 335-336.

[25] Poter Koltalo F, Desbene P L, C Treiner. Self-desorption of mixtures of anionic and nonionic surfactants from a silica/water interface. Langmuir, 2001, 17: 3858-3862.

[26] Kitaev V, Seo M, Mcgovern M E, et al. Significant efficiency improvement of the black dye-sensitized solar cell through protonation of TiO2 films. Langmuir, 2001, 17: 4272-4276.

[27] Cheng L, Niu L, Gong J, et al. Electrochemical growth and characterization of polyoxometalate-containing monolayers and multilayers on alkanethiol monolayers self-assembled on gold electrodes. Chem. Mater., 1999, 11: 1465-1475.

[28] Liu S Q, Tang Z Y, Wang Z X, et al. Oriented polyoxometalate-polycation multilayers on a carbon substrate. J. Mater. Chem., 2000, 10: 2727-2733.

[29] Liu S Q, Tang Z Y, Bo A L, et al. Electrochemistry of heteropolyanions in coulombically linked self-assembled monolayers. J. Electroanal. Chem., 1998, 458: 87-97.

[30] Lu X Q, Hu L N, Wang X Q. Thin-layer cyclic voltammetric and scanning electrochemical microscopic study of antioxidant activity of ascorbic acid at liquid/liquid interface. Electroanalysis, 2005, 17(11): 953-958.

[31] Walczak M M, Popenoe D D, Deinhammer R S, et al. Reductive desorption of alkanethiolate monolayers at gold: a measure of surface coverage. Langmuir, 1991, 7: 2687-2693.

[32] Lee L Y S, Lennox R B. Electrochemical desorption of n-alkylthiol sams on polycrystalline gold: Studies using a ferrocenylalkylthiol probe. Langmuir, 2007, 23: 292-296.

[33] 程志亮, 杨秀荣. 电化学交流阻抗技术表征自组装多层膜. 分析化学, 2001, 29: 6-9.

[34] Inoue References and further reading may be available for this article. To view references and further reading you must purchase this article. K Y, Ino K, Shiku H, et al. Electrochemical detection of endotoxin using recombinant factor C zymogen. Electrochemistry Communications, 2010, 12(8): 1066-1069.

[35] Pardo-Yissar V, Katz E, Lioubashevski O, et al. Layered polyelectrolyte films on Au-electrodes: Characterization of electron-transfer features at the charged-polymer-interface and application for selective redox reactions. Langmuir, 2001, 17: 1110-1118.

[36] Harris J J, Bruening M L. Electrochemical and in situ ellipsometric investigation of the permeability and stability of layered polyelectrolyte films. Langmuir, 2000, 16: 2006-2013.

[37] Lu X Q, Yuan H Q, Zuo G F, et al. Study of the size and separation of pinholes in the self-assembled thiol-porphyrin monolayers on gold electrodes. Thin Solid Films, 2008, 516: 6476-6482.

[38] Bollo S C, Yáez J, Sturm L, et al. Cyclic voltammetric and scanning electrochemical microscopic study of thiolated β-cyclodextrin adsorbed on a gold electrode. Langmuir, 2003, 19: 3365-3370.

[39] Chlistunoff J, Cliffel D, Bard A. Electrochemistry of fullerene films. Thin Solid Films, 1995, 257: 166-184.

[40] Lou X H, Lin H. Does cardiopulmonary bypass still represent a good investment? The biomaterials perspective. Sensors Actuators B: Chemical, 2008, 129: 225-231.

[41] 朱梓华，盛晓霞，张智军，等. 表面增强喇曼光谱检测卟啉单分子膜. 高等学校化学学报，2001，22(8)：1368-1372.

[42] Asanuma H, Bishop E M, Yu H Z. Electrochemical impedance and solid-state electrical characterization of silicon (111) modified with ω-functionalized alkyl monolayers. Electrochimi. Acta, 2007, 52: 2913-2919.

[43] Tominaga M, Ohira A, Yamaguchi Y, et al. Electrochemical, AFM and QCM studies on ferritin immobilized onto a self-assembled monolayer-modified gold electrode. J. Electroanal. Chem., 2004, 566: 323-329.

[44] Schoenfisch M H, Pemberton J E. Air stability of alkanethiol self-assembled monolayers on Ag and Au surfaces. J. Am. Chem. Soc., 1998, 120: 4502-4513.

[45] 张 群. 原子力显微镜. 上海计量测试，2002，29：38-39.

[46] Stranick S J, Parikh A N, Tao Y T, et al. Phase separation of mixed-composition self-assembled monolayers into nanometer scale molecular domains. J. Phys. Chem., 1994, 98: 7636-7646.

[47] Stranick S J, Atre S V, Parikh A N, et al. Nanometer-scale phase separation in mixed composition self-assembled monolayers. Nanotechnology, 1996, 7: 438-442.

[48] Viana A S, Leupold S, Eberle C, et al. A novel fullerene lipoic acid derivative: synthesis and preparation of self-assembled monolayers on gold. Surface Science, 2007, 601: 5062-5068.

[49] Liang S, Chen M, Xue Q, et al. Site selective micro-patterned rutile TiO2 film through a seed layer deposition. J. Colloid. Interf. Sci., 2007, 311: 194-202.

[50] Janssen D, Palma R D, Verlaak S, et al. Static solvent contact angle measurements, surface free energy and wettability determination of various self-assembled monolayers on silicon dioxide. Thin Solid Films, 2006, 515: 1433-1438.

[51] Yokota S, Matsuyama K, Kitaoka T, et al. Thermally responsive wettability of self-assembled methylcellulose nanolayers. Appl. Surf. Sci., 2007, 253: 5149-5154.

[52] Yao X, Yang M L, Wang Y F, et al. Au-NPs enhanced SPR biosensor based on hairpin DNA without the effect of nonspecific adsorption. Sensor Actuat B: Chem., 2007, 122: 351-356.

[53] Ulman A. An introduction to ultrathin organic films from L-B to SAM. Boston: Academic Press, 1991.

[54] Finklea H O. in: Bard A J, Rubinstein I (Eds.). Electroanalytical chemistry. New York: Marcel Dekker, 1996.

[55] Murray R W. Molecular design of electrode surface; techniques of chemistry series. New York: Wiley, 1992.

[56] Swalen J D, Allara D L, Andrade J D, et al. Molecular monolayers and films. Langmuir, 1987, 3: 932-950.

[57] Karpovich D S, Blanchard G J. Direct measurement of the adsorption kinetics of alkanethiolate self-assembled monolayers on a microcrystalline gold surface. Langmuir, 1994, 10: 3315-3332.

[58] Pan W, Durning C J, Turro N J. Kinetics of alkanethiol adsorption on gold. Langmuir, 1996, 12: 4469-4473.

[59] Hu K, Bard A J. In situ monitoring of kinetics of charged thiol adsorption on gold using an atomic force microscope. Langmuir, 1998, 14: 4790–4794.

[60] Bensebaa F, Voicu R, Huron L, et al. Chain n-alkanethiolate monolayers on polycrystalline gold. Langmuir, 1997, 13: 5335–5340.

[61] De Beno R F, Loucks G D, Manna D D, et al. Study using surface plasmon resonance techniques. Can. J. Chem., 1996, 74: 677–688.

[62] Dannenberger O, Buck M, Grunze M. Self-assembly of n-alkanethiols: A kinetic study by second harmonic generation. J. Phys. Chem. B., 1999, 103: 2202–2213.

[63] Forouzan F, Bard A J, Mirkin M V. Israel. Journal of Chemistry, 1997, 37: 155–163.

[64] Tirado J D, Acevedo D, Bretz R L, et al. Adsorption dynamics of electroactive self-assembling molecules. Langmuir, 1994, 10: 1971–1979.

[65] Bretz R L, Abruna H D. Synthesis of redox-active self-assembling osmium thiol monolayers and the effects of the working electrode's potential during deposition. J. Electroanal. Chem., 1995, 388: 123–128.

[66] Acevedo D, Bretz R L, Tirado J D, et al. Thermodynamics of adsorption of redox-active self-assembling monolayers of transition-metal complexes. Langmuir, 1994, 10: 1300–1305.

[67] Lorenzo E, Sanchez L, Pariente F, et al. Thermodynamics and kinetics of adsorption and electrocatalysis of NADH oxidation with a self-assembling quinone derivative. Anal. Chim. Acta., 1995, 309: 79–88.

[68] Shao H B, Yu H Z, Cheng G J, et al. Formation kinetics and electrochemical behavior of azobenzene self-assembled monolayers on gold electrodes. J. Phys. Chem. B., 1998, 102: 111–117.

[69] Bain C D, Troughton E B, Tao Y T, et al. Formation of monolayer films by the spontaneous assembly of organic thiols from solution onto gold. J. Am. Chem. Soc., 1989, 111: 321–335.

[70] Shimazu K, Yagi I, Sato Y, et al. In situ and dynamic monitoring of the self-assembling and redox processes of a ferrocenylundecanethiol monolayer by electrochemical quartz crystal microbalance. Langmuir, 1992, 8: 1385–1387.

[71] Hong H G, Park W. A study of adsorption kinetics and thermodynamics of ω-mercaptoalkylhydroquinone self-assembled monolayer on a gold electrode. Electrochimica Acta, 2005, 51: 579.

[72] Bendall D S. Protein Electron Transfer, Bios. Sci., Oxford, 1996.

[73] Li T T-T, Liu H Y, Weaver M J. Intramolecular electron transfer at metal surfaces. Ⅲ. Influence of bond conjugation on reduction kinetics of cobalt(Ⅲ) anchored to metal surfaces via thiophenecarboxylate ligands. J. Am. Chem. Soc., 1984, 106: 1233–1239.

[74] Li T T-T, Weaver M J. Intramolecular electron transfer at metal surfaces. 4. Depen dence of tunneling probability upon donor-acceptor separation distance. J. Am. Chem. Soc., 1984, 106: 6107–6108.

[75] Chidsey C E D. Free energy and temperature dependence of electron transfer at the metal-electrolyte interface. Science, 1990, 251: 919–922.

[76] Creager S E, Rowe G K. Normal-phase liquid chromatographic separation of stratum corneum ceramides with detection by evaporative light scattering and atmospheric pressure chemical ionization mass spectrometry. Anal. Chim. Acta., 1991, 246: 233-239.

[77] Finklea H O, Hanshew D D. Electron-transfer kinetics in organized thiol monolayers with attached pentaammine (pyridine)ruthenium redox centers. J. Am. Chem. Soc., 1992, 114: 3173-3181.

[78] Finklea H O, Ravenscroft M S, Snider D A. Electrolyte and temperature effects on long range electron transfer across self-assembled monolayers. Langmuir, 1993, 9: 223-227.

[79] Rowe G K, Creager S E. Interfacial solvation and double-layer effects on redox reactions in organized assemblies. J. Phys. Chem., 1994, 98: 5500-5507.

[80] Richardson J N, Peck S R, Curtin L S, et al. Electron-transfer kinetics of self-assembled ferrocene octanethiol monolayers on gold and silver electrodes from 115 to 170 K. J. Phys. Chem., 1995, 99: 766-772.

[81] Smalley J F, Feldberg S W, Chidsey C E D, et al. The kinetics of electron transfer through ferrocene-terminated alkanethiol monolayers on gold. J. Phys. Chem., 1995, 99: 13141-13149.

[82] Sachs S B, Dudek S P, Hsung R P, et al. Rates of interfacial electron transfer through pi-conjugated spacers. J. Am. Chem. Soc., 1997, 119: 10563-13564.

[83] Creager S E, Yu C J, Bamdad C, et al. Electron transfer at electrodes through conjugated "molecular wire" bridges. J. Am. Chem. Soc., 1999, 121: 1059-1064.

[84] Sek S, Misicka A, Bilewicz R. Effect of inter-chain hydrogen bonding on electron transfer through alkanethiol monolayers containing amide bonds J. Phys. Chem. B., 2000, 104: 5399-5402.

[85] Miller C J, Cuendet P, Gratzel M. Adsorbed omega-hydroxy thiol monolahers on gold electrodes-evidebce for electron-tunneling to redox species in solution. J. Phys. Chem., 1991, 95: 877-886.

[86] Miller C J, Gratzel M. Electrochemistry at omega -hydroxythiol coated electrodes. 2. Measurement of the density of electronic states distributions for several outer-sphere redox couples. J. Phys. Chem., 1991, 95: 5225-5233.

[87] Becka A M, Miller C J. Electrochemistry at omega-hydroxy thiol coated electrodes. 3. Voltage independence of the electron tunneling barrier and measurements of redox kinetics at large overpotentials. J. Phys. Chem., 1992, 96: 2657-2668.

[88] Becka A M, Miller C J. Electrochemistry at omega-hydroxy thiol coated electrodes. 4. Comparison of the double layer at omega-hydroxy thiol and alkanethiol monolayer coated Au electrodes. J. Phys. Chem., 1993, 97: 6233-6239.

[89] Xu J, Li H-L, Zhang Y. Relationship between electronic tunneling coefficient and electrode potential investigated by using self-assembled alkanethiol monolayers on gold electrodes. J. Phys. Chem., 1993, 97: 11497-11500.

[90] Terrettaz S, Becka A M, Traub M J, et al. omega-Hydroxythiol monolayers at au electrodes. 5. insulated electrode voltammetric studies of cyano/bipyridyl iron complexes. J. Phys. Chem., 1995, 99: 11216-11224.

[91] Cheng J, Saghi-Szabo G, Tossel J A, et al. Modulation of electronic coupling through

SAMs via internal chemical modification. J. Am. Chem. Soc., 1996, 118: 680–684.

[92]Slowinski K, Chamberlain R V, Bilewicz R, et al. Evidence of inefficient chain-to-chain coupling in electron tunneling through liquid alkanethiol monolayer films on mercury. J. Am. Chem. Soc., 1996, 118: 4709–4710.

[93]Cheng J, Miller C J. Ruthenium(II) hydrido complexes of S-methyldithiocarbazates. J. Phys. Chem., 1997, 101: 1058–1062.

[94]Slowinski K, Chamberlain R V, Miller C J, et al. Through-bond and chain-to-chain coupling. two pathways in electron tunneling through liquid alkanethiol monolayers on mercury electrodes. J. Am. Chem. Soc., 1997, 119: 11910–11919.

[95]Slowinski K, Slowinska K U, Majda M. Electron tunneling across hexadecanethiolate monolayers on mercury electrodes. reorganization energy, structure and permeability of the alkane/water interface. J. Phys. Chem., 1999, 103: 8544–8551.

[96]Portsailo L V, Fawcett W R. Studies of electron transfer through self-assembled monolayers using impedance spectroscopy. Electrochim. Acta., 2000, 45: 3497–3505.

[97]Sek S, Bilewicz R. Kinetics of long-range electron transfer through alkanethiolate monolayers containing amide bonds. J. Electroanal. Chem., 2001, 509: 11–18.

[98]Bilewicz R, Sek S, Zawisza I. Electron transport through composite monolayers. Elektrokhimiya, 2002, 38: 29–38.

[99]Sek S, Palys B, Bilewicz R. Contribution of intermolecular interactions to electron transfer through monolayers of alkanethiols containing amide groups. J. Phys. Chem. B., 2002, 106: 5907–5914.

[100]Xing Y F, O'Shea S J, Li S F Y. Electron transfer kinetics across a dodecanethiol monolayer self assembled on gold. J. Electroanal. Chem., 2003, 542: 7–11.

[101]Chidsey C E D, Bertozzi C R, Putvinski T M, et al. Coadsorption of ferrocene-terminated and unsubstituted alkanethiols on gold: electroactive self-assembled monolayers. J. Am. Chem. Soc., 1990, 112: 4301.

[102]Weber K, Hockett L, Creager S. Long-range electronic coupling between ferrocene and gold in alkanethiolate–based monolayers on electrodes. J. Phys. Chem. B., 1997, 101: 8286–8291.

[103]Caldwell W B, Campbell D J, Chen K, et al. A highly ordered self-assembled monolayer film of an azobenzenealkanethiol on Au(111)-electrochemical properties and structural characterization by synchrotron in-plane X-ray diffraction, atomic force microscopy, and surface-enhanced raman spectroscopy. J. Am. Chem. Soc., 1995, 117: 6071–6082.

[104]Campbell D J, Herr B R, Hulteen J C, et al. Ion-gated electron transfer in self-assembled monolayer films. J. Am. Chem. Soc., 1996, 118: 10211–10219.

[105]Ye Q, Fang J, Sun L. Distance dependence of enhanced raman scattering from an azobenzene terminal group. J. Phys. Chem. B., 1997, 101: 8221–8224.

[106]Zak J, Yuan H, Ho M, et al. Thiol–derivatized metalloporphyrins: monomolecular films for the electrocatalytic reduction of dioxygen at gold electrodes. Langmuir, 1993, 9: 2772–2774.

[107]Uosaki K, Kondo T, Zhang X-Q, et al. Very efficient visible-light-induced uphill electron transfer at a self-assembled monolayer with a porphyrin-ferrocene-thiol linked molecule.

J. Am. Chem. Soc., 1997, 119: 8367-8368.

[108] Imahori H, Norieda H, Ozawa S, et al. Chain length effect on photocurrent from polymethylene-linked porphyrins in self-assembled monolayers. Langmuir, 1998, 14: 5335-5338.

[109] Mirkin C A, Caldwell W B. Thin Film, Fullerene-based materials. Tetrahedron, 1996, 52: 5113-5130.

[110] Imahori H, Azuma T, Ajavakom A, et al. An investigation of photocurrent generation by gold electrodes modified with self-assembled monolayers of C60. J. Phys. Chem. B., 1999, 103: 7233-7237.

[111] Imahori H, Sakata Y. Fullerenes as novel acceptors in photosynthetic electron transfer. Eur. J. Org. Chem., 1999, 2445-2457.

[112] Doron A, Portnoy M, Lion-Dagan M, et al. Amperometric transduction and amplification of optical signals recorded by a phenoxynaphthacenequinone monolayer electrode: photochemical and pH-gated electron transfer. J. Am. Chem. Soc., 1996, 118: 8937-8944.

[113] Fox M A. Fundamentals in the design of molecular electronic devices: long-range charge carrier transport and electronic coupling. Acc. Chem. Res., 1999, 32: 201-207.

[114] Liu X H, Hu L N, Zhang L M, et al. Electron transfer between reactants ferric ion and decamethyferrocene located on NB/H_2O interfaces by thin layer method. Electrochim Acta, 2005, 51(3): 467-473.

[115] Lu X Q, Jin J, Kang J W, et al. Electrochemical behavior of unitary or binary self-assembled monolayers with thiol-derivatized cobaltous porphyrin. Mater. Chem. Phys., 2002, 77: 952-957.

[116] Lu X Q, Lv B Q, Xue Z H, et al. Self-assembled monolayers of a thiol-derivatized porphyrin on gold electrode: Film formation and electrocatalytic dioxygen reaction. Thin Solid Films, 2005, 488: 230-235.

[117] Lu X Q, Li M R, Yang C H, et al. Electron transport through self-assembled monolayer of thiol-end functionalized tetraphenylporphines and metal tetraphenylporphines. Langmuir, 2006, 22: 3035-3039.

第十一章 卟啉自组装膜电子传递性质研究

第一节 卟啉自组装膜与分子信息存储

自从 1974 年 Aviram 和 Ratner[1]首次提出有机分子具有半导体电子学的功能之后,关于分子二极管、分子开关、分子电容器以及具有光电类型记忆和逻辑体系的分子电子器件已有大量研究报道[2,3]。而关于记忆和逻辑功能的分子电子器件通常是将具有氧化还原活性的分子固定于作为活性存储介质的导电物表面,信息存储于分子分离状的氧化态中。卟啉类化合物作为一种光电活性物质广泛存在于生物界,它被选作分子信息存储的研究对象是基于它重要的特殊性质[4-9]:

(1)形成的 π 阳离子激发态在室温下相对稳定,可用于实际装置的制备和研究;

(2)卟啉分子在相对较低的电位下,即可产生多重氧化态,预示着它在能量消耗较低的情况下,可以提供分子的多点信息存储;

(3)在工作电势断开的状态下,卟啉氧化态储存的电荷可保留一定时间(长达数分钟),这进一步减少了能量消耗,并且在记忆装置所要求的更新速率方面,也可以得到有效的改善。

卟啉自组装膜是以自组装技术为基础,将功能化的卟啉或金属卟啉直接组装于基底表面或间接将卟啉化合物作用于自组装膜上。卟啉化合物在电极表面形成自组装膜之后,可以方便、灵活地研究卟啉本体的性质以及其他物质对卟啉性质的影响。卟啉化合物在电极表面具有多步电子转移过程,而且其氧化还原电位清晰、可逆。这预示着卟啉自组装膜在信息存储领域具有广阔的研究应用前景[10,11]。

1. 卟啉自组装膜分子信息存储的装置设计

在信息存储的应用研究中,不仅要证实电荷储存的存在,更要精确测量电荷的有效保留时间。目前,固定在电极表面电活性物质的氧化还原性质已经通过各种电化学方法进行了研究[12-17]。然而,这些方法由于是在极短的时间量程内(微秒或更小)进行的测量,背景电流的存在往往掩盖了法拉第电流信号。因此,必须发展一种新的方法把法拉第电流从其电荷电流中分离出来并进行测量,这种方法称为 OCP(Open Circuit Potential)法。卟啉自组装膜氧化态电荷保留时间 OCP 法是 Lindsey 等在考察卟啉自组装膜中卟啉氧化态电荷的存储时,应用快速循环伏安法在电解池开路状态下,以卟啉氧化态荧光特性的恢复来独立测量卟啉自组装膜氧化态的电荷保留时间。这种测量证明了当电化学池电位快速衰减到开路电位(OCP)时,卟啉自组装膜仍然保留其氧化态形式并延至一定时间(数百秒)。基于以上原理,设计的开环电路电势分析法就被应用于检测在电路断开状态下,当充电电流已经衰减时自组装膜氧化态电荷的输出。而且,伴随半波电位表征,开路电势产生了一个 S 形的响应,表明这是一个能斯特过程。这种实验建立了用来定量检测自组装膜中电荷存储量的方法体系[18](图 11-1)。

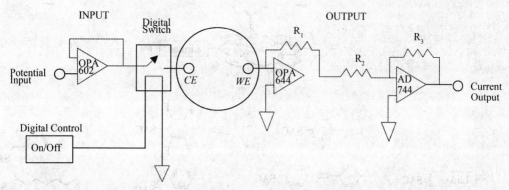

一个双电极的电流分析系统由一个高带宽的电流放大器组成,这个放大器使用了两个高带宽的运算放大器,R_1、R_2 和 R_3 分别是 2 kΩ、50 Ω 和 1 kΩ。工作电势经由一个快速的水银—继电器开关与一个对电极的电压输出器相连。

图 11-1　具有高带宽的电流分析装置图[18]

2. 卟啉自组装膜分子信息存储的影响因素

影响卟啉自组装膜信息存储的因素主要有:卟啉自组装膜的表面覆盖、存储单元的分子结构、卟啉自组装膜的表面空间定位以及自组装膜基底的材料等。

基于具有氧化还原活性的卟啉分子被作用于电极表面对分子信息存储系统研究提供的诱人途径,Lindsey 等详细考察了影响信息存储效率的各种因素。如研究者在考察自组装膜上金属卟啉基团—电极表面间的电子传递速率与卟啉氧化态电荷保留值的关系时,考察了各种卟啉自组装膜氧化还原反应动力学特性,并测量了在工作电势存在下电子传递速率常数以及在工作电势断开情况下电荷随时间($t_{1/2}$)衰减的速率常数。结果表明,所有卟啉自组装膜的电子转移常数值介于 $10^4 \sim 10^5$ s^{-1}。炔键连接卟啉自组装膜的电子转移常数值大于烷基链连接的卟啉结构,后者的速率常数值随亚甲基数的增加而减小。而三层三明治形卟啉结构的电子转移速率通常更慢。电子转移常数值的变化趋势与 $t_{1/2}$ 相平行,电荷衰减速率值比电子传递速率约小 6 个数量级,即卟啉自组装膜展示了相对较快的电子传递速率和一个缓慢的电荷衰减速率($t_{1/2}$ 更长)。另外,研究也发现,卟啉自组装膜的电子传递速率和电荷衰减速率对卟啉分子的表面覆盖非常敏感,都随自组装膜的致密而减小,分析认为这种行为是因为排除了溶剂、相反电荷离子和空间电荷的影响。卟啉表面覆盖量对速率的影响掩盖了不同连接基团和不同链长卟啉自组装膜测量时存在的差异。这些研究对设计分子信息存储装置中的分子元件具有重要意义[19-22]。

自组装膜上卟啉信息存储单元的分子结构一般包括卟啉与电极界面间的连接体结构;卟啉本体的分子结构以及不同金属离子中心的卟啉结构等。

为了考察卟啉自组装膜上金属卟啉基团与电极间的连接体结构对其信息存储的影响,研究者合成了 100 余种金属卟啉基团—尾端巯基基团间不同连接结构的卟啉分子,并将它们组装于电极表面。通过对这类自组装膜电化学性质的研究,发现卟啉基团和电活性表面间的电子传递速率是影响分子信息存储系统设计的关键因素。因此,探索连接体的结构及长度对卟啉分子信息存储的研究具有重要意义[23-32](图 11-2)。

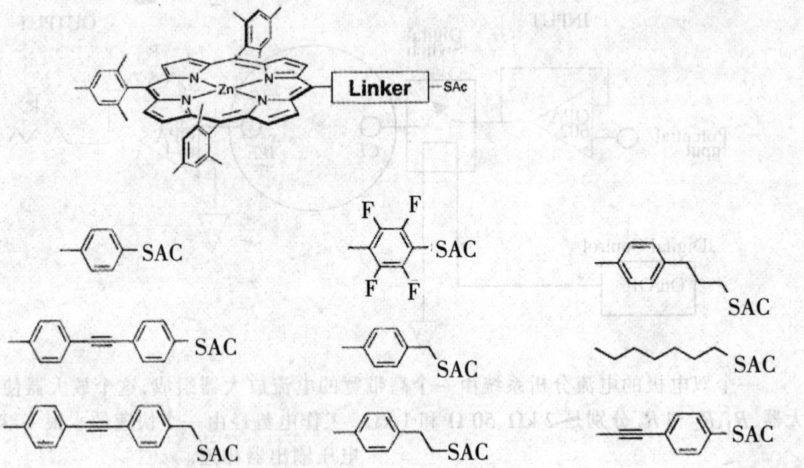

图 11-2　不同卟啉—巯基连接体的卟啉分子结构[23]

早期对卟啉自组装膜信息存储的研究，重点是设计和合成可以形成更加稳定自组装膜活性的卟啉结构，要求形成的自组装膜在长时间的电化学过程中没有损失，且具有较高的电荷储存效率。研究的对象包括二茂铁、锌卟啉、二茂铁—锌卟啉、镁酞菁以及三层三明治结构铕卟啉等活性分子的相应巯基化合物。研究者分别将这些活性化合物自组装于金电极表面，探索工作电势存在下这些活性分子自组装膜的电子传递速率以及开路状态下电荷的衰减速率。结果表明自组装膜上的三明治结构卟啉分子较其他分子更稳定。而且，三明治结构卟啉分子自组装膜的电子传递和电荷衰减速率值基本相近，表明三明治结构分子自组装膜在电化学行为过程中没有损失，这为分子信息存储电化学研究提供了一个超稳定的结构模型[33,34]。

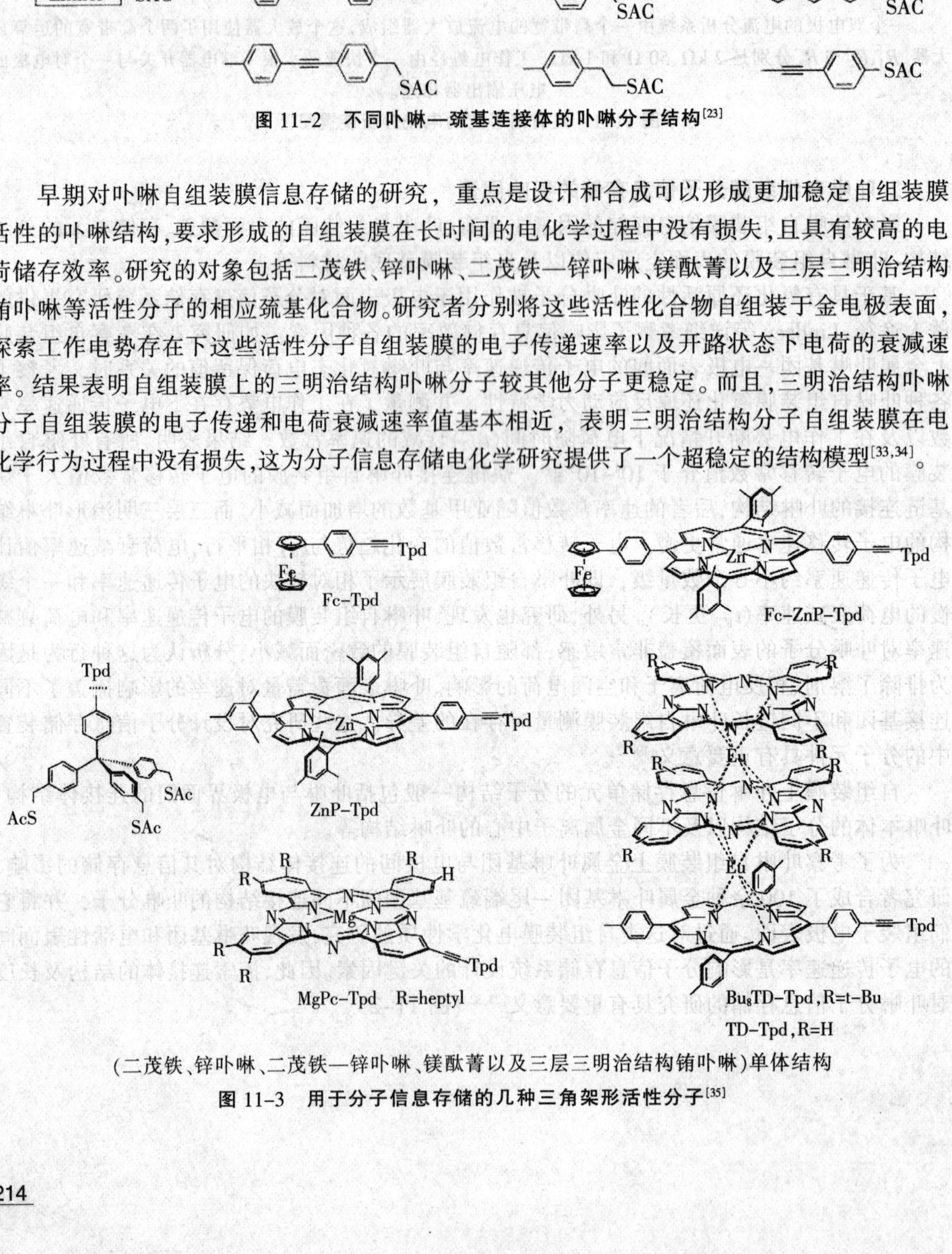

（二茂铁、锌卟啉、二茂铁—锌卟啉、镁酞菁以及三层三明治结构铕卟啉）单体结构

图 11-3　用于分子信息存储的几种三角架形活性分子[35]

为了更好地控制分子在基底表面的空间定位,加强分子膜的稳定性,研究者合成了三脚架形巯基尾式金属卟啉(图 11-3),考察了相对较低的电活性分子表面覆盖对分子信息存储的影响。研究表明:这种膜上卟啉环与电极表面呈垂直定位,卟啉分子之间的相互作用减弱,分子膜的稳定性增强。另外,研究者还考察了不同金属中心卟啉自组装膜的氧化还原活性,并以此卟啉分子结构为基础,合成了大量三脚架形活性分子用于分子信息存储研究[35-38]。

不同的卟啉结构对自组装膜的信息存储产生着影响,为了进一步考察卟啉及其自组装膜结构与其信息存储间的相互关系,研究者进行了大量的研究工作。如在多巯基尾式刚性连接的卟啉自组装膜的研究中,Lindsey 等首先合成了一系列巯基刚性连接的尾式锌卟啉并将其自组装于金膜表面(图 11-4),并使用 X-光电能谱、傅立叶反射红外光谱以及各种电化学方法对其进行了表征。结果表明:由于刚性连接基团——炔键的存在,$ZnPS_n$ 分子经由单一的巯基与金结合,而不必考虑其他巯基的存在对自组装的影响;$ZnPS_3$ 和 $ZnPS_4$ 在金表面的空间定位较其他卟啉更直立,形成的自组装膜也更为致密,其表面覆盖量与 $ZnPS_1$ 自组装膜相似,而与其较大的分子尺寸无关;所有自组装膜的热力学和电子传递过程动力学性质都相似,这归因于它们在自组装膜上具有相似的结合模式[39,40]。

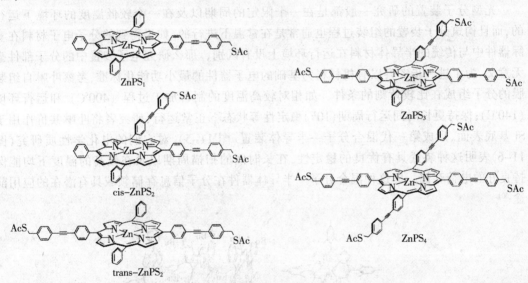

图 11-4 系列多巯基刚性连接的尾式锌卟啉结构[39]

除了合成相关的巯基尾式金属卟啉并将其自组装于金表面来研究分子信息存储之外,研究者还合成了一些羟基、巯基、巯基保护基、硒保护基的卟啉分子结构,考察了这类卟啉在硅、锗及金表面形成的单分子膜的氧化还原活性与分子的信息存储[41,42]。

如研究者将两种电活性分子 Fc—BzOH 和 Por—BzOH 通过苄羟基以 Si—O 键的形式结合于 Si(100)表面,测量了在工作电势下的氧化电子转移速率常数以及在工作电势断开情况下的电荷衰减速率(电荷保留时间的一半 $t_{1/2}$)。两种类型的 $t_{1/2}$ 具有平行变化规律。电荷衰减速率约小于电子转移速率 6 个数量级,它们强烈依赖于电活性物质在界面的浓度。对于 Por—BzOH,其在 Si(100)表面的 k_0 和 $t_{1/2}$ 以及表面覆盖值与其巯基卟啉衍生物在 Au(111)的值基本相同;而对于 Fc—BzOH,其在 Si(100)表面比相应的巯基衍生物在 Au(111)表面表现出更慢的电子转移和电荷衰减速率。电子转移和电荷衰减速率的减小与表面覆盖的增加以及对应于其衍生物在金表面的行为,一种解释是单层膜—溶液界面上可能存在空间电荷效应,第二种解释是连接链的空间定位可能改变了氧化还原中心与表面间的距离。总之,分子信息存储系统制备的简单化和稳定性决定了电活性分子在 Si(100)表面可能用于混合分子—半导体分子电子器件的研制和发展[43]。

另外,Lindsey 等将合成的炔键尾式苯基卟啉分别自组装于 Si(100)、SiO$_2$、Au(111)和导电玻璃表面,应用电化学方法研究了 Si(100)表面形成的炔键尾式苯基卟啉自组装膜的电化学特性。结果表明在 Si(100)表面可以形成 50 倍于饱和单分子膜厚的多层膜。膜的层数可以通过改变实验条件来进行控制。多层炔键尾式苯基卟啉膜的电化学行为表明其氧化还原态的热力学特性与单层膜基本相同。FTIR、激光拉曼光谱和 SEM 也分别用来考察了卟啉基团间缺失刚性炔键时膜结构的特征以及膜厚在数十到数百纳米时的情况。这种在线聚合技术不仅能够控制卟啉的膜厚,而且可以有效避免炔键尾式苯基卟啉在溶液中发生聚合时可能存在的溶解性问题。相比较于单分子膜,这种膜上卟啉氧化激发态增加了表面电荷的密度,表明这种卟啉膜可能作为电子寻址和分子信息存储的材料[44-49]。

光型分子装置的研究一般都是在一个限定的周期以及在一个较低温度的环境下运行的,而且构筑分子装置的组装过程也通常是在常温下进行的。如果得到的分子电子材料在实际器件中与传统的半导体材料在运行环境上没有区别,那么研究电子装置中的分子部件毫无意义。为了进一步考察以有机材料为基础的电子器件的最小功能化标准,考察卟啉自组装膜的分子组成在比较苛刻的条件,如相对较高温度的制备加工过程(400℃)和运行环境(140℃);保持更长时间运行周期(10^{12})稳定性等状态下正常运行,研究者将卟啉共价作用于 Si 基底表面,形成第一代混合分子—半导体装置(图 11-5),对它们的电化学性质研究(图 11-6)表明这种装置具有优良的稳定性,在长时间的扫描周期以及在较高的温度下均能保持良好的电荷存储,即这种混合分子—半导体器件在分子信息存储领域具有潜在的应用前途[50]。

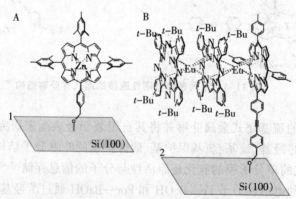

A 为卟啉单体;B 为三层三明治形卟啉/酞菁结构
图 11-5　以 Si 为基底形成自组装膜的卟啉分子结构图[50]

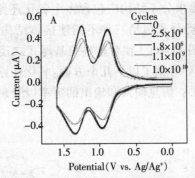

图 11-6 卟啉单体在 Si 表面形成的自组装单层膜在不同运行周期下的伏安响应[50]

3. 卟啉自组装膜分子信息存储研究现状

金属卟啉自组装膜在分子信息存储材料领域中良好的应用前景激发了研究者更大的研究兴趣。目前，对分子信息存储的研究已开始由单点信息存储向超高密度多点信息存储发展。分子水平上的超高密度多点信息存储要求设计的分子结构应具有较多的、明晰的氧化态。例如，为了达到多点分子信息存储研究的目的，研究者首先探索了比较简单的多点分子信息存储装置结构，设计的电活性卟啉单元通过一个 σ 键直接相连，而不是通过一个基团（或分子）连接（图 11-7）。研究表明：巯基尾式卟啉自组装膜显示了 4 个可逆的氧化还原峰，但是由于形成的卟啉 π 阳离子激发态使其氧化态发生部分重叠。而且，随着电化学扫描时间的增加，极易造成 SAM 上 meso—meso 直接连接的卟啉稳定性减弱，发生 meso—meso σ 键的断裂，表现为其卟啉单体的性质。研究表明这种分子结构在分子信息存储应用性方面受到一定限制[51]。

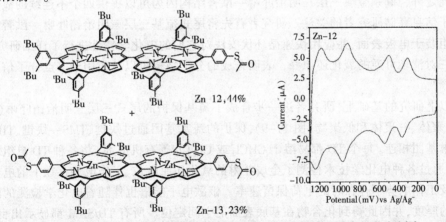

图 11-7 通过一个 σ 键直接相连的双卟啉分子结构及其电化学响应[51]

为了探索更加稳定的卟啉自组装膜分子结构单元，研究者设计合成了多种多活性中心卟啉化合物。其中二聚和三聚体的卟啉单元间由苯炔键相连。对于卟啉三聚体，中心卟啉及两端卟啉基团连接有不同的取代基或金属中心，因此，这种卟啉结构有可能提供不同的氧化还原性能。将此卟啉化合物自组装于电极表面，表征得到卟啉三聚体的中心金属卟啉基团与电极表面呈平行定位，其单层膜在电极表面上的排列更为致密、有序[52-54]。基于多点信息存储材料在分子水平上的设计要求分子结构应包含多个氧化还原活性单元，Lindsey 等将二茂铁基引入卟啉分子结构，这类分子包括三种不同连接体结构的二茂铁基卟啉分子和一种二-二茂铁基苯基卟啉分子(图11-8)。在对这些巯基尾式二茂铁基苯基卟啉自组装膜多点信息存储的研究中，研究者主要考察了以下几个方面：氧化还原活性分子的尺寸对分子信息存储系统设计空间结构的要求；信息输入和输出的速率以及氧化还原活性基团在信息分子不同位置时的电荷保留等[55-57]。

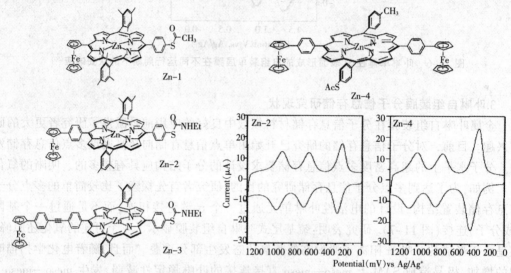

图 11-8　几种不同连接体的二茂铁基卟啉分子结构及其电化学响应[55]

除此之外，镧系金属三层三明治卟啉—酞菁结构因为可以提供四个不连续的氧化态而受到分子信息存储研究者的关注。研究者首先将尾式巯基三层三明治铈卟啉—酞菁结构化合物自组装于电极表面，并应用快速循环伏安法对它们的电化学特性进行了详细研究，可以观察到三对清晰可逆的氧化还原峰，表明合成的分子结构可以被应用于多点分子信息存储的研究[58-60]。

在以上研究的基础上，研究者进一步合成了巯基保护的尾式三层三明治铈卟啉(TD)的单体、二聚体、三聚体和低聚物(图11-9)，保护的巯基基团通过炔键、TMS—炔键、TIPS—炔键与卟啉基团相连。每个 TD 都易溶于 $CHCl_3$ 或 CH_2Cl_2 等有机溶剂。将各种 TD 自组装于金表面，并通过各种电化学技术检测了金属卟啉的氧化还原电位、工作电势存在下的电子传递速率以及工作电势不存在时的电荷保留速率。俄歇电子能谱也伴随着电化学检测的同时测量了膜的厚度，并因此得到化合物在基底表面的空间定位。所有 TD/SAM 都显示出良好、可逆的伏安响应，在 0 到+1.6 V 电势扫描范围内表现出 4 对明显的氧化还原峰(分别对应于三层三明治铈卟啉—酞菁分子结构的单体、二聚体、三聚体和低聚物)。所有自组装膜的电子传递速率常数都相似，范围为 $10^4\sim10^5\ s^{-1}$。电荷扩散速率(根据电荷保留寿命的一半)也相似，范围为 10~60 s。研究表明，这些速率受到分子聚集的密度以及电极表面的分子定位影响。而且所有的实验数据均支持了以下观点：假设所有巯基尾式 TD 化合物在电极表面具有相似的几何形状。特别是这些化合物都通过它们的连接体——大环与表面呈平行定位，这与单巯基尾式相似化合物与电极表面呈垂直定位不同，多巯基尾式 TD 化合物与金表面通过双巯基以共价键结合形成高质量的平行定位(图 11-10)[61-63]。

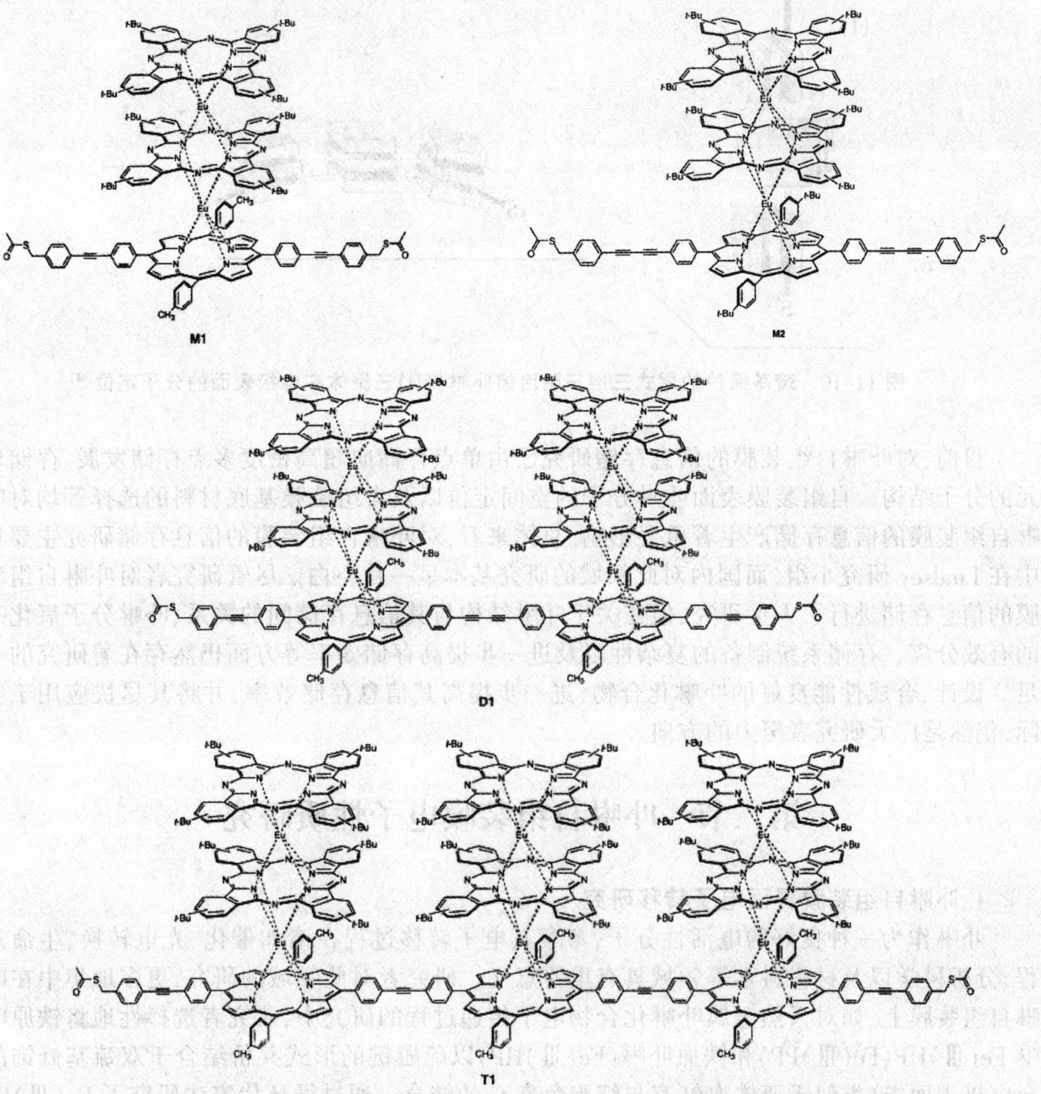

图 11-9 巯基保护的尾式三层三明治铕卟啉(TD)的单体、二聚体、三聚体和低聚物分子结构[58]

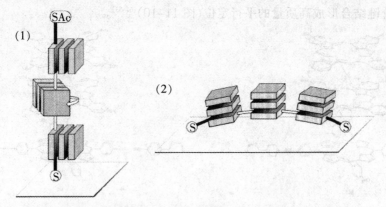

图 11-10　巯基保护的尾式三层三明治铕卟啉(TD)三聚体在电极表面的分子定位[61]

目前,对卟啉自组装膜的信息存储研究已由单点存储向超高密度多点存储发展。存储单元的分子结构、自组装膜表面卟啉分子的空间定位以及自组装膜基底材料的选择等均对卟啉自组装膜的信息存储产生着重要影响。总体来看,对卟啉自组装膜的信息存储研究主要集中在 Lindsey 研究小组,而国内对此领域的研究基本呈一片空白。尽管研究者对卟啉自组装膜的信息存储进行了大量研究,但在关于卟啉结构与其信息存储间的关系、卟啉分子氧化态的有效分离、存储系统制备的复杂性以及进一步提高存储效率等方面仍然存在着研究的不足。设计、合成性能良好的卟啉化合物,进一步提高其信息存储效率,并将其尽快应用于实际,仍然是广大研究者努力的方向。

第二节　卟啉自组装膜电子性质研究

1. 卟啉自组装膜界面电子转移研究

卟啉作为一种良好的电活性分子,考察其电子转移过程在诸如催化、光电转换、生命过程、分析科学以及材料科学等领域具有重要意义。研究者对此领域的研究,更多地集中在卟啉自组装膜上。如对天然金属卟啉化合物电子传递过程的研究中,研究者选择性地将铁原卟啉 $Fe(III)HP(Fe(III)PP)$ 和铁血卟啉($Fe(III)HP$)以硫醚键的形式共价结合于双巯基修饰的金电极表面,这类似于亚铁血红素与细胞色素 C 的结合。通过循环伏安法研究了 $Fe(III)PP$ 和 $Fe(III)HP$ 自组装膜的电化学性质,并从电荷的转移估计金属卟啉的表面覆盖约为单层膜的 30%。$Fe(III)PP$ 和金电极间的异相电子转移常数随着空间链长的增加而呈指数减小,其电子隧穿系数值为 $1.0\ \text{Å}^{-1}$,而且,在 pH=8 的溶液中,氧化还原基团的形式电位随链长增加呈线性增长。这表明在膜上的烷基链和卟啉基团呈密集排列,氧化还原电对的形式电位受到电极到烷基链自组装膜表面的静电降以及膜上 $Fe(III)PP$ 和溶液中的阴离子之间发生强烈的离子电对相互作用的影响。自组装膜上氧化还原物质 $Fe(III)PP$ 的形式电位与溶液的 pH 值呈现明显、复杂的依赖关系,这可以通过加入氢氧根离子、水以及分子氧等来改变铁的配位进行解释:铁既与基底表面的功能基团存在着相互作用,也与卟啉环上的丙酸酯基团存在着相互作用,所以并没有表现出电子—质子转移的结合机制[64]。

Imahori 等合成了几种双硫键连接的不同空间链长的锰卤代四苯基卟啉衍生物:$(MnPFPP—C_n—S)_2$,$(MnDCPP—C_n—S)_2$ 和 $(MnTTP—C_n—S)_2$($n=2,6,12$)(图 11-11),用于考察空间链长和卟啉结构对其电子传递的影响。在水和 DMSO 溶液中,通过电化学和光学技术对各种卟啉自组装膜的形成及氧化还原响应进行了研究。自组装膜的结构依赖于空间烷基链的亚甲基数和卟啉环上卤素的种类,氟代的和烷基链较长的卟啉自组装膜更为致密。从循环伏安法和 CCV 电位调制反射法(ER)(即在溶液中的电子反射)可以看到电极表面单层膜的电子转移发生在 Mn(Ⅲ)和 Mn(Ⅱ)间从而掩蔽了卟啉环(电子转移)。ER 法具有比 CV 法更高的灵敏性,尤其是在水溶液中。除了较大的表面浓度,长链卟啉的形式电位发生正移,形式电位的移动也依赖于卤素的种类;MnTTPC$_n$—S/SAMs>MnDCPP—C$_n$—S/SAMs>MnPF—PP—C$_n$—S/SAMs。在 DMSO 和水溶液中,从单分子膜到电极表面的电子转移速率常数可以通过电位调制反射法(ER)与调制频率的关系计算得到,它随着空间链长的减少而增加,并且依 MnDCPP—C$_n$—S/SAMs>MnTTPC$_n$—S/SAMs>MnPFPP—C$_n$—S/SAMs 的顺序递减。卤素种类的影响可以通过膜上卟啉基团间空间位阻的存在使其很难发生聚集来进行解释,尤其是在水溶液中。根据 Marcus 相关理论,研究者还计算得到卟啉自组装膜的电子隧穿系数为 0.070 Å^{-1}。获得的隧穿系数远小于文献报道的其他相关电活性物质自组装膜的值,这可能是由于锰卟啉与电极间电子发生了相互作用的结果[65-67]。

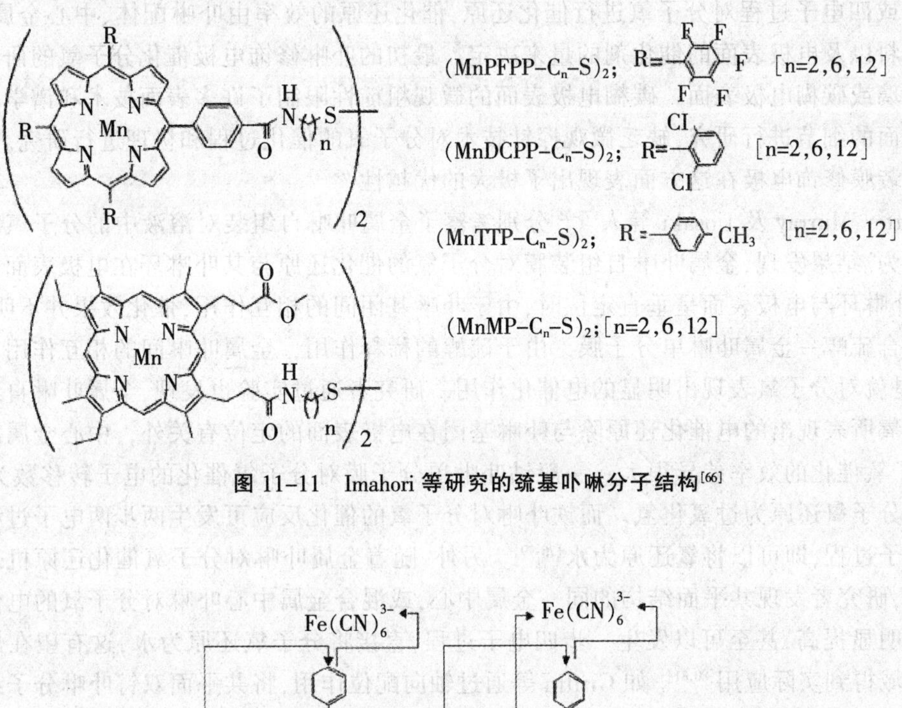

图 11-11 Imahori 等研究的巯基卟啉分子结构[66]

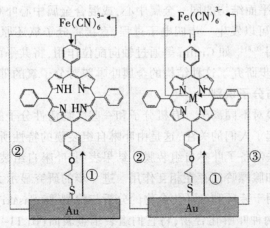

图 11-12 溶液中氧化还原电对跨越卟啉、金属卟啉自组装膜的电子传递模式[68]

另外，Lu 等人[68]也通过溶液中的分子探针 $Fe(CN)_6^{3-}$—$Fe(CN)_6^{4-}$ 考察了金属卟啉自组装膜对电子传递的阻滞。研究表明，$Fe(CN)_6^{3-}$—$Fe(CN)_6^{4-}$ 与电极表面间的电子传递除了通过隧穿和针孔机制外，自组装膜上卟啉金属中心对溶液中的电活性与电极间的电子传递也可能起到了中介作用（图11-12）。

2. 金属卟啉自组装膜对分子氧催化还原的电化学行为研究

自组装膜的发展使我们可以创造性地设计和控制电极表面，这种表面引起的关注之一是可以利用它们作为理想的电催化平台，因为只有在电极表面被很好控制的前提下，才能够获得电极表面结构和催化剂活性点之间的确定关系。在众多电催化剂选择的过程中，卟啉因其在生命系统以及在仿生学中良好的功能，备受研究者的关注。从卟啉化合物的结构来看，研究者特别感兴趣的是其卟啉环可以在电极表面被很好地控制。基于此，金属卟啉自组装膜被广泛地应用于基底物质的电催化氧化还原反应。

金属卟啉对分子氧的还原表现出良好的催化活性，电极表面结合金属卟啉后可以通过两电子或四电子过程对分子氧进行催化还原，催化还原的效率由卟啉配体、中心金属离子、电极材料以及电极表面的催化剂的量来决定。最初的卟啉修饰电极催化分子氧的研究多集中在玻璃或碳糊电极表面，碳糊电极表面的微观粗糙性限制了许多表面技术和谱学手段对电极表面的细节进行研究，缺乏微观探针技术对分子氧的催化过程和机理进行研究。金属表面自组装膜修饰电极在这方面表现出了极大的优越性[69-72]。

Porter、Murray 及 Uosaki 等人[73-75]分别考察了金属卟啉自组装对溶液中的分子氧的催化还原行为。结果发现，金属卟啉自组装膜对分子氧的催化还原与其卟啉环在电极表面的定位有关。卟啉环与电极表面呈垂直定位时，由于卟啉基团间的相互作用，催化效果并不明显，但对于混合硫醇—金属卟啉单分子膜，由于硫醇的稀释作用，金属卟啉间的相互作用大大减弱，于是就对分子氧表现出明显的电催化作用。研究者通过实验也发现，金属卟啉自组装膜对分子氧所表现出的电催化还原除与卟啉基团在电极表面的定位有关外，中心金属离子不同，分子氧催化的效率差异很大。一般钴卟啉单分子膜对分子氧催化的电子转移数为2，即只能将分子氧还原为过氧化氢。而铁卟啉对分子氧的催化反应可发生两步两电子过程或一步四电子过程，即可以将氧还原为水[76-79]。另外，随着金属卟啉对分子氧催化还原机理研究的深入，研究者发现共平面结构的同一金属中心，或混合金属中心卟啉对分子氧的电催化还原效率明显提高，甚至可以发生一步四电子过程，直接将分子氧还原为水，这有望在燃料电池等领域得到实际应用[80,81]。如 Griffin 等通过轴向配位作用，将共平面双钌卟啉分子结合于硫醇自组装膜表面，初步研究了这种结构的金属卟啉膜对分子氧的催化还原性质[82]。

3. 卟啉自组装膜与分子识别

利用卟啉自组装膜对不同离子、有机分子和一些生物活性分子的渗透性和运输差异进行选择性的测定也引起了人们的关注，这是由卟啉自组装膜的特性所决定的[83-85]。如有研究者通过表面合成的方法制备了卟啉自组装膜。结果表明，卟啉自组装膜与模型化合物，如荧蒽、1,10-邻二氮杂菲和腺嘌呤间存在相互作用。进一步的研究显示大环低聚吡咯功能基团的自组装膜可以被应用于芳环和杂环化合物的分析与检测[86]。Sasaki 等人的工作更具意义，他们创造性地合成了两种卟啉化合物，将它们组装于金表面（图11-13），并使用 X-光电能谱、紫外—可见光谱以及扫描隧道显微镜对膜进行了表征。XPS 结合能的位移表明卟啉通过硫—金键化学吸附于金的表面。吸收光谱 Soret 带的红移说明卟啉分子排列在金表面呈肩并肩定位。围绕 STM 的卟啉分子形貌，其直径约为 2 nm，这与相应的四苯基卟啉的尺寸接近（1.8 nm）。综合实验数据表明 TPP—P—二硫化物和 TPP—P—硫醇化合物在金表面均形成了均一的单分子膜，而且，卟啉环平行定位于金基底表面。在室温条件下，卟啉自组装膜的稳定性至少可保留一周。进一步的研究表明，这些膜可以作为固定生物分子如蛋白、细胞识别的理想空间，成为新的生物识别表面的适合模型[87]。

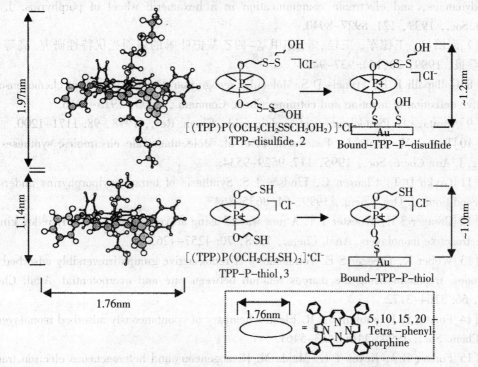

图 11-13 Sasaki 等人研究的卟啉分子结构及其自组装膜[87]

聚合物和表面间的氢键特定相互作用也可以被应用去选择性地修饰金表面。界面识别过程可以应用石英晶体微天平从一个非极性溶剂中的二胺基吡啶功能基（DAP）的聚苯乙烯的吸附过程中，通过观察胸腺嘧啶自组装膜频率的改变来进行研究。石英晶体微天平同时也结合了XPS、接触角以及椭圆偏振光测量了聚合物修饰表面，证明了聚合物—表面氢键相互作用的选择性。这些研究表明聚合物和表面间识别元素功能在速率、选择性以及聚合物的表面覆盖的检测中是至关重要的[88]。除此之外，利用杯芳烃、冠醚以及卟啉空穴的尺寸差异，也有进行卟啉自组装膜超分子固定的相关研究报道[89]。

参考文献

[1] Aviram A, Ratner M A. Molecular rectifiers. Chem. Phys. Lett., 1974, 29: 277-283.

[2] Kwok K S, Ellenbogen J C. Moletronics: Future. Electronics. Mater. Today, 2002, 2002: 28-37.

[3] Carroll R L, Gorman C B. The genesis of molecular electronics. Angew. Chem. Int. Ed., 2002, 41: 4378-4440.

[4] Moore T A. Photodriven charge separation in a carotenoporphyrin-quinone triad.

Nature, 1984, 307: 630-632.

[5] Collman J P, Hutchison J E, Wagenknecht P S, et al. Unprecedented, bridged dihydrogen complex of a cofacial metallodiporphyrin and its relevance to the bimolecular reductive elimination of hydrogen. J. Am. Chem. Soc., 1990, 112: 8206-8208.

[6] Li J Z, Ambroise A, Yang S I, et al. Template-directed synthesis, excited-state photodynamics, and electronic communication in a hexameric wheel of porphyrins. J. Am. Chem. Soc., 1999, 121: 8927-8940.

[7] 谢腾峰, 王德军, 王瑛, 等. 四甲基-四乙基钯卟啉的表面光伏特性研究. 高等学校化学学报, 1999, 20(16): 937-940.

[8] Gollapalli R D, Francis D S. Molecular recognition directed porphyrin chemosensor for selective detection of nicotine and cotinine. Chem. Commun., 2000, 1915-1916.

[9] Armitage B. Photocleavage of nucleic acids. Chem. Rev., 1998, 98: 1171-1200.

[10] Tour J M, Jones L, Pearson D L. et al. Molecular scale electronics: syntheses and testing. J. Am. Chem. Soc., 1995, 117: 9529-9534.

[11] Gryko D T, Clausen C, Lindsey J S. Synthesis of tetraphenylporphyrins under very mild conditions. J. Org. Chem., 1999, 64: 8635-8647.

[12] Creager S E, Wooster T T. A new way of using ac voltammetry to study redox kinetics in electroactive monolayers. Anal. Chem., 1998, 70: 4257-4263.

[13] Weber K, Creager S E. Voltammetry of redox-active groups irreversibly adsorbed onto electrodes. treatment using the marcus relation between rate and overpotential. Anal. Chem., 1994, 66: 3164-3172.

[14] Forster R J, Faulkner L R. Electrochemistry of spontaneously adsorbed monolayers. J. Am. Chem. Soc., 1994, 116: 5453-5461.

[15] Forster R J, Keyes T E, Majda M. Homogeneous and heterogeneous electron transfer dynamics of osmium-containing monolayers at the air/water interface. J. Phys. Chem. B, 2000, 104: 4425-4432.

[16] Kertesz V, Chambers J Q, Mullenix A N. Chronoamperometry of surface-confined redox couples. Application to daunomycin adsorbed on hanging mercury drop electrodes. Electrochim. Acta, 1999, 45: 1095-1104.

[17] Palecek E, Tomschik M, Stankova V, et al. Chronopotentiometric stripping of DNA at mercury electrodes. Electroanalysis, 1997, 9: 990-997.

[18] Roth K M, Lindsey J S, Bocian D F, et al. Characterization of charge storage in redox-active self-assembled monolayers. Langmuir, 2002, 18: 4030-4040.

[19] Roth K M, Gryko D T, Clausen C, et al. Comparison of electron-transfer and charge-retention characteristics of porphyrin-containing self-assembled monolayers designed for molecular information storage. J. Phys. Chem. B., 2002, 106: 8639-8648.

[20] Carcel C M, Laha J K, Loewe R S, et al. Porphyrin architectures tailored for studies of molecular information storage. J. Org. Chem., 2004, 69: 6739-6750.

[21] Padmaja K, Wei L, Lindsey J S, et al. A Compact all-carbon tripodal tether affords high coverage of porphyrins on silicon surfaces. J. Org. Chem., 2005, 70: 7972-7978.

[22] Thamyongkit P, Yu L, Padmaja K, et al. Porphyrin dyads bearing carbon tethers for studies of high-density molecular charge storage on silicon surfaces. J. Org. Chem., 2006, 71:

1156-1171.

[23] Gryko D T, Clausen C, Roth K M, et al. Synthesis of "porphyrin-linker-thiol" molecules with diverse linkers for studies of molecular-based information storage. J. Org. Chem., 2000, 65: 7345-7355.

[24] Roth K M, Dontha N, Dabke R B, et al. Molecular approach toward information storage based on the redox properties of porphyrins in self-assembled monolayers. J. Voc. Sci. Technol. B., 2000, 18(5): 2359-2364.

[25] Gryko D T, Li J, Diers J R, et al. Studies related to the design and synthesis of a molecular octal counter. J. Mater. Chem., 2001, 11: 1162-1180.

[26] Roth K M, Liu Z, Gryko D T, et al. Molecules as Components of Electronic Devices; ACS Symposium Series 844. Washington DC: American Chemical Society, 2003.

[27] Liu Z, Schmidt I, Thamyongkit P, et al. Synthesis and film-forming properties of ethynyl porphyrins. Chem. Mater., 2005, 17: 3728-3742.

[28] Zou Z-Q, Wei L, Chen F, et al. Solution STM Images of porphyrins on HOPG reveal that subtle differences in molecular structure dramatically alter packing geometry. J. Porphyrins Phthalocyanines, 2005, 9: 387-392.

[29] Zaidi S H H, Loewe R S, Clark B, et al. Nearly chromatography-free synthesis of the A3B-porphyrin 5-(4-hydroxymethylphenyl)-10,15,20-tri-p-tolylporphinatozinc (II). Org. Process Res. Dev., 2006, 10: 304-314.

[30] Schmidt I, Jiao J, Thamyongkit P, et al. Investigation of stepwise covalent synthesis on a surface yielding porphyrin-based multicomponent architectures. J. Org. Chem., 2006, 71: 3033-3050.

[31] Zuo G F, Liu X H, Yang J D, et al. Study of the adsorption kinetics of thiol-derivatized porphyrin on the surface of gold electrode. J. Electroanal. Chem., 2007, 605: 81-88.

[32] 左国防, 雷新有, 杜捷, 等. 系列卟啉的合成及其光谱及电化学性质研究. 光谱实验室, 2008, 25(3): 290-296.

[33] Wei L Y, Tiznado H, Liu G M, et al. Adsorption characteristics of tripodal-thiol-functionalized porphyrins on gold. J. Phys. Chem. B, 2005, 109: 23963-23971.

[34] Tomizaki K-Y, Yu L, Wei L, et al. Synthesis of cyclic hexameric porphyrin arrays. anchors for surface immobilization and columnar self-assembly. J. Org. Chem., 2003, 68: 8199-8207.

[35] Wei L Y, Padmaja K, Youngblood W J, et al. Diverse redox-active molecules bearing identical thiol-terminated tripodal tethers for studies of molecular information storage. J. Org. Chem., 2004, 69: 1461-1469.

[36] Wei L, Syomin D, Loewe R S, et al. Structural and electron-transfer characteristics of carbon-tethered porphyrin monolayers on Si (100). J. Phys. Chem. B, 2005, 109: 6323-6330.

[37] Liu Z, A Yasseri A, Loewe R S, et al. Synthesis of porphyrins bearing hydrocarbon tethers and facile covalent attachment to Si(100). J. Org. Chem., 2004, 69: 5568-5577.

[38] Loewe R S, Ambroise A, Muthukumaran K, et al. Porphyrins bearing mono or tripodal benzylphosphonic acid tethers for attachment to oxide surfaces. J. Org. Chem., 2004, 69: 1453-

1460.

[39] Yasseri A A, Syomin D, Malinovskii V L, et al. Characterization of self-assembled monolayers of porphyrins bearing multiple thiol−derivatized rigid−rod tethers. J. Am. Chem. Soc., 2004, 126: 11944−11953.

[40] Zuo G F, Yuan H Q, Yang J D, et al. Study of orientation mode of cobalt-porphyrin on the surface of gold electrode by electrocatalytic dioxygen reduction. J. Mol. Catal. A., 2007, 269: 46−52.

[41] Balakumar A, Lysenko A B, Carcel C, et al. diverse redox-active molecules bearing O-, S-, or Se-terminated tethers for attachment to silicon in studies of molecular information storage. J. Org. Chem., 2004, 69: 1435−1443.

[42] Yasseri A A, Syomin D, Loewe R S, et al. Structural and electron-transfer characteristics of O−, S−, and Se-tethered porphyrin monolayers on Si (100). J. Am. Chem. Soc., 2004, 126: 15603−15612.

[43] Roth K M, Yasseri A A, Liu Z M, et al. Measurements of electron-transfer rates of charge-storage molecular monolayers on si(100). Toward hybrid molecular/semiconductor information storage devices. J. Am. Chem. Soc., 2003, 125: 505−517.

[44] Liu Z M, Schmidt I, Thamyongkit P, et al. Synthesis and film-forming properties of ethynyl porphyrins. Chem. Mater., 2005, 17: 3728−3742.

[45] Li Q, Surthi S, Mathur G, et al. Electrical characterization of redox-active molecular monolayers on SiO_2 for memory applications. Appl. Phys. Lett., 2003, 83: 198−200.

[46] Muthukumaran K, Loewe R S, Ambroise A, et al. Porphyrins bearing arylphosphonic acid tethers for attachment to oxide surfaces. J. Org. Chem., 2004, 69: 1444−1452.

[47] Gowda S, Mathur G, Li Q, et al. Hybrid silicon/molecular memories: co−engineering for novel functionality. Technical Digest−IEEE Meeting on Electron Devices, 2003, 537−540.

[48] Gowda S, Mathur G, Li Q, et al. Modulation of drain current by redox-active molecules incorporated in Si MOSFETS. IEEE Tech. Dig., 2004, 707−710.

[49] Jiao J, Anariba F, Tiznado H, et al. Stepwise formation and characterization of covalently linked multiporphyrin-imide architectures on Si(100). J. Am. Chem. Soc., 2006, 128: 6965−6974.

[50] Liu Z M, Yasseri A A, Lindsey J S, et al. Molecular memories that survive silicon device processing and real-world operation. Science, 2003, 302: 1543−1545.

[51] Clausen C, Gryko D T, Yasseri A A, et al. Investigation of tightly coupled porphyrin arrays comprised of identical monomers for multibit information storage. J. Org. Chem., 2000, 65: 7371−7378.

[52] Gryko D T, Zhao F, Yasseri A A, et al. Synthesis of thiol-derivatized ferrocene-porphyrins for studies of multibit information storage. J. Org. Chem., 2000, 65: 7356−7362.

[53] Schweikart K-H, Malinovskii V L, Diers J R, et al. Design, synthesis, and characterization of prototypical multistate counters in three distinct architectures. J. Mater. Chem., 2002, 12: 808−828.

[54] Li Q, Surthi S, Mathur G, et al. Multiple-bit storage properties of porphyrin monolayers on SiO2. Appl. Phys. Lett., 2004, 85: 1829−1831.

[55] Clausen C, Gryko D T, Dabke R B, et al. Synthesis of thiol-derivatized porphyrin

dimers and trimers for studies of architectural effects on multibit information storage. J. Org. Chem., 2000, 65: 7363-7370.

[56] Li Q, Mathur G, Homsi M, et al. Capacitance and conductance characterization of self-assembled ferrocene monolayers on silicon surfaces for memory applications. Appl. Phys. Lett., 2002, 81: 1494-1496.

[57] Li Q, Mathur G, Gowda S, et al. Multibit memory using self-assembly of mixed ferrocene/porphyrin monolayers on silicon. Adv. Mater., 2004, 16: 133-137.

[58] Li J Z, Gryko D, Dabke R B, et al. Synthesis of thiol-derivatized europium porphyrinic triple-decker sandwich complexes for multibit molecular information storage. J. Org. Chem., 2000, 65: 7379-7390.

[59] Gross T, Chevalier F, Lindsey J S. Investigation of rational syntheses of heteroleptic porphyrinic lanthanide (europium, cerium) triple-decker sandwich complexes. Inorg. Chem., 2001, 40: 4762-4774.

[60] Lysenko A B, Malinovskii V L, Kisari P, et al. Multistate molecular information storage using s-acetylthio-derivatized dyads of triple-decker sandwich coordination compounds. J. Porphyrins Phthalocyanines, 2005, 9: 491-508.

[61] Schweikart K-H, Malinovskii V L, Yasseri A A, et al. Synthesis and characterization of bis(S-acetylthio)-derivatized europium triple-decker monomers and oligomers. Inorg. Chem., 2003, 42: 7431-7446.

[62] Lysenko A B, Thamyongkit P, Schmidt I, et al. Diverse porphyrin dimers as candidates for high-density charge-storage molecules. J. Porphyrins Phthalocyanines, 2006, 10: 22-32.

[63] Padmaja K, Youngblood W J, Wei L, et al. Triple-decker sandwich compounds bearing compact triallyl tripods for molecular information storage applications. Inorg. Chem., 2006, 45: 5479-5492.

[64] Pilloud D L, Chen X X, Dutton P L, et al. Electrochemistry of self-assembled monolayers. J. Phys. Chem. B., 2000, 104: 2868-2877.

[65] Yamada T, Hashimoto T, Kikushima S, et al. Electron transfer of manganese halogenated tetraphenylporphyrin derivatives assembled on gold electrodes. Langmuir, 2001, 17: 4634-4640.

[66] Yamada T, Kikushima S, Hikita T, et al. Molecular assembly of manganese mesoporphyrin derivatives on a gold electrode and their electron transfer activity Thin Solid Films, 2005, 474: 310-321.

[67] Yamada T, Nango M, Ohtsuk T. Potential modulation reflectance of manganese halogenated tetraphenylporphyrin derivatives assembled on gold electrodes. J. Electroanal. Chem., 2002, 528: 93-102.

[68] Lu X Q, Li M R, Yang C H, et al. Electron transport through self-assembled monolayer of thiol-end functionalized tetraphenylporphines and metal tetraphenylporphines. Langmuir, 2006, 22: 3035-3040.

[69] Shi C, Anson F C. Electrocatalysis of the reduction of molecular oxygen to water by tetraruthenated cobalt meso-tetrakis(4-pyridyl)porphyrin adsorbed on graphite electrodes. Inorg. Chem., 1992, 31: 5078-5083.

[70] Choi E M, Jeong H, Park D H, et al. Electrocatalytic reduction of dioxygen by new

water soluble cobalt (ii) tetrakis-(1,2,5,6-tetrafluoro-4-NN'N" -trimethylanilinium)-b-octabromoporphyrin in aqueous solutions. Bull. Korean Chem. Soc., 1999, 20(9): 1256-1260.

[71] Liu M H, Su Y O. A Comparison of the reaction products at different oxidation states. J Electroanal. Chem., 1997, 426: 197-202.

[72] Ishito A, Majima T. Photocurrent generation of a porphyrin self-asembly monolayer on a gold film electrode by surface plasmon excitation using near-IR light. Chem. Phys. Lett., 2000, 322: 242-246.

[73] Zak J, Yuan H, Ho M, et al. Thiol-derivatized metalloporphyrins: monomolecular films for the electrocatalytic reduction of dioxygen at gold electrodes. Langmuir, 1993, 9: 2772-2774.

[74] Postlethwaite T A, Hutchison J E, Hathcock K W, et al. Optical, electrochemical, and electrocatalytic properties of self-assembled thiol-derivatized porphyrins on transparent gold films. Langmuir, 1995, 11: 4109-4116.

[75] Nishimura N, Ooi M, Shimazu K, et al. Post-assembly insertion of metal ions into thiol-derivatized porphyrin monolayers on gold. J. Electroanal. Chem., 1999, 473: 75-84.

[76] Cordas C M, Viana A S, Leupold S, et al. Self-assembled monolayer of an iron (III) porphyrin disulphide derivative on gold. Electrochemistry Communications, 2003, 5: 36-41.

[77] Viana A S, Leupold S, Montforts F-P, et al. Self-assembled monolayers of a disulphide-derivatised cobalt-porphyrin on gold. Electrochimica Acta., 2005, 50: 2807-2813.

[78] Fuerte A, Cormab A, Iglesias M, et al. Approaches to the synthesis of heterogenised metalloporphyrins-Application of new materials as electrocatalysts for oxygen reduction. J. Mol. Cat. A,. 2006, 246: 109-117.

[79] Groves J T. High-valent iron in chemical and biological oxidations. Journal of Inorganic Biochemistry, 2006, 100: 434-447.

[80] Kadish K M, Fre'mond L, Burdet F, et al. Cobalt (IV) corroles as catalysts for the electroreduction of O2. reactions of heterobimetallic dyads containing a face-to-face linked Fe (iii) or mn(iii) porphyrin. J. Inorg. Biochem., 2006, 100: 858-868.

[81] Herbert W, Vésper Y O, Sergio D, et al. Supramolecular tetracluster-cobalt porphyrin: A four-electron transfer catalyst for dioxygen reduction. Electrochim. Acta., 2004, 49: 3711-3718.

[82] Hassan S A, Hassan H A, Hashem K M, et al. Catalytic and physical characteristics of axially modified cobalt (III) tetraphenyl porphyrin immobilized on chemically different supports. Applied Catalysis A: General, 2006, 300: 14-23.

[83] Trevin B, Bedioui F, Devynck J. New electropolymerized nickel porphyrin films. Application to the detection of nitric oxide in aqueous solution. J. Electroanal. Chem., 1996, 408: 261-265.

[84] Dobson D J, Saini S. Porphyrin-modified electrodes as biomimetic sensors for the determination of organohalide pollutants in aqueous samples. Anal. Chem., 1997, 69: 3532-3538.

[85] Araki K, Anganes L, Azevedo C M, et al. Electrochemistry of a tetraruthenated cobalt porphyrin and its use in modified electrodes as sensors of reducing analytes. J. Electroanal. Chem., 1995, 397: 205-210.

[86] Zaruba K, Matejka P, Volf R, et al. Formation of porphyrin-and sapphyrin-containing monolayers on electrochemically prepared gold substrates: A FT Raman spectroscopic study. Langmuir, 2002, 18: 6896–6906.

[87] Boeckl M S, Bramblett A L, Hauch K D, et al. Self-assembly of tetraphenylporphyrin monolayers on gold substrates. Langmuir, 2000, 16: 5644–5653.

[88] Norsten T B, Jeoung E, Thibault R J, et al. Specific hydrogen bond-mediated recognition and modification of surfaces using complimentary functionalized polymers. Langmuir, 2003, 19: 7089–7093.

[89] Zhang S, Echegoyen L. Non-covalent immobilization of C_{60} on gold surfaces by SAMs of porphyrin derivatives. Tetrahedron, 2006, 62: 1947–1954.

第十二章 卟啉光诱导电子转移和能量传递以及卟啉自组装膜光电转换研究

卟啉类化合物是非常理想的 D—A 体系中的供电子部分。多年来国内外化学工作者竭力合成各种具有不同结构特点的卟啉化合物。模拟植物体内光合反应过程或者作为分子光电器件的模拟系统,已成为国内外十分活跃的研究领域[1]。本章重点对卟啉自组装膜近几年来的光致电荷转移、能量传递以及光电转换等方面的研究进行述评。

第一节 卟啉光诱导电子转移和能量传递

1. 金属卟啉及其配合物的光诱导性质

研究者[2]采用稳态吸收光谱、荧光光谱及皮秒时间分辨技术研究了轴向配体 4-N,N-二甲氨基吡啶(DMAP)对锌卟啉(Zn(p-OCH$_3$)TPP)的电子激发态性质的影响。发现轴向配体的引入导致热荧光寿命缩短,并使 S_1 态振动模式发生变化,分子内振动能量再分配以及振动驰豫过程在新的振动模式间进行得更快。该实验结果及相关的机理探讨对加深理解金属卟啉衍生物在化学和生物体系中的反应机制有一定的意义。

Fukuzumi 等[3]合成了包括金属锌和金的三卟啉配合物(ZnPQ—2HPQ—AuPQ$^+$)。瞬时吸收光谱研究表明:开始能量从 ZnPQ 传递到 2HPQ,电子从 2HPQ 转移到 AuPQ$^+$,同时电子从 ZnPQ 转移到 2HPQ$^+$,产生了电荷分离态 ZnPQ$^{\cdot+}$—2HPQ—AuPQ。与其他三卟啉相比,在此配合物中由于电荷分离态的距离较远,所产生的电荷分离态有了最长的停留寿命(7.7 μs)。该系统主要是模拟了光合反应中心的电子转移特征。

2. 周边基团修饰卟啉衍生物的光诱导性质

研究者[4]采用荧光光谱法研究了通过控制卟啉周边不同活性基团(如羟基等)的数量和位置来获得不同荧光强度的卟啉化合物。研究结果显示:随着活性基团数量的增加,卟啉分子的共平面性降低,这增加了卟啉的 $S_1 \rightarrow T_1$ 系间窜跃和 $S_1 \rightarrow S_0$ 之间的内转化,因此导致荧光量子产率减小,S_1 态时间寿命降低。研究者[5]采用相同方法对系列不同取代基(R=2H,CH$_3$,OCH$_3$,NO$_2$,NH$_2$ 等)卟啉化合物的荧光性质进行了研究,结果发现:(1)卟啉类化合物浓度增大时,分子间碰撞增多,引起分子间荧光猝灭;(2)卟啉环上吸电子基导致荧光强度降低,是因为卟啉环上吸电子基的 n 电子到 π-键跃迁属于禁阻跃迁,产生的激发态分子数较少。同时,吸电子基使$S_1 \rightarrow T_1$(三重激发态)的系间跨越程度增大,$S_1 \rightarrow S_0$ 放出光子的数量大大减少,致使荧光减弱。

Li 等[6]合成了包含四羧酸二酰亚胺(PIm)分子的单卟啉化合物和双卟啉化合物,这些化合物在可见光区的吸收光谱特征匹配于太阳光谱。在极性溶剂苄腈中,通过稳态法和时间分辨光谱法对激发态电子转移过程进行检测,发现当卟啉环分别受到光激发后,从卟啉环到类物质间均会出现激发单线态产生的电荷分离现象,电荷分离态的存在寿命为 7~14 ns。该类化合物在光传感电子调谐系统中具有潜在的应用价值。

3. 卟啉和富勒烯组装体的光诱导性质

富勒烯分子有较低的还原电势和重组能，其单重激发态可以诱发卟啉分子单线态—单线态的激发吸收。因此，卟啉—富勒烯化合物分子具有优良的光学性能，有望在光转换器、信号转换和数据存储等光电子领域获得应用。

Moore 等[7]研究了一种由自由卟啉(P)、C_{60} 电子受体和一个二氢吲嗪发色团(DHI)构成的化合物(DHI—P—C_{60})。在此卟啉化合物中，光诱导电子转移过程的时间为 2.3 ns，产生的电荷分离态 DHI—$P^{·+}$—$C_{60}^{·-}$ 的量子产率为 82%。通过时间常数为 56 ps 的光诱导电子转移过程，BT—P—C_{60} 中的卟啉激发会产生 99% 的电荷分离态 BT—$P^{·+}$—$C_{60}^{·-}$。再通过光或热使异构化的 BT—P—C_{60} 返回 DHI—P—C_{60}。通过二氢吲嗪的光异构化，就可以使这种开关控制光诱导电子转移的开或关，利用此原理可设计分子光电器件。

在可见区和近红外区，Kamat 等[8]通过增加 α-聚缩氨酸结构中的卟啉单元的数量来研究光反应的幅度和光电化学效应。结果显示：使用由卟啉—多肽—富勒烯构成的超分子化合物所形成的有机光电池在光电化学性质方面有明显的增强，光能转化效率为 1.3%，光子—光电流的产生效率为 42%。

这些实验数据进一步说明：在富勒烯与连有多肽骨架的多卟啉分子间，分子组装体的形成可以有效地控制带有富勒烯的卟啉—缩氨酸低聚体的超分子化合物中的电子转移效率。

同样，Gust 等[9]研究在由二氢芘(DHP)—卟啉(P)—富勒烯(C_{60})组成的化合物中，卟啉环受到激发产生电荷分离态 DHP—$P^{·+}$—$C_{60}^{·-}$，进而转化为 $DHP^{·+}$—P—$C_{60}^{·-}$。用可见光(300 nm)辐射化合物会使 DHP 部分光异构化变为 CPD，而 CPD—P—C_{60} 同样也可产生电荷分离态 CPD—$P^{·+}$—$C_{60}^{·-}$。因此，光被用于控制可长时间存在的光诱导电荷分离的开或关，该实验原理可用于光电分子系统的设计。

Valentin 等[10]通过对胡萝卜素(C)—卟啉(P)—富勒烯(C_{60})化合物的时间分辨光谱研究，发现在不同介质中，化合物经历了两个步骤的光诱导电子转移过程，从开始 C—$P^{·+}$—$C_{60}^{·-}$ 态产生了一个长时间存在的电荷分离态 $C^{·+}$—P—$C_{60}^{·-}$，电荷重组三线态定位于胡萝卜素部分。

Guldi 等[11]合成了具有新颖结构特点的卟啉化合物：Fc（二茂铁）—ZnP—ZnP—C_{60} 和 Fc—ZnP—H_2P—C_{60}。通过瞬时吸收光谱研究发现，在化合物 Fc—ZnP—ZnP—C_{60} 中，光诱导电子转移产生了电荷分离态 Fc^+—ZnP—ZnP—$C_{60}^{·-}$，在四氢呋喃溶剂中其存在寿命为 1.6 s，这在已报道的人工光合反应中心模拟系统研究中是最长的电荷分离态停留寿命。与 Fc—ZnP—H_2P—C_{60} 相比，Fc—ZnP—ZnP—C_{60} 有更小的静电耦合常数，这是因为具有更高能量 LUMO 轨道的 ZnP 导致在 Fc^+—ZnP—ZnP—$C_{60}^{·-}$ 中的 ZnP 与 $C_{60}^{·-}$ 之间比在 Fc^+—ZnP—H_2P—$C_{60}^{·-}$ 中的 H_2P—$C_{60}^{·-}$ 之间有更小的轨道重叠。此外，光激发 Fc—ZnP—ZnP—C_{60} 产生 Fc^+—ZnP—ZnP—$C_{60}^{·-}$ 的过程有较高的量子产率(40%)。

4. 卟啉自组装体的光诱导性质

Scandola 等[12]设计合成了一种发色团以超分子形式排列的自组装体，采用皮秒和飞秒的时间分辨光谱技术进行研究。结果显示：在两个钌卟啉上连接有分子的组装体中，

DPyPBI[Ru(TPP)(CO)$_2$]受到激发后,组装体分子所表现出的强烈的荧光会完全猝灭。时间分辨荧光光谱显示,从钌卟啉到分子通过电荷的重组会发生光诱导电子转移过程,这种超分子排列的光物理性质为以下内容提供了明显的例证:(1)较大变化的激发波长就能产生从电子到能量传递的快速转换;(2)在超分子排列中存在着三重态的能量传递。

在 Nocera 等[13]对由氢键形成卟啉供—受电子组装体的研究中,在四氢呋喃做溶剂时,皮秒荧光瞬时吸收光谱显示,S_1 单重态和 T_1 状态之间存在一个激发状态下的等吸光点 (λ = 650 nm),由此可进一步观察质子耦合电子转移的过程,因此可建立一个质子网络来控制电荷的转运。该过程类似于血红蛋白—蛋白质化合物中的质子转移过程。质子网络控制电子转移模拟系统作为仿生系统机理方面的研究,其应用将越来越广泛。

同时利用氢键和配位键两点作用,El-Khouly 等[14]组装成富勒烯—卟啉—DMA(N,N-二甲替苯胺)化合物。通过时间分辨发射实验证实,当锌卟啉接受光子后,产生了阳离子部分 ZnTPP$^{·+}$和阴离子部分 $C_{60}^{·-}$,之后 DMA 将电子转移给 ZnTPP$^{·+}$,该机理发生过程有效地降低了电荷的重组,增加了电荷分离态的停留寿命。Souza 等[15]通过轴向反应将带有富勒烯吡咯的咪唑配位到共价键相连的锌硼卟啉二吡咯基的中心金属锌上,组装出超分子组装体。利用超分子组装体的电子转移引起的能量传递来作为光合成的触角反应中心化合物的模型中,硼二吡咯相当于触角叶绿素,用来吸收光能并空间转运至光合反应的中心,而从激发的锌卟啉到富勒烯间的电子转移模拟最初光合反应中心以电荷分离形式发生的电子激发能量到化学能的转化过程。该模型的重要特征是利用相对简单的超分子方法模拟相对复杂的光合成的触角反应中心。

5. 卟啉—纳米薄膜层的光诱导性质

为了提高光电流的量子产率,Sereno 等[16]设计合成了一种不对称的双卟啉,用来修饰 SnO$_2$ 薄膜电极作为光接收系统。通过光谱分析,发现在 PZn—P 体系中,从 PZn 到 P 有单重态—单重态的激发能量传递。光电流光谱和吸收光谱说明双卟啉激发态的产生导致电子进入 SnO$_2$ 薄膜,SnO$_2$ 薄膜阻止了电荷分离态 PZn$^{·+}$—P—SnO$_2$ 的重组,增强了光电流。该装置有潜力用作半导体太阳能电池。

Kamat 等[17]采用电泳的方法将卟啉修饰的 TiO$_2$ 纳米微粒和 C$_{60}$ 分子沉积到具有纳米结构的 SnO$_2$ 薄膜上研究光电诱导性质。在此装置中,TiO$_2$ 纳米微粒和 C$_{60}$ 分子作为电子受体,导致光电流增强。在可见光的激发下,该薄膜有光敏性反应,光子—光电流的产生效率为 41%。研究结果说明该薄膜在光能转换方面起到了重要的作用。

Vuorinen 等[18]将卟啉—富勒烯体系和多聚 3-己基噻吩(PHT)制作成 L-B 薄膜体系,通过电化学和光谱学研究得知该薄膜体系进行了多步电子转移过程。当卟啉环受到光激发时,从卟啉环到富勒烯分子之间发生了光诱导电子转移,随后从多聚物 PHT 到卟啉阳离子进行第二步电子转移,如此循环可以得到长时间存在的电荷分离态。

总之,利用人工合成的卟啉类化合物模拟植物光合反应中心光诱导电子转移和能量传递的研究是当前卟啉仿生化学研究的重要方向。到目前为止,科研工作者已掌握和了解了光合反应中的一些基本原理,但是,还存在许多需进一步探索的基本问题,如激发态的电子结构和性质,激发态的形成与弛豫机制以及激发态的调控等。对这些机理问题的深入研究,也会为人类设计和制备性能优越的分子光电材料(如:分子传感器、分子开关、生物传感器等)

提供重要的理论基础。

第二节 卟啉自组装膜光电转换研究

1. 卟啉自组装膜光电转换研究

卟啉化合物是自然界光合作用得以发生的重要物质,共轭的大环π电子体系使得分子的 HOMO 和 LUMO 之间能量差降低,因此具有广泛的光谱响应范围,尤其在可见光区的发光量子产率较高;同时,卟啉及其衍生物结构多样、易于修饰,人们可以根据需要通过改变周边官能团改变其荧光发射波长,从而实现制备各种光电分子器件的目的。

在对卟啉及其衍生物光电性能的研究方面,Imahori 小组[19]做了大量重要且有意义的工作。他们将不同烷基链长的尾式双硫基苯基卟啉自组装于金电极表面,通过紫外—可见吸收光谱法、循环伏安法、荧光光谱法等分析手段详细考察了烷基链长和膜结构对光电转化的影响(图 12-1)。研究结果表明,随着烷基链长的增加,自组装膜更加有序、致密,但同时单层膜结构依赖于尾端烷基链亚甲基数的奇偶性变化,这是由于亚甲基数目奇偶性变化会影响卟啉分子二聚体形成的难易程度,从而改变单层膜的结构。据此他们首次提出了链长数目以及奇偶性的变化与卟啉组装体的电子转移存在关系。对自组装膜性能的电化学研究表明,其量子产率开始呈一个锯齿形增加趋势,当链长数超过 6 时略微减少,并趋于恒定。对于光电流的这种变化趋势可解释为:链长数目减少,电子在卟啉与电极之间更容易发生耦合而通过能量转移使卟啉激发单线态发生猝灭去活;同时,卟啉二聚体的形成也增强了激发态卟啉去活的速率,随着链长的进一步增加,电子从金电极表面到卟啉阳离子基的转移速率减小,导致光电流减小。

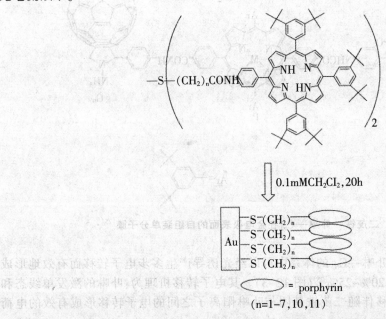

图 12-1 不同链长卟啉分子自组装膜[19]

另外,他们[20]利用十一烷基硫醇调节卟啉在单层薄膜中的浓度,从而控制卟啉分子之间的相互作用对电子转移的影响(图 12-2)。结果表明,随着烷基硫醇比例的增加,卟啉分子之间的相互作用减弱,但光电流产生的量子产率基本保持不变。这些研究为在表面构建光活性分子的排列提供了基本信息。

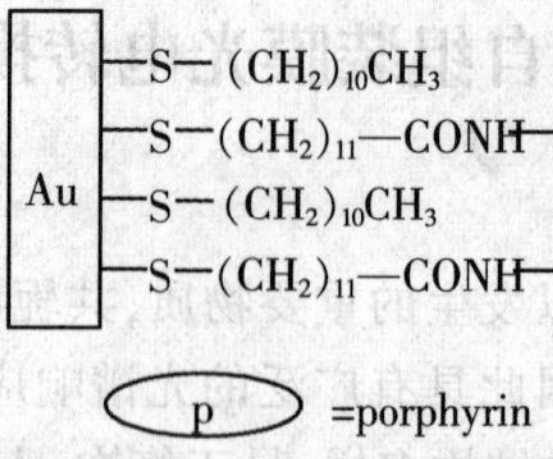

图 12-2 卟啉—烷基硫醇混合自组装膜[20]

2. 卟啉—C_{60}以及二茂铁—卟啉—C_{60}复合系统自组装膜光电转换研究

在光电转换器件方面的研究中,Imahori 等[21]首先发现了富勒烯作为受体具有可加快电荷分离、延迟电荷重组的优势,并将其用来构筑卟啉—C_{60}体系以实现较高的光电转化效率。起初,他们通过 L—B 成膜技术得到了有序的分子组装体,但是较差的稳定性和较多的缺陷使得薄膜内部的量子产率不到 10%,从而极大地限制了其在光电转化方面的应用。后来,他们利用自组装制备单层膜的方法将卟啉—C_{60}体系通过化学键固定在电极表面,得到了一系列作为模拟光合成电荷分离的分子光电器件[22-24]。

研究结果表明:电极表面的卟啉连接 C_{60} 作为电子受体,光电流产生的量子产率(6.4%)比相应没有 C_{60} 的系统高出 30 倍。其光电流的增强归因于电子转移过程中从卟啉激发单线态到 C_{60} 之间存在一个有效的光诱导电子转移机制。

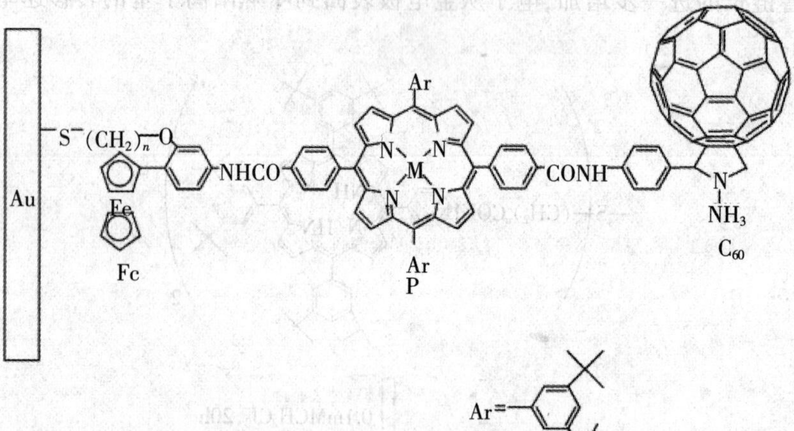

图 12-3 二茂铁—卟啉—C_{60} 在金电极表面的自组装单分子膜[25]

此外,将二茂铁引入卟啉—C_{60}的体系中可以经光诱导产生多步电子转移而有效地形成电荷分离,量子产率达到 20%~25%[25](图 12-3)。其电子转移机理为:卟啉的激发单线态和 C_{60} 之间的光诱导电子转移伴随二茂铁基团与卟啉阳离子之间的电子转移形成有效的电荷分离态(Fc^+—P—C_{60}^-),金电极与二茂铁阳离子之间的电子转移使得体系中产生了整体的电子流动(图 12-4)。一方面,由于较小的电荷重组能,电子在卟啉与 C_{60} 之间发生快速转移,可以与电极产生的非理想能量转移猝灭相竞争,电子从 C_{60}^- 转移至电子载体如氧气分子(O_2/O_2^+)

和甲基紫($MV^{2+}/MV^{·+}$),使得电子载体的能量状态较低而最终将电子传给对电极。另一方面,电子从金电极到二茂铁的转移速率受控于应用在电极上的电压,因此随着电压的降低电子转移速率增大从而导致光电流的增大。

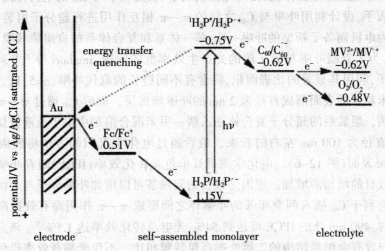

图 12-4　金电极表面二茂铁—卟啉—C_{60} 的自组装单分子膜光电流产生的能量分布示意图[25]

利用光诱导多步电子转移的机理,研究者[26]还成功地在金电极表面构建了完整的二茂铁基—卟啉—C_{60}人工光合成系统(图 12-5),同时模拟光捕获和反应中心的电荷分离,该体系表现出了优异的光电流产生效率。在组装中引入硼—联吡咯烷基硫醇可以增强体系在绿光及蓝光区对光的吸收。除此之外,硼—联吡咯的发射光谱与卟啉的吸收光谱有一定程度的重叠,因此,混合组装体系可以实现硼—联吡咯与卟啉之间最大的能量转移。当硼—联吡咯烷基硫醇与二茂铁基—卟啉—C_{60}的比例为 37:63 时,本系统在极易发生光子—电流转换的最佳波长 510 nm 处的阴极光电流量子产率高达 50%±8%,430 nm 处的量子产率达到 20%±3%,总转化效率达到 1.6%±0.3%,比没有引入硼—联吡咯的非混合组装在波长为 510 nm 处的量子产率高 30 倍。在所有单层膜修饰金属电极和使用电子给体接受器连接分子的人工膜光电流产生的报道中,这一组装体系具有最高的量子产率,它不仅在人工光合成中模拟了光捕获和电荷分离过程,而且也是分子器件中一个高效的光电转换器。

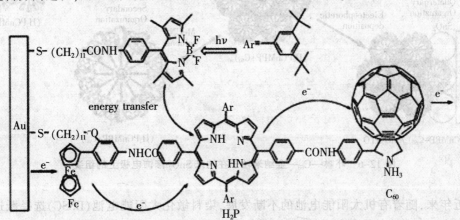

图 12-5　二茂铁—卟啉—C_{60} 人工光合成系统[26]

卟啉自组装膜电化学

尽管上述系统量子产率有了明显的提高。但由于较低的光捕获效率,光电转化效率仍然得不到有效的提高(最大 1%~2%)。为了克服这一缺点,提高光电转化效率使其达到应用的要求,Imahori 与 Kamat 等合作,从提高光捕获效率、加速卟啉与 C_{60} 的电荷分离、减少不合理的电荷重组入手,设计利用卟啉与 C_{60} 之间的 π—π 相互作用进行超分子组装,并应用合适的半导体作为电极制备了新型的卟啉—C_{60} 单一体系和复合体系的自组装薄膜。其中,最值得一提的是,为了克服卟啉单分子膜的光产生效率低的问题,Imahori 等[29-30]采用三维结构的金纳米粒子,利用其显著的比表面积,将含有不同链长的取代卟啉($n=5,11,15$)通过硫基组装在金纳米粒子的表面形成直径为 2 nm 的卟啉修饰层,而后 C_{60} 通过 π—π 相互作用与卟啉进行组装,组装后的超分子复合体在乙腈—甲苯混合溶剂中由于疏液胶体之间的相互作用形成了直径为 100 nm 左右的胶束,最后通过电化学沉积的方法将胶束沉积于纳米 SnO_2 修饰电极表面(图 12-6)。电化学测得其单色光转化效率(IPCE)随着卟啉与金纳米粒子间亚甲基数目的增加而增加,原因是较长的间隔基可以增加卟啉分子与金纳米粒子之间的距离,更有利于 C_{60} 插入两个相邻的卟啉环之间形成 π—π 作用而有效提高光电流的产生。当 $n=15$ 时,490 nm 处的 IPCE 可达到 54%,光电总转化效率达 1.5%[31]。该方法制得的光电转换器件与没有金纳米结构的二维卟啉自组装膜相比,不仅光捕获效率得到了极大的促进,而且转化效率也有了明显的提高,这一结果充分证明纳米 SnO_2 修饰电极上的三维卟啉—C_{60} 网络结构有助于将分离的电荷注入半导体的导带从而极大地提高光电转化效率,该研究为发展光电化学电池提供了一个可借鉴的方法。

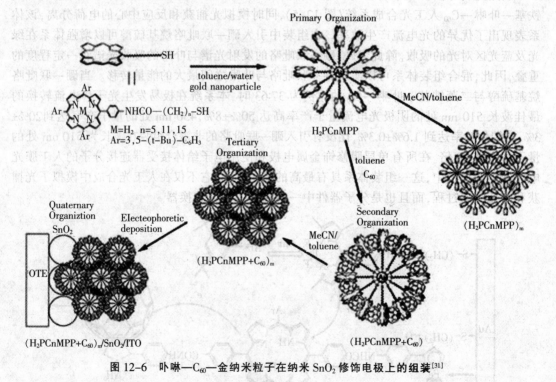

图 12-6 卟啉—C_{60}—金纳米粒子在纳米 SnO_2 修饰电极上的组装[31]

近年来,随着有机太阳能电池的不断发展,染料敏化太阳能电池(DSSC)选择跃迁能量和太阳光匹配的有机染料敏化半导体,使体系的光谱响应延伸到可见光区,是目前太阳能电池研究中的热点。卟啉作为出色的光捕获剂,在这方面有了广泛的应用。人们合成各种具有特殊官能团的卟啉衍生物并将其通过共价键或非共价键修饰纳米半导体如 SnO_2、TiO_2 等,目的是提高其对光的捕获效率,从而提高染料敏化太阳能电池的光电转化效率。Seunghun 等[32]合成了一系列新型、含有噻吩和呋喃取代基的金属锌卟啉络合物,并通过羧基键合在 TiO_2 的表面,达到敏化的目的,其最高 IPCE(420 nm)可达 65%,光电总转化效率达 3.1%。随后,他们[33]又合成了分别含有单羧基和二羧基喹喔啉结构的金属锌卟啉络合物,同样用于敏化 TiO_2,用二羧基喹喔啉卟啉锌敏化的 TiO_2 光电总转化率可达 4%,而经单羧基喹喔啉卟啉锌敏化的 TiO_2 光电总转化率可达 5.2%[33]。利用碳纳米管良好的 π 共轭结构与卟啉进行组装,开发新型光电化学太阳能电池是近几年广受关注的重要研究方向。Hasobe 等[34]将卟啉组装在长度不同的多层杯状的碳纳米结构(SCCNT)表面,然后通过电化学将其沉积在纳米 SnO_2 修饰的光透电极(OTEPSnO_2)组成光电池。实验证明具有三维有序结构的卟啉—碳纳米结构有效地增加了薄膜在可见光区的光电化学响应,其最大 IPCE(440 nm)可达 32%,比单纯的 OTEPSnO_2PSCCNT(IPCE=19%)有了很大的提高。Tagmatarchis 等与 Hasobe 等[35,36]合作先利用带有氨基的卟啉对碳纳米角冠结构的表面进行功能化修饰,再通过电泳沉积的方法使其在纳米 SnO_2 改性电极表面形成自组装单层膜(图 12-7)。电化学测得其 IPCE(440 nm)可达 5.8%,大大超过了单纯将碳纳米角冠材料和卟啉通过物理方法混合后所测得的光电转化效率。

除了在光电转换方面基于卟啉自组装功能膜显示卓越性能外,自组装技术还被广泛用于制作其他纳米电子器件,例如光子导线[37-39]、光致分子开关[40-42]、分子存储器件[43-45]等。由于篇幅的限制,这里就不再展开论述了。

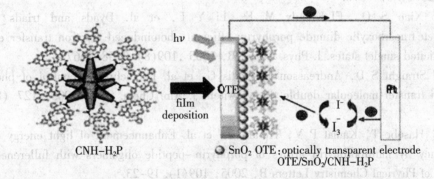

图 12-7 卟啉修饰角冠碳纳米结构及用其组成的光电化学太阳能电池结构示意图[35]

鉴于对卟啉及其衍生物自组装的研究已经比较深入,研究者选择金属酞菁络合物为自组装膜构筑基元,基于配位键合作用,实现了层—层自组装。而且,利用全共轭 4-(2-(4-吡啶基)乙炔基)苯基重氮盐在紫外光照射下,能够使重氮基正离子与基片表面负离子之间的离子键作用转化为苯环与基片之间的共价键连接的特点,从而保证苯乙炔基吡啶分子轴向与基片表面形成近垂直角度,有效提高了自组装膜与基底之间载流子传输效率和稳定性。

通过自组装制备超薄膜的技术得到了科学界的广泛关注,科学家对自组装膜的结构和影响组装的因素进行了较为深入的研究[46]。随着合成技术和相关表征技术的不断发展,卟啉自组装膜在基础研究和应用研究方面均取得了较大发展。同时,认识卟啉在光合作用中的地位及卟啉给—受体分子的光诱导电子转移和电荷分离的能力，在开发光化学合成方面也取得了一定的进展。但是，真正模拟生物体系实现分子水平的电子转移，还需要科学家在纳米尺度内对电子转移的机理做大量详细的研究。另外，设计和合成具有特殊功能的卟啉分子依旧面临很多困难，尤其是作为敏化太阳能电池的光敏剂，如何通过分子设计和组装形式进一步提高太阳能电池的转化效率成为当前有机太阳能电池的研究重点。近年来，对组装功能化的卟啉自组装膜用于纳米光电转化器件的研究也刚刚起步，如何使它在实际电路中发挥作用还有很多问题需要解决，因此要实现功能化的自组装卟啉膜在化学、材料、生物等领域的广泛应用，仍需要多学科科研工作者的不断努力。

参考文献

[1] 王树军,彭玉苓.卟啉光诱导电子转移和能量传递的研究进展.西华大学学报:自然科学版,2007, 26(3): 102-104.

[2] 冯娟,张慧娟,向俊锋,等.锌卟啉的轴向配位对其激发态光物理性质的影响.中国科学,B, 2003, 33(1): 8-13.

[3] Ohkubo K, Sintic P Jm, Tkachenko N V, et al. Photoinduced electron transfer dynamics and long lived CS states of donor-acceptor linked dyads and a triad containing a gold porphyrin in nonpolar solvents. Chemical Physics, 2006, 326: 3-14.

[4] 郭喜明,师宇华,张忆华,等.系列羟基苯基卟啉的合成及其荧光光谱研究.有机化学, 2006, 26(2): 247-251.

[5] 陈新斌,孙咏芬,陈俊琛,等.卟啉化合物荧光性能的取代基效应研究.光谱实验室, 2006, 23(2): 374-376.

[6] Xiao S Q, EI-Khouly M E, Li Y L, et al. Dyads and triads containing perylenete tracarboxylic diimide porphyrin: efficient photoinduced electron transfer elicited via both exicited singlet states. J. Phys. Chem. B, 2005, 109(8): 3658-3667.

[7] Straight S D, Andreasson J, Kodis G, et al. Photochromic control of photoinduced electron transfer molecular double throw switch. J. Am. Chem. Soc., 2005, 127 (8): 2717-2724.

[8] Hasobe T, Kamat P V, Troiani V, et al. Enhanncement of light energy conversion efficiency by multiporphyrin arrays of porphyrin-peptide oligomers with fullerence clusters. Journal of Phyrical Chemistry Letters B, 2005, 109(1): 19-23.

[9] Liddell P A, Kodis G, Andreasson J, et al. Photonic switching of photoinduced electron transfer in a dihydropyrene porphyrin fullerene molecular triad. J. Am. Chem. Soc., 2004, 126(15): 4803-4811.

[10] Valentin M D, Bisol A, Agostini G, et al. Electronic coupling effects on photoinduced electron transfer in carotene porphyrin fullerene triads detected time resolved EPR. J. Chem. Inf. Model, 2005, 45(6): 1581-1590.

[11] Guldi D M, Imahori H, Tamaki K, et al. A molecular tetrad allowing efficient energy storage for 116s at 163K. J. Phys. Chem. A, 2004, 108(4): 541-548.

[12] Prodi A, Chiorboli C, Scandola F, et al. Wavelength dependent electron and energy transfer pathways in a side-to-face ruthenium porphyrin/perylene bisimide assembly. J. Am. Chem. Soc., 2005, 127(5): 1454-1462.

[13] Damrauer N H, Hodgkiss J M, Rosenthal J, et al. Observation of proton-coupled electron transfer by transient absorption spectroscopy in a hydrogen-bonded, porphyrin donor-acceptor assembly. J. Phys. Chem. B, 2004, 108(20): 6315-6321.

[14] EI-Khouly M E, Ito O, Smith P M, et al. Intermolecular and supramolecular photoinduced electron transfer processes of fullerene porphyrin/phthalocyanine systems. J. Photochem. Photobio. C, 2004, 5: 79-104.

[15] Souza F D, Smith P M, Zandler M E, et al. Energy transfer followed by electron transfer in a supramolecular triad composed of boron dipyrrin zinc porphyrin and fullerene: a model for the photosynthetic antenna reaction center complex. J. Am. Chem. Soc, 2004, 126(25): 7898-7907.

[16] Gervaldo M, Otero L, Milanesio M E, et al. Photosensitization of thin SnO_2 nanocrystalline semiconductor film electrodes with electron donor-acceptor metallodiporphyrin dyad. Chemical Physics, 2005, 312: 97-109.

[17] Hasobe T, Hattori S, Kamat P V, et al. Supramolecular nanostructured assemblies of different types of porphyrins with fullerene using TiO_2 nanoparticles for light energy conversion. Tetrahedron, 2006, 62: 1937-1946.

[18] Vuorinen T, Kauniso K, Tkachenko N V, et al. Photoinduced interlayer electron transfer in alternating porphyrin-fullerene dyad and regioregular poly (3-exylthiophene) Langmuir-Blodgett films. J. Photochem. Photobio. A, 2006, 178: 185-191.

[19] Imahori H, Noried H, Nishimura Y, et al. Chain length effect on the structure and photoelectrochemical properties of self-assembled monolayers of porphyrins on gold electrodes. J. Phys. Chem. B, 2000, 104(6): 1253-1260.

[20] Imahori H, Hasobe T, Yamada H, et al. Concentration effects of porphyrin monolayers on the structure and photoelectrochemical properties of mixed self-assembled monolayers of porphyrin and alkanethiol on gold electrodes. Langmuir, 2001, 17(16): 4925-4931.

[21] Imahori H, Tanakia K, Yamada H. Photosynthetic electron transfer using fullerenes as novel acceptors. Carbon, 2000, 38: 1599-1605.

[22] Yamada H, Imahori H, Nishimura Y, et al. Enhancement of photocurrent generation by ITO electrodes modified chemically with self-assembled monolayers of porphyrin-fullerene dyads. Adv. Mater., 2002, 14(12): 892-895.

[23] Yamada H, Imahori H, Fukuzumi S. Comparison of reorganization energies for intra- and intermolecular electron transfer. J. Mater. Chem., 2002, 12(7): 2034-2040.

[24] Yamada H, Imahori H, Nishimura Y, et al. Photovoltaic properties of self-assembled monolayers of porphyrins and porphyrin-fullerene dyads on ITO and gold surfaces. J. Am. Chem. Soc., 2003, 125(30): 9129-9139.

[25] Imahori H, Yamada H, Nishimura Y, et al. Vectorial multistep electron transfer at the gold electrodes modified with self-assembled monolayers of ferrocene-porphyrin-fullerene triads. J. Phys. Chem. B, 2000, 104(9): 2099-2018.

[26] Imahori H, Norieda H, Yamada H, et al. Light-harvesting and photocurrent generation by gold electrodes modified with mixed self-assembled monolayers of boron-dipyrrin and ferrocene-porphyrin-fullerene triad. J. Am. Chem. Soc., 2001, 123(1): 100-110.

[27] Imahori H, Nishimura Y, Yamazaki I, et al. Metal and size effects on structures and

photophysical properties of porphyrin-modified metal nanoclusters. J. Mater. Chem., 2003, 13 (12): 2890-2898.

[28] Imahori H, Seki S, Ueda M. Effects of porphyrin substituents on film structure and photoelectrochemical properties of porphyrin/fullerene composite clusters electrophoretically deposited on nanostructured SnO2 electrodes. Chem. Eur. J., 2007, 13: 10182-10193.

[29] Imahori H, Kashiwagi Y, Hanada T, et al. Electrophoretic deposition of donor-acceptor nanostructures on electrodes for molecular photovoltaics. J. Mater. Chem., 2007, 17(1): 31-41.

[30] Imahori H. Creation of fullerene-based artificial photosynthetic systems. Bull. Chem. Soc. Jpn., 2007, 80(4): 621-636.

[31] Hasobe T, Imahori H, Kamat P V, et al. Photovoltaic cells using composite nanoclusters of porphyrins and fullerenes with gold nanoparticles. J. Am. Chem. Soc., 2005, 127(4): 1216-1228.

[32] Seunghun E, Shinya H, Tomokazu U, et al. Effects of 5-membered heteroaromatic spacers on structures of porphyrin films and photovoltaic properties of porphyrin-sensitized TiO_2 cells. J. Phys. Chem. C, 2007, 111(8): 3528-3537.

[33] Seunghun E, Shinya H, Tomokazu U, et al. Carboxyquinoxaline-fused porphyrins for dye-sensitized solar cells. J. Phys. Chem. C, 2008, 112(11): 4396-4405.

[34] Hasobe T, Murata H, Kamat P V. Tube-length dependence and charge transfer with excited porphyrin. J. Phys. Chem. C, 2007, 111(44): 16626-16634.

[35] Georgia P, Atula S D S, Tagmatarchis N, et al. Covalent functionalization of carbon nanohorns with porphyrins: nanohybrid formation and photoinduced electron and energy transfer. Adv. Funct. Mater., 2007, 17: 1705-1711.

[36] Pagona G, Hasobe T, Tagmatarchis N, et al. Characterization and photoelectrochemical properties of nanostructured thin film composed of carbon nanohorns covalently functionalized with porphyrins. J. Phys. Chem. C, 2008, 112(40): 15735-15741.

[37] Arounaguiry A, Christine K, Richard W W, et al. Weakly coupled molecular photonic wires: synthesis and excited-state energy-transfer dynamics J. Org. Chem., 2002, 67(11): 3811-3826.

[38] Zin S Y, Shanmugam E, Dongho K. Screening for tyrosinase inhibitors among extracts of seashore plants and identification of potent inhibitors from Garcinia subelliptica. Bull. Korean Chem. Soc., 2008, 29(1): 197-201.

[39] Fortage J, Boixel J, Blart E, et al. Single-step electron transfer on the nanometer scale: ultra-fast charge shift in strongly coupled zinc porphyrin-gold porphyrin dyads. Chem. Eur. J., 2008, 14(11): 3467-3480.

[40] Reddy D R, Maiya B G. A molecular photoswitch based on an "axial-bonding" type phosphorus(V) porphyrin. Chem. Commun., 2001, (1): 117-118.

[41] Wang Z, Nygard A M, Cook M J, et al. an evanescent-field-driven self-assembled molecular photoswitch for macrocycle coordination and release. Langmuir, 2004, 20(14): 5850-5857.

[42] Peters M V, Goddard R, Hecht S. Asymmetric iminium ion catalysis with a novel bifunctional primary amine thiourea: controlling adjacent quaternary and tertiary stereocenters.

J. Org. Chem., 2006, 71(20): 7846-7849.

[43] Li C L, Lei J, Fan B, et al. Binary memories based on nanowire / molecular wire heterostructures. J. Phys. Chem. B, 2004, 108(28): 9646-9649.

[44] Kulikov O V, Schmidt I, Muresan A Z, et al. Synthesis of porphyrins for metal deposition studies in molecular information storage applications. J. Porphyrins Phthalocyanines, 2007, 11(10): 699-712.

[45] Satake A, Tanihara J, Kobuke Y. Ligand-induced interconversion of a coordination-organized porphyrin dimer: a potential fluorescence-based molecular memory monitor. Inorg. Chem., 2007, 46: 9700-9707.

[46] 赵玮, 佟斌, 支俊格, 等. 卟啉自组装超薄膜的制备及其在光电转换方面的研究. 化学进展, 2009, 21(12): 2625-2634.